FIRE ALARM SYSTEMS

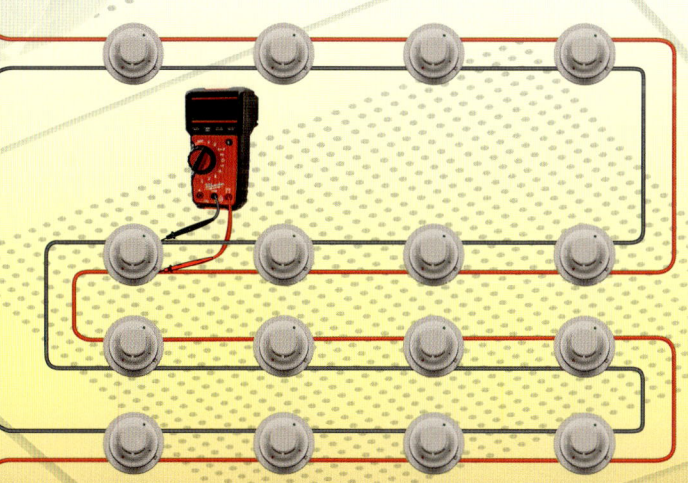

Based on the 2013 *NFPA 72®* and the 2014 *NEC®*

NJATC Fire Alarm Systems is intended to be an educational resource for the user and contains procedures commonly practiced in industry and the trade. Specific procedures vary with each task and must be performed by a qualified person. For maximum safety, always refer to specific manufacturer recommendations, insurance regulations, specific job site and plant procedures, applicable federal, state, and local regulations, and any authority having jurisdiction. The electrical training ALLIANCE assumes no responsibility or liability in connection with this material or its use by any individual or organization.

© 2014, 2011, 2008 National Joint Apprenticeship and Training Committee for the Electrical Industry

This material is for the exclusive use by the IBEW-NECA JATCs and programs approved by the NJATC. Possession and/or use by others is strictly prohibited as this proprietary material is for exclusive use by the NJATC and programs approved by the NJATC.

All rights reserved. No part of this material shall be reproduced, stored in a retrieval system, or transmitted by any means whether electronic, mechanical, photocopying, recording, or otherwise without the express written permission of the NJATC.

5 6 7 8 9 – 14 – 9 8 7 6 5

Printed in the United States of America

M43431

Contents

Chapter 1 — INTRODUCTION TO FIRE ALARM SYSTEMS ... 1
Overview of Fire Alarm Systems ... 2
History of Fire and Fire Alarm Signaling Systems ... 3
Basic Types of Fire Alarm Systems ... 8
Other Emergency Communications Systems ... 12
Circuit Types ... 16
Suppression Systems ... 19
Codes and Standards ... 21

Chapter 2 — FUNDAMENTALS AND SYSTEM REQUIREMENTS ... 31
General Requirements ... 32
Signals and Signal Types ... 37
Monitoring for Integrity ... 38
Power Supplies ... 50

Chapter 3 — INITIATING DEVICES ... 57
Types of Initiating Devices ... 58
Initiating Device Operating Theory ... 59
Installation and Spacing Requirements for Initiating Devices ... 65

Chapter 4 — NOTIFICATION APPLIANCES ... 85
General Requirements ... 86
Audible Signaling ... 88
Visible Signaling and the Americans with Disabilities Act ... 94

Chapter 5 — WIRING AND WIRING METHODS ... 105
Wiring Requirements for Fire Alarm Systems ... 106
Non–Power-Limited and Power-Limited Fire Alarm Circuits ... 110
Circuit Integrity Cable ... 116
Class A and Class X Circuit Separation ... 117
Calculations ... 117
Firestopping Concepts ... 120

Contents

Chapter 6 SYSTEM INTERFACES AND SAFETY CONTROL FUNCTIONS 127

Combination Systems and Interconnected Fire Alarm Systems 128
Connection Methods 131
Interconnected Suppression Systems 133
Emergency Control Function Interfaces 139

Chapter 7 ADVANCED DETECTION TOPICS 157

Basic Fire Science 158
Advanced Detector Applications 163
Smoke Control Applications 174
Smoke Detectors for Door Release Service 179

Chapter 8 EMERGENCY COMMUNICATIONS SYSTEMS (ECS) 185

Types of Emergency Communications Systems 186
Power Supplies for Emergency Communications Systems 191
Fire Command Centers 191
Types of Evacuation Signals 192
Intelligibility 194
Circuits 196

Chapter 9 PUBLIC EMERGENCY SYSTEMS AND SUPERVISING STATIONS 205

Public Emergency Alarm Reporting Systems 206
Supervising Stations 209

Contents

Chapter 10 HOUSEHOLD FIRE ALARM SYSTEMS 227
Basic Requirements 228
Required Detection 231
Workmanship and Wiring 236
Occupant Notification 236
Power Supplies 238
Household Fire Alarm Systems 240

Chapter 11 PLANS AND SPECIFICATIONS 245
Specifications 246
Plans and Drawings 248
Standardized Fire Protection Plans and Symbols 256
Contracting Methods 257

Chapter 12 INSPECTION, TESTING, AND MAINTENANCE 261
General Requirements 262
Inspections and Tests 268
Initial/Acceptance Tests 268
Periodic Inspections and Tests 268
Reacceptance Tests 269
Testing Methods and Frequency 270
Maintenance 274
Troubleshooting 274

Appendix 287
Appendix A — Certification 289
Appendix B — One Prosperity Place Specifications 299
Appendix C — One Prosperity Place Drawings .... Enclosed

Index 319

Features

Facts offer additional information related to Fire Alarm Systems.

Red Headers and **Subheaders** organize information within the text.

Figures, including photographs and artwork, clearly illustrate concepts from the text.

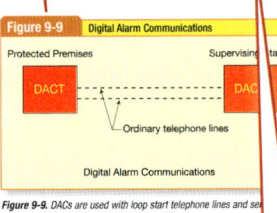

Code Excerpts are "ripped" from *NFPA 72* or other sources.

For additional information related to QR Codes, visit qr.njatcdb.org Item #1079

Quick Response Codes (QR Codes) create a link between the textbook and the Internet. They can be scanned using Smartphone applications to obtain additional information online. (To access the information without using a Smartphone, visit qr.njatc.org and enter the referenced Item #.)

NJATC Fire Alarm Systems

Features

Clear, easy-to-read **Contents** pages in the front of the textbook and inside each Chapter enable the reader to quickly find important fire alarm content.

The **Introduction** and **Objectives** at the beginning of each chapter introduce readers to the concepts to be learned in each chapter.

At the end of each chapter, a concise chapter **Summary** and **Review Questions** reinforce the important concepts included in the text.

The **Appendix** includes information related to Certification and sample Specifications and Drawings.

IMSA and NICET certification exams can be accessed at qr.njatcdb.org Item #1083

Introduction

Fire alarm systems are designed and installed to protect people and property. Because these systems involve life safety, fire alarm systems must be treated much differently than other building systems, such as Heating, Ventilating, and Air Conditioning (HVAC) or burglar alarm systems. Fire alarm systems have special installation codes and design criteria. In fact, there are many other things that make fire alarm systems unique from other building systems.

The major difference between fire alarm systems and other building systems is that fire alarm systems are intended to protect people and property from fires and other similar emergencies. In addition, fire alarm systems can actually manage other fire protection (life safety) systems such as fire pumps, smoke control, HVAC, access control, and smoke doors. Fire alarms systems, the foundation system for building occupants, must be installed with the greatest of care to ensure reliability.

Fire alarm installers must understand the essential code installation requirements for fire alarm systems and appreciate the relationship between good workmanship and reliable systems. One of the primary objectives of this text is to provide a working knowledge of fire alarm systems and the required installation requirements. Equipped with this knowledge, you will be well prepared to perform high-quality installations.

Many modern fire alarm systems on the market today utilize microprocessor-based controls. Fire alarm systems are essentially computers with peripheral circuits extending throughout the building. Modern systems require installers to have good working knowledge of programming, power supplies, wiring, detectors, and notification appliances. Installers must also have an understanding of other life safety systems that interact with fire alarm systems.

The study of fire alarm systems begins by reviewing the interesting history of fire alarm signaling then developing an understanding of the state-of-the-art technology available today. Topics also covered include basic requirements such as fire protection system types, power supplies, monitoring for integrity, and wiring. Each chapter also contains the basic operating principles of devices or equipment. Understanding how equipment works reinforces *Code* intent and helps the student develop better installation techniques.

About This Book

This fourth edition includes the changes related to nationally recognized certification programs. The curriculum has been revised and re-organized to streamline the materials and make it easier to use. The course materials have been updated to reflect changes to the 2013 edition of *NFPA 72®*, *National Fire Alarm and Signaling Code®*, and the 2014 edition of *NFPA 70®*, *National Electrical Code®*.

This guide is specifically designed to assist the reader in understanding fire alarm system requirements from design to installation to testing. The primary focus is to encourage students to apply for and obtain certification in the International Municipal Signal Association (IMSA) or National Institute for Certification in Engineering Technologies (NICET) programs. This edition of *Fire Alarm Systems* reflects changes that have been made pertaining to these certifications.

Finally, other revisions and changes were made to this curriculum based upon user input. The reader is encouraged to assist in the development of a better curriculum by submitting ideas and recommended changes to the NJATC so we can continuously improve our products.

Acknowledgments

Acknowledgments

ASME
Fire-Lite Alarms by Honeywell
Honeywell Security Group
National Fire Protection Association (NFPA)
System Sensor

QR Codes

3M
Ansul, a Tyco International Company
Edwards Signaling
Fire-Lite Alarms by Honeywell
Gold-Line
Home Safeguard Industries
Honeywell System Sensor
National Fire Protection Association (NFPA)

NFPA 70®, National Electrical Code, and NEC® are registered trademarks of the National Fire Protection Association, Quincy, MA.

National Fire Alarm Code® and NFPA 72® are registered trademarks of the National Fire Protection Association, Quincy, MA.

About the Author

Merton Bunker is the President of Merton Bunker & Associates in Stafford, Virginia. He has a Bachelor of Science in Electrical Engineering from the University of Maine, a Master of Science in Engineering Management from Western New England College, and is a registered professional engineer. He is also a certified fire and explosion investigator and a Master Electrical Inspector (IAEI).

Merton Bunker has twenty-five years of engineering experience, sixteen of which are in fire protection. Some of his experiences include seven years at the National Fire Protection Association, where he served as Senior Signaling Systems Engineer and Chief Electrical Engineer. During his tenure at NFPA, he co-authored the 1996 and 1999 editions of the NFPA 72 Handbook. He also co-authored the NFPA Pocket Guide to Fire Alarm Systems in 2001, 2004, and 2009. Merton has written dozens of trade publication articles and continues to develop training materials for the NJATC.

Merton currently serves as the Chair of the NFPA Technical Committee on Protected Premises Fire Alarm and Signaling Systems and is a principal member of the National Electrical Code Technical Correlating Committee. He also served as a principal member of the NFPA Technical Committee on Testing and Maintenance of Fire Alarm Systems.

Merton is a member of:
National Fire Protection Association
Institute of Electrical & Electronics Engineers
National Association of Fire Investigators
Society of Fire Protection Engineers
Automatic Fire Alarm Association

Introduction to Fire Alarm Systems

The history of fire and fire alarm systems dates to the Middle Ages, if not earlier. The systems have ranged from simple town criers patrolling the streets looking for fires or smoke to Boston's early telegraph-enabled street boxes. Modern systems employ complex circuitry to allow alarm devices to be monitored remotely and to control various forms of fire suppression systems. A sophisticated system of standards has evolved to make buildings safer and fire alarm and prevention systems more reliable. But as technology advances, codes and standards strive to keep pace. This chapter explores fire alarm system technology and the codes and standards that regulate its use.

Objectives

- » Describe the history of fire and fire alarm systems
- » Explain the basic types of fire alarm systems
- » Describe fire alarm system components and signals
- » Explain circuit types important to fire alarm systems
- » Explain suppression systems
- » Locate and refer to codes and standards relevant to fire alarm systems

Chapter 1

Table of Contents

Overview of Fire Alarm Systems..................2
    Overview of Fire Statistics 2
    Purpose and Importance of Fire Alarm Systems..2
    Goals of a Fire Alarm System 2
    Fire Alarm System Functions.................3
    Installation and Maintenance.................3

History of Fire and Fire Alarm Signaling Systems 3
    Colonial America 4
    Early Fire Alarm Systems......................4
    Modern Fire Alarm Equipment...............6
    Formation of Standards Developing Organizations (SDO) 7

Basic Types of Fire Alarm Systems 8
    Protected Premises (Local) Systems 8
    In-Building Fire Emergency Voice/Alarm Communications Systems 10
    Supervising Station Alarm Systems 10
    Public Emergency Alarm Reporting Systems / Auxiliary Alarm Systems...... 11
    Single and Multiple-Station Alarms and Household Fire Alarm Systems 12

Other Emergency Communications Systems..12
    Control Units 12
    Initiating Devices 14
    Signals 15
    Notification Appliances.........................15

Circuit Types...16
    Initiating Device Circuits 16
    Signaling Line Circuits.........................18
    Notification Appliance Circuits 19

Suppression Systems.................................19

Codes and Standards 21
    Codes and Standards Development in the U.S..21
    Using Codes and Standards 23

Summary...28

Review Questions.......................................28

OVERVIEW OF FIRE ALARM SYSTEMS

A fire in a building will likely cause a significant amount of damage. In addition, people in a building during a fire are at risk of being seriously injured or even dying as a result of the fire.

Overview of Fire Statistics

Fire is one of the most destructive and costly elements known to humanity. The National Fire Protection Association (NFPA) reports that, in the year 2011, there were 1,389,500 fires in the United States alone. This figure includes all types of reported fires, such as structure fires, industrial fires, transportation-related fires, and wildfires. That means a fire occurs about every 23 seconds in the United States. Total civilian deaths in 2011 were 3,005, and there were 17,500 civilian injuries. There were 484,500 structure fires in 2011 that resulted in 2,640 civilian deaths and 15,635 civilian injuries. Structure fires in the United States accounted for $9.7 billion in property damage in the year 2011. While the number of civilian deaths has slowly declined, the amount of property damage has increased significantly. On average, there is one structure fire reported every 65 seconds.

Purpose and Importance of Fire Alarm Systems

Fire alarm systems are designed, installed, and maintained to protect people and property from the devastating effects of fire, whether the fire is accidental or set on purpose by an arsonist. Different systems may have different features, but every system is intended to provide warning time to react to the fire so appropriate action may be taken.

Fire protection is an important element in any facility, regardless of occupancy type or use group. It comprises three basic parts:

1. Detection and alarm systems
2. Suppression systems
3. Containment

Detection and suppression systems are more active in nature, while containment is passive. Alarm and signaling systems supervise suppression systems and are an important part of the overall fire protection approach. The requirement to install a fire alarm system in any given building is generally found in building codes. Some codes define buildings or areas by "use group," while others define buildings or areas by "occupancy type." These terms are similar but do not always carry the same meaning in the different codes in which they are used. Depending on the use group or occupancy type, fire alarm systems may be required. Most building owners would rather spend money on aesthetically pleasing features like carpet and paint. However, codes and standards may require the owner to install a fire alarm system, depending on building size, the number of occupants, building height, occupancy type, use group, and many other variables.

Goals of a Fire Alarm System

Smoke and heat are deadly and can cause significant property damage. The contents of a building and the building itself will contribute to the production of smoke and heat when they are burned. For the purposes of fire protection, room contents (furnishings, commodities, stored items, etc.) are often referred to as "fuels." The fuels being burned will determine the amount of smoke and heat produced. Some fires produce very little smoke but produce large amounts of heat. Other fires may smolder for long periods, producing large amounts of smoke but little heat.

Occupants are adversely affected by smoke and heat; so, it is natural to assume that fire alarm systems are provided to detect fires before conditions become untenable. This use of fire alarm systems is commonly referred to as "life-safety protection." In this use, fire alarm systems are designed to alert occupants to the existence of a fire through the use of audible and visible signals. But smoke and heat can also damage or destroy buildings and their contents. Fire alarms are also installed to provide early warning for the purposes of property protection. Many insurance providers require fire alarm systems solely for this purpose and place little emphasis on life-safety protection.

Fact

NFPA 72, National Fire Alarm and Signaling Code, is the document which provides installation, performance, inspection, testing and maintenance requirements for fire alarm systems. Wiring requirements for fire alarm systems are provided by NFPA 70, National Electrical Code.

Another use of a fire alarm system may be to provide mission continuity. For example, a financial institution cannot be expected to earn a profit if a computer server room containing transaction records is destroyed by fire. In this case, the fire alarm system can provide early warning to notify system operators, shut down the system, and actuate a suppression system. This type of system is rarely required by any code but is often required by insurance carriers.

Fire alarm systems can also provide environmental protection. For example, rubber tires produce large amounts of toxic smoke when they are burned. A tire warehouse with an early detection system to actuate a suppression system will help limit losses and mitigate the impact on the environment. This type of system also provides property protection, even though the primary goal is environmental protection.

Finally, heritage protection is usually afforded by installing systems in culturally significant buildings. These can include libraries, museums, and architecturally important buildings. Property protection is an intertwined goal with heritage protection.

Fire Alarm System Functions

In addition to signaling an alarm, fire alarm systems sometimes actuate emergency control functions to make the building safer in the event of a fire. Examples of emergency control functions include smoke control, HVAC shutdown, elevator recall, door unlocking, and smoke door closure. Fire alarms also supervise other fire protection systems, such as automatic sprinkler systems, fire pumps, and other suppression systems.

Installation and Maintenance

Because fire alarm systems are used to protect people and property, buildings and their contents, they must be designed and installed with great care to ensure that they are extremely reliable. Although many features of fire alarm designs are redundant, improper field installations can render them useless. Failure of fire alarm systems in the past has all too often led to tragic losses of property and life.

There are four major components involved in fire alarm system reliability: design, equipment, installation and maintenance. This book focuses on the benefits of properly installing and maintaining a fire alarm system. Following applicable codes, standards, specifications, and the manufacturer's recommendations for installation and maintenance is the best way to ensure that a new fire alarm system will properly operate and achieve design objectives. Proper maintenance and testing of the system will generally prevent or quickly discover problems or failures. Proper and timely maintenance and testing will also extend the useful life of a system, giving better value to the owner.

HISTORY OF FIRE AND FIRE ALARM SIGNALING SYSTEMS

The need to signal a fire, to raise an alarm, has existed since the discovery of fire. The approaches to signaling or raising an alarm have evolved with technology and architectural practices.

During the Middle Ages in Europe, most houses were constructed of wood with thatched roofs. Chimneys were frequently made of wood and grout, and in some cases were nonexistent. Cooking and heating were the usual causes of deadly, fast-moving fires. The only effective way to alert others of a fire was through verbal communications or by ringing church bells. Not until 1400 did the King of England institute organized fire service and outlaw wooden chimneys. The Great Fire of London in 1666 destroyed five-sixths of the city but, surprisingly, caused only six deaths. London alehouse fires in the late 1600s were, however, responsible for thousands of deaths. These events lead to the birth of the first fire codes of modern history. These codes required new buildings to be constructed from brick and limited their height.

 Fact

The National Fire Protection Association (NFPA) sponsors Fire Prevention Week (FPW), which is held annually on the anniversary of the Great Chicago Fire of 1871.

Early firefighting equipment was primitive at best. Leather buckets and metal hooks were used to combat fires. Leather hoses and pumps were developed later; these were still largely ineffective because many buildings were wooden and burned very quickly. Water mains were made from wood and buried deep in the ground to prevent freezing. When a fire was discovered, a hole was dug to the water main and a hole was drilled in the pipe to permit the surrounding hole to fill with water, which could be pumped or carried to the fire. After the fire, the main was plugged with a wooden plug, which is where the term "fire plug" originated.

Colonial America

The first settlers in North America had little protection from fires. They, too, built thatched-roof huts and used open fires as had been done in England. However, they did not import any of the King's fire codes when they immigrated. Bricks were expensive, and wood was in ready supply because of the virgin forests nearby. Most dwellings were made of wood for these reasons. While cost effective and readily available, wood burns very easily.

Fires in Colonial America were all too common. Careless use of cooking and heating fires and a lack of codes regulating building construction led to many fire losses and countless deaths. Fires continued well into the nineteenth and twentieth centuries. For example, the Great Chicago Fire of 1871 destroyed more than $192,000,000 (adjusted to today's value) of property and took 300 lives. Fire departments were formed, but they often lacked good equipment and training. The best detection available at the time was primarily limited to human senses. The problem with humans is that they must be awake to be effective. Smoke inhalation tends to put sleeping humans deeper into sleep and is a leading cause of deaths. Clearly, there was a need to have quicker detection and warning to allow firefighters more time.

The history of American fire alarm signaling has its roots in the fire service. During colonial times, roving night watchmen would patrol cities and towns looking for fires or smoke. The human sense of smell is excellent at detecting smoke; however, the human voice is not loud enough to carry very far. Watchmen were given a wooden rattle, which consisted of a reed and ratcheted handle. When the rattle was twirled, it made a loud noise to awaken volunteer firefighters. Church bells were also commonly used to alert others of a fire. These practices continued well into the middle of the nineteenth century until electric bell strikers and other technologies were used.

When the watchman's rattle was twirled, the loud noise would awaken volunteer firefighters.

Early Fire Alarm Systems

Samuel Morse developed and built the first telegraph in 1844. The telegraph was the basis for fire alarm signaling. Moses Farmer and William Channing, two inventors from New Hampshire, enhanced telegraph technology to begin development of the world's first public fire alarm reporting system in 1845.

At noon on April 12, 1852, the first public fire alarm reporting system became operational in Boston, Massachusetts. This system used a combination of bell strikers and boxes outfitted with hand cranks, which operated an elaborate telegraph system capable of tapping out an individual box number. The box number was transmitted to a central location from which firefighters could be dispatched. The box number corresponded to a specific street address. A dispatcher would look up the box number and telegraph the closest fire station. There were 40 boxes and 19 electric tower bell strikers dispersed around the city, all of which were connected to a central reporting station through three circuits.

Unfortunately, many excited responders cranked the box handle too quickly, which often garbled the signal to the central station. A new spring-wound, hook-type box, which eliminated the problem of excited responders using the manual crank, was designed in 1864 but did not see service until 1868. In fact, the original Boston telegraph system remained in service for over 20 years before being replaced with a new system in 1874.

Many more public fire reporting systems were installed in cities throughout the Northeast. For example, the City of Baltimore installed a Gamewell public fire reporting system in 1858. The original system had 30 boxes and 30 miles of conductors. It quickly grew to 1,100 boxes and was deemed the most complete of any in the world. Many Gamewell public fire reporting systems are still in service today, mostly throughout the northeastern United States. However, many of these systems are becoming difficult and costly to maintain. Additionally, cellular telephone technology has rendered public fire reporting systems somewhat obsolete.

Street boxes used in public fire reporting systems were once a common sight in major cities of the U.S.

Public fire reporting systems were a vast improvement, but they still relied on humans to detect and report a fire. Automatic systems were clearly needed to eliminate the fallible human element and make fire alarm systems more reliable. The first automatic fire alarm detection device was a heat detector developed around 1849. It used a fusible link element, which uses the same technology as many automatic sprinklers in use today. Other early heat detectors used bimetallic strips that bent and closed contacts when heated. Heat detectors represented a huge advance in technology, but they had limitations. Heat detectors will not actuate until there is sufficient heat present to operate the element, thereby causing an alarm. Eventually, heat detectors within individual buildings were connected to the public fire reporting system through the use of master boxes located near the building. Occupant notification was also needed. Moses Farmer also developed the electric bell in 1848, and this technology is still used on fire alarm systems today.

The loss of life was a great problem, but the burgeoning insurance industry also recognized significant property loss was also a problem. The northeastern United States was a center for manufacturing during the Industrial Revolution of the late nineteenth century. Many factories started using the new electrical distribution system, but poor workmanship practices in the rapidly growing electrical and fire sprinkler system industries caused devastating losses to industry during this period. Insurance interests saw a need to regulate the installation of these systems. Underwriters Laboratories (UL) was founded in 1895 to address product safety. The National Fire Protection Association (NFPA) was founded in 1896 and immediately began developing codes and standards for fire protection. The first fire alarm standard dates to 1899. These standards developing organizations (SDO) are still producing standards today.

Fire alarm system technology changed very little from the turn of the century until the first pneumatic heat detection devices were developed in the 1940s. These systems used a sealed copper capillary tube routed throughout the protected spaces. A fire would heat the tube, causing expansion of the air inside the tube, which could be detected by pressure switches located at one end of the tube. These detectors were slightly more sensitive than fusible link heat detectors but still had limitations. Leaks would sometimes render the system useless. Line-type (cable) heat detectors were also developed in the 1940s and are still in use today.

Smoke detectors became commercially available in the 1960s. They operated on a 220-volts-direct-current (VDC) supply, provided by a control unit, and offered earlier warning than heat detectors. This achievement paved the way for significant advances in detection technology. However, these system-powered smoke detectors were still not practical or affordable for household use. At the time, average annual fire deaths in dwellings exceeded 8,000 persons in the U.S. A solution was clearly needed for the household fire problem.

Modern Fire Alarm Equipment

Smoke alarms for dwellings became commercially available in the mid-1970s. Early detectors operated on large, expensive batteries and were very costly. They also used an ionized chamber technology. But poor placement of detectors and an extremely sensitive design caused many nuisance alarms, usually from cooking vapors. Underwriters Laboratories revised the product standard for dwelling smoke detectors in the middle 1980s, requiring a lower sensitivity and standardized battery types. The revised product standard caused an improved product life span and increased reliability, saving countless lives.

A series of costly computer room fires in Australia and New Zealand in the late 1970s spurred the development of the world's first active sampling smoke detector. An Australian company developed this technology to provide significantly earlier warning of smoke in a protected space. This detector is known as the Very Early Smoke Detection Apparatus, or VESDA. The detector was located in a housing that contained a fan to draw air samples into the detector through sampling pipes that extended into the protected space. Other manufacturers developed similar systems. These systems are hundreds of times more sensitive than ordinary smoke detectors and are known collectively as air-sampling smoke detectors. The first VESDA used a Xenon strobe technology to detect smoke particles, but newer models use laser particle counters.

Before the development of microprocessors in the early 1980s, all fire alarm system controls used relay or transistor logic. Alarms were created by contact closure on a circuit. If any device on a circuit actuated, the controls were not capable of knowing or annunciating which device was in alarm. These systems are referred to as conventional systems. This technology is still in use today, and many of these older systems are still in service.

With the development of personal computers during the 1980s, the microprocessor-based fire alarm controller became possible. Modern addressable systems are capable of annunciating exactly which device is in alarm, resulting in improved response times. This advancement has further led to the development of "intelligent" detectors, which can report the quantity of smoke present in the protected area. Electronic heat detectors are capable of transmitting the actual temperature in a room. In addition, these newer technologies can perform self-test and calibration routines to improve their reliability and accuracy.

Fire alarm systems also take advantage of wireless technology. These systems do not require wires between the controls and field devices. Wireless systems are especially useful in retrofit applications, particularly in historic buildings where wiring would not be practical or attractive. In 1981, the world's first wireless system was installed at a condominium complex in Boca Raton, Florida. There are many wireless products on the mar-

> **Fact**
> The first smoke detector was actually developed in the late 1930's by Swiss physicist Walter Jaeger. He was actually trying to develop a poison gas detector. The idea was deemed a failure at the time.

Fact

Microprocessor-based fire alarm systems are capable of providing much more information to the user than conventional (zoned) systems. This results in faster response times, which helps mitigate damage.

ket today. Some are only listed for household use and cannot be used in commercial applications.

Advancements in detection have spurred the latest changes in fire alarm technology. Detectors are now more sensitive, but they are less susceptible to nuisance alarms than ever before. Computer algorithms have enabled manufacturers to develop smart detectors that build a database of information over time that can be compared to the conditions being sensed. Smoke detectors containing laser-based detection were developed in the mid-1990s. But like the VESDA system, these are primarily developed for cleanroom applications. Additionally, multi-criteria detectors were also developed at that time and have helped to reduce nuisance alarms while providing quick responses. Multi-criteria detectors primarily use heat and smoke signatures to discriminate between real fires and transient conditions. Still, newer technologies may use smoke, heat, carbon monoxide, carbon dioxide, infrared, and other sensing technologies to ensure that the detector quickly responds to fires.

Another type of detection (projected beam) developed in the late 1980s solved the problem of protecting large open areas. Projected beam smoke detection involves the use of a transmitter that sends an infrared light beam across the pro-tected space to a receiver located up to 300 feet (100 m) away. Newer models can reach distances of 500 feet (140 m) and use multiple receivers. Projected beam detection is perfect for large open spaces like gymnasiums, atria, transportation centers, convention halls, and trans-portation centers. Newer models have eliminated the need for a separate receiver through the use of mirrors. Further advances have even incorporated motorized test lenses that eliminate the need for manual testing from a ladder or scaffolding. More recent developments in communications technologies have also led to the development of optical fiber cable communications between control panels. Optical fiber connections have increased bandwidth, which decreases the time required to transmit signals between panels, and have the added advantage of lightning resistance. Many manufacturers also provide remote monitoring or diagnostics through the Internet. These gateways can be installed in the controls, allowing licensed users to access the panel from outside the building.

Formation of Standards Developing Organizations (SDO)

The National Fire Protection Association (NFPA) was founded in 1896. By 1899, NFPA committees developed *NFPA 71-D, General Rules and Requirements for the Installation of Wiring and Apparatus for Automatic Fire Alarms; Hatch Closers, Sprinkler Alarms, and Other Automatic Alarm Systems and Their Manual Auxiliaries*. NFPA 71-D later became the basis for *NFPA 72, National Fire Alarm and Signaling Code*. The Underwriters Association of the Middle Department developed the first rules for electrical wiring in 1897. This standard was called *Installation of Wiring and Apparatus for Light, Heat, and Power*. This document became known as the *National Electrical Code®*

For additional information, visit qr.njatcdb.org Item #1003

Projected beam (linear) smoke detection is widely used to protect large open areas with high ceilings.

(NEC) and would later be published by NFPA. These documents and the advancement of organized training programs caused immense improvements in safety and reliability of building safety systems.

The Americans with Disabilities Act (ADA) became law in 1992 and was intended to provide equal access to buildings and facilities for all Americans. Where a building is equipped with an audible fire alarm system, the ADA generally requires visible signals to alert hearing-impaired occupants. It also requires manual fire alarm boxes to be accessible to all occupants, including those with physical handicaps. Until the late 1980s, visible signaling was achieved through the use of notification appliances that contained low-wattage incandescent flashing lights. However, these lights were largely ineffective because they were not sufficiently bright to alert an occupant unless he or she was directly viewing the appliance. Underwriters Laboratories conducted a series of tests in the late 1980s using hearing-impaired students to determine how much flashing light was required to alert occupants who were not viewing appliances. Those findings were incorporated into *NFPA 72-G, Recommended Practice for the Installation of Notification Appliances*.

By 1989, the original signaling document, *NFPA 71-D*, had grown into 11 documents, each covering a different type of technology. In 1990, the NFPA committees responsible for the development of standards relating to fire alarm systems decided to incorporate the 11 signaling standards into a single document. The consolidated document is known as *NFPA 72, National Fire Alarm Code*. This document has been revised several times since then, but the basic organization remains largely the same.

Changes to the 2010 edition of the *NFPA72* expanded the scope of the document to include all signaling for the protection of life and property. New technologies, such as directional audible signaling and narrow band signaling, have improved the capabilities of systems to alert building occupants to danger. Mass notification systems (MNS) can be used not only to alert people to the dangers of fire but can be used to warn of other life threatening events, such as a terrorist attack. With these additions, the title of the *Code* changed to the *National Fire Alarm and Signaling Code*. Other changes also permit new methods of communications, such as through the Internet. The 2013 edition of *NFPA 72* incorporates a number of changes, such as the relocation and consolidation of all requirements for circuits and pathways in Chapter 12. Other changes are discussed throughout this textbook.

BASIC TYPES OF FIRE ALARM SYSTEMS

There are five basic classifications (types) of fire alarm systems. These are:
1. Protected premises (local) systems
2. In-building fire emergency voice/alarm communications systems
3. Supervising station alarm systems (central station, proprietary station, and remote station)
4. Public fire emergency reporting systems/auxiliary fire alarm systems
5. Single- and multiple-station alarms and household fire alarm systems

Some of the individual classifications are broken down into sub-categories. **See Figure 1-1.**

Protected Premises (Local) Systems

Protected premises (local) systems are installed to protect commercial buildings. These systems can provide life safety, property protection, mission continuity, or environmental protection. They can also manage other building systems, such as suppression systems. Protected premises systems can be integrated with security and access systems; heating, ventilating, and air

Fact

The electrical safety system in the United States comprises installation codes and standards, product standards, and enforcement.

Figure 1-1	The Basic Types of Fire Alarm Systems			
System Type	Sub-category	Description	Typical Uses	Features
Protected Premises (Local) System		A system, using controls; system powered initiating devices; notification appliances; and control elements. Not connected to supervising station.	Commercial buildings such as hotels, warehouses, assembly occupancies, and office buildings	Provides life safety, property protection, and mission continuity.
In-Building Fire Emergency Voice/Alarm Communications System		A fire alarm paging system that uses amplifiers, speakers, and messages to direct occupants to safety.	High-rise, health care, assembly	Provides more information than a single-stroke notification system.
Supervising Station Alarm Systems	Central Station Alarm System	A centralized system that monitors fire alarm systems at protected premises, through communications with the property. Alarms are re-transmitted to fire department.	Used to monitor property conditions on a 24/7 basis. Often used for high cost assets, often required by insurers. Strict requirements for construction and operation.	Includes a contract that provides maintenance and testing. Somewhat expensive. Required to monitor all signals.
	Proprietary Supervising Station Alarm System	A centralized system that monitors fire alarm systems at protected premises, under a single ownership, through communications with the property. Alarms are re-transmitted to fire department.	Used to monitor property conditions on a 24/7 basis. Often used for high cost assets, often required by insurers.	Can monitor only properties under a single ownership. Required to monitor all signals.
	Remote Supervising Station Alarm System	A centralized system that monitors fire alarm systems at protected premises, through communications with the property. Alarms are re-transmitted to fire department.	Used to monitor property conditions on a 24/7 basis. Often used for high cost assets, often required by insurers. Reduced requirements for construction and operation.	Does not include maintenance and testing. Less expensive. Not required to monitor all signals.
Public Emergency Reporting System	Public Emergency Reporting System	A series of street boxes and conductors throughout the city. Master boxes are used to connect protected premises systems directly to the public fire dispatcher through the system.	Used for 24/7 coverage of street and master boxes.	Fast response to alarm signals. Trouble and supervisory not transmitted. Maintenance is expensive.
	Auxiliary Fire Alarm System	Connected to a public fire reporting system. Alarms are transmitted to the public fire dispatcher.	Used where immediate response is desired or required, such as a high-rise or health care facility. Cannot have an auxiliary system without a public fire reporting system.	Uses protected premises system tied to existing public fire reporting system. Immediate response. Trouble and supervisory not transmitted. Relatively inexpensive.
Single- and Multiple-Station Alarms and Household Fire Alarm Systems		Stand-alone units used for dwelling units only. Often battery powered or 120 VAC powered.	Provide life safety only. Not intended for property protection.	Inexpensive and easy to install. Low maintenance.

Figure 1-1. There are five basic types of fire alarm systems. Some types are further divided into sub-categories.

Fact

In-building fire emergency/voice alarm communications systems are one type of emergency communications system (ECS). These systems are used to direct recorded or live instructions to occupants during a fire or other emergency.

conditioning (HVAC) systems; or other systems. However, fire alarm systems are generally stand-alone fire alarm systems that may be connected to other systems through contacts or digital connections. Protected premises systems do not transmit signals outside the building served by the system.

In-Building Fire Emergency Voice/Alarm Communications Systems

In-building fire emergency voice/alarm communications systems (EVACS) are intended to provide voice instruction, either live or recorded, to aid occupants in the event of a fire or other emergency. Recorded messages are stored in the control unit memory and responders can also use a microphone to broadcast live messages. These systems are often integral to protected premises fire alarm systems controls. EVACS have large amplifiers in the control cabinet or distributed amplifiers connected to speakers located throughout the protected space. In-building emergency voice/alarm communications systems are commonly found in high-rise buildings, assembly occupancy (theatre or auditorium), transportation centers, health care facilities, and other occupancies, utilizing partial or selective evacuation, where total evacuation is impractical or impossible.

Large buildings, such as high-rises, are also frequently provided with two-way, in-building emergency communications systems, formerly called firefighters' telephone systems. As the name implies, they are used for fire department or responder communications. They are usually provided whenever an emergency voice/alarm communications system is provided, but building codes dictate their use based upon occupancy type or use group. Two-way in-building emergency communications systems are generally integrated with the system controls. These systems provide a means of communication between the main controls and remote points in the building. Telephone jacks or handset cabinets are located in or near stairwells, elevator lobbies, fire pump rooms, and elevator machine rooms. They are very useful when responder (fire department) radio traffic or building steel prevents effective radio communications.

Supervising Station Alarm Systems

Supervising station alarm systems are designed to monitor or supervise fire alarm systems from a central point. These systems are intended to provide 24-hour monitoring of a building system so that the appropriate authorities are notified in case of fire alarms or other signals.

Figure 1-2. Central Station

Figure 1-2. A central station monitors many facilities owned by separate entities, and may include subsidiary stations.

Supervising station alarm systems fall into one of three types:
1. Central station systems
2. Remote station systems
3. Proprietary supervising station systems

These systems all have similar architectures but have very different requirements, particularly in the operation of the system.

Central station systems can be classified as providing all elements of fire alarm coverage. These systems begin at the last device located at the protected premises and extend to the central station facilities. Central station systems are required to include a maintenance contract; they have the most restrictive requirements. Central station systems are the least common type of supervising station and typically include a binding contract between the building owner, the fire alarm installation company, and the central station. **See Figure 1-2.**

Remote station systems have lesser requirements and differ in many ways. For example, remote supervising stations are not required to have the extensive security measures required for central stations. Remote supervising stations are the most common type of supervising station. The remote supervising station uses a local facility (such as a 911 call center or fire department in a small community) as the remote station. **See Figure 1-3.**

Proprietary supervising station systems are owned and operated by the property owner. The proprietary supervising station and all buildings monitored by the system must be under a single ownership. For example, a college campus may choose to monitor the fire alarm systems throughout the campus from the security offices that are on-site and staffed 24/7. **See Figure 1-4.**

Supervising station systems often use telephone lines or other means to transmit signals between the protected premises and the central station.

Public Emergency Alarm Reporting Systems / Auxiliary Alarm Systems

Public emergency alarm reporting systems are commonly found in cities throughout the northeastern U.S. These systems were formerly called "public fire alarm reporting systems" and "auxiliary fire alarm systems." These systems were renamed in the 2010 edition of *NFPA 72* because of the addition of emergency communications systems to the *Code*.

Public emergency alarm reporting systems use miles of conductors and street boxes throughout the protected areas of a

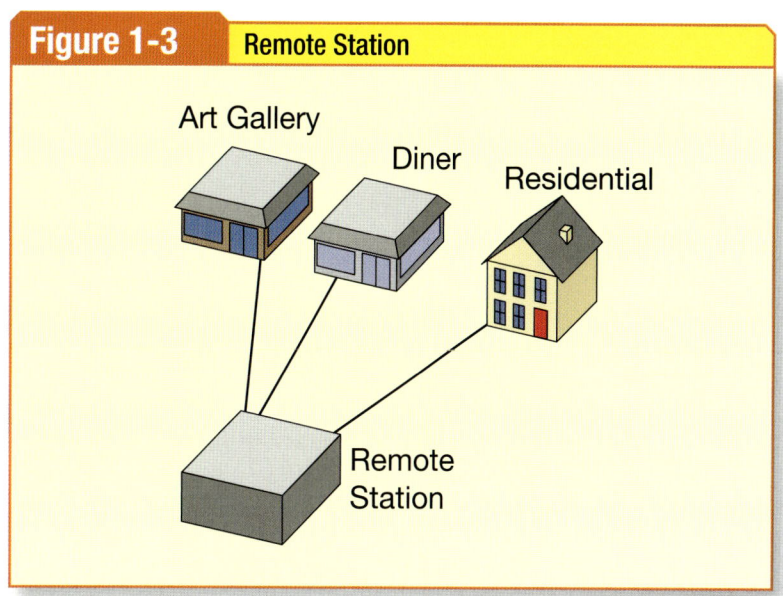

Figure 1-3. *Remote stations have fewer requirements than a central station.*

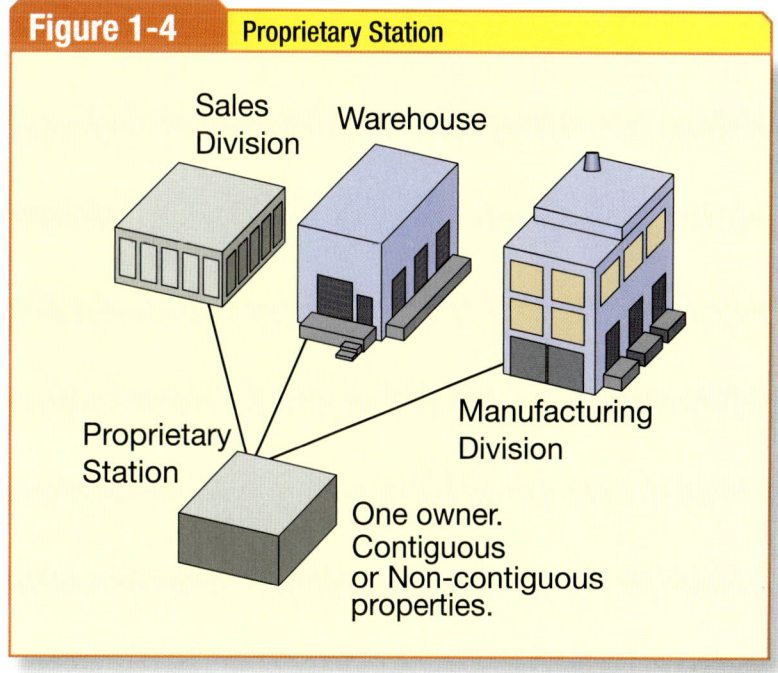

Figure 1-4. *A proprietary station monitors the facilities of a single organization.*

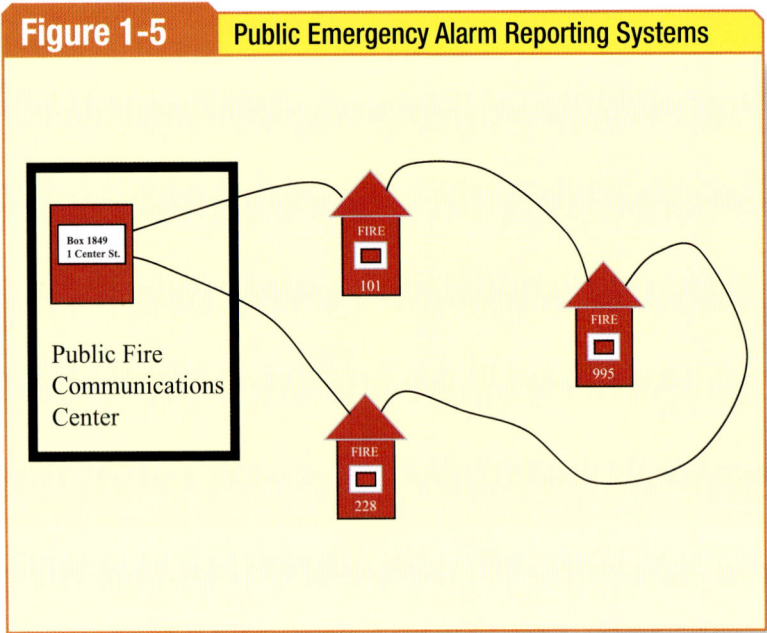

Figure 1-5. Public fire reporting systems use miles of conductors connected to numerous street boxes.

community. Master boxes have contacts that allow a protected premises system to communicate with the public emergency alarm reporting system. These connections are sometimes referred to as auxiliary alarm systems. **See Figure 1-5.**

Single and Multiple-Station Alarms and Household Fire Alarm Systems

Single- and multiple-station smoke and heat alarms are for use only within dwelling units and are intended to provide only life-safety protection. Property protection is not implied or intended by their use. In multi-dwelling properties, such as apartment buildings, the dwelling units are protected with single-station or multiple-station smoke and heat alarms. Common areas, such as corridors, are protected by protected-premises systems. Single-station smoke and heat alarms are self-contained stand-alone units, which are not intended to connect to system controls. Multiple-station smoke and heat alarms can be interconnected so that actuation of one unit causes all connected units to operate within a single dwelling unit. Smoke alarms can be battery powered or a combination of 120 volts alternating current (VAC) with battery backup.

Arrangements using system-powered detectors and fire alarm control units in dwellings are permitted in place of smoke and heat alarms, but cost usually precludes their use in this application. The common spaces of an apartment building are generally protected by a protected premises system or a supervising station system, using system-powered detection devices. Dwelling systems are not interconnected between individual dwelling units because nuisance alarms could cause frequent evacuations and because such interconnection is not permitted by *NFPA 72*. Dwelling and protected-premises systems are not usually interconnected for the same reasons.

OTHER EMERGENCY COMMUNICATIONS SYSTEMS

There are other types of emergency communications systems (ECS) permitted by Chapter 24 of *NFPA 72*. These include in-building mass notification systems, outdoor (Giant Voice) mass notification systems, and distributed recipient mass notification systems.

In-building mass notification systems (MNS) may be stand-alone or integrated with the fire alarm system. Where they are stand-alone systems, they may look much like a standard paging system. However, there are many new *Code* requirements that apply to these systems. Outdoor MNS may include large, high-power speaker arrays (HPSA) to broadcast signals to large numbers of people in an outdoor or urban setting. Many areas near Washington, DC, have outdoor MNS to alert large numbers of government workers in an emergency.

Distributed recipient MNS systems may include reverse 9-1-1, pagers, e-mail, text messaging, and variable highway signs. These systems are becoming widely used on college campuses following the Virginia Tech shootings. All of these systems are now under the scope of *NFPA 72*.

Control Units

All protected premises and supervising station systems utilize some type of control unit. Small properties may use a single control unit, but large properties may employ dozens of control units. The control unit is the brain of the system. The controls are generally provided

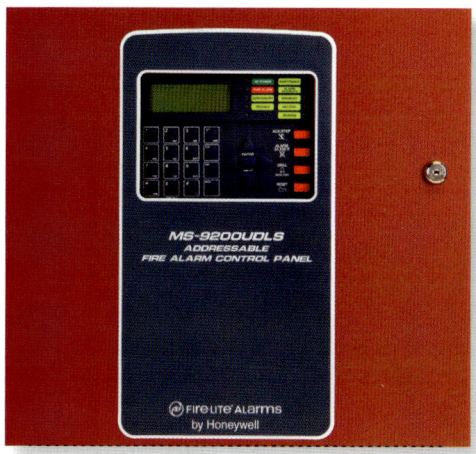

The fire alarm control unit is the brain of the system.

with 120- or 240-VAC power, but field circuits usually operate at either 12 or 24 VDC, nominal, though some field circuits may operate as high as 70 VDC. The control unit provides power to the field devices, monitors the wiring for opens and grounds, monitors its power supply, and provides control of fire-safety outputs. The control unit also provides a user interface to provide operators with system information. The user interface may consist of touch pads, switches, liquid crystal displays (LCD), printers, light-emitting diodes (LED), horns, buzzers, and lamps.

The fire alarm system controls are connected to field devices that provide both input and output for the system. The controls are also connected to transmitters, annunciators, and sometimes are connected to other fire alarm system controls. **See Figure 1-6.**

Microprocessor Controlled (Addressable) Controls. Addressable controls use microprocessors and software to operate the system. Every field device has a unique address that is identified at the panel. This address may be set by switches on the device base or on the device itself, depending on the manufacturer. Some manufacturers require a separate programmer to set the address device before installing it. The addressable controls will monitor the status of the field devices and wiring by periodically interrogating or polling each device. Once an input

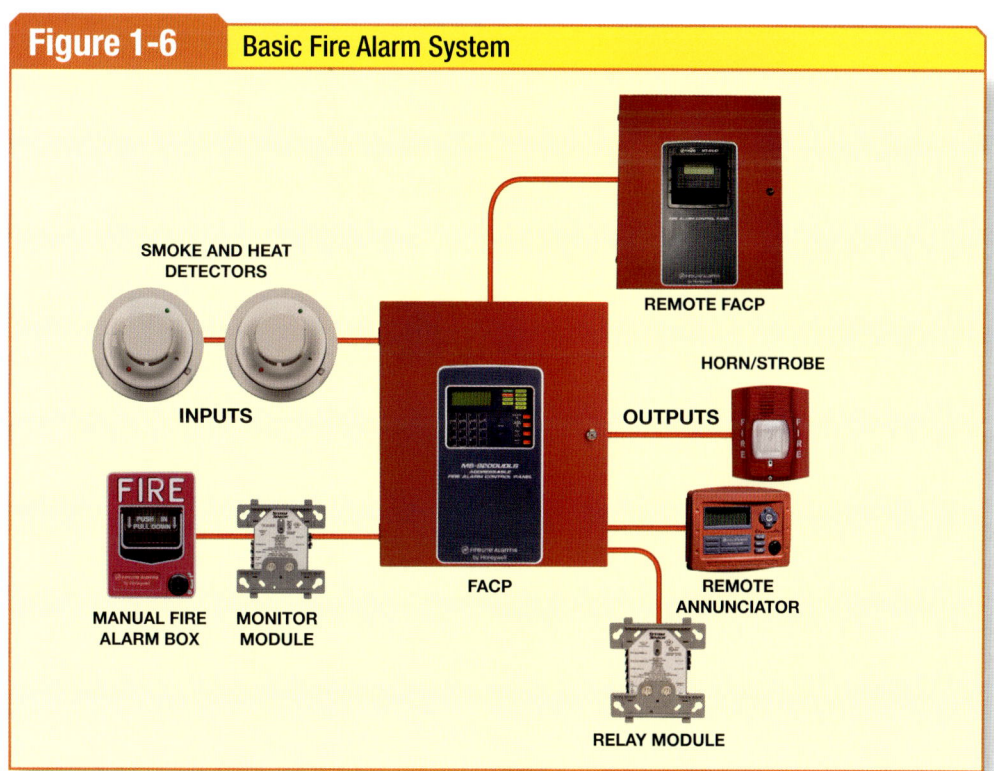

Figure 1-6. *Fire alarm controls connect to various input and output field devices, and sometimes connect to other control units.*

device reaches threshold level, the device will report its status and the software in the control unit will cause an alarm signal. The control unit knows what device has operated and usually displays the device address, physical location, and device type on an alphanumeric display. This feature greatly shortens response times. Addressable panels also record events in memory, which can be useful for troubleshooting. Paper printers are frequently used on large systems to permanently record system events.

Conventional Controls. Conventional controls contain no software, and the input (initiating) devices, rather than the controls, cause all signals. Devices contain a pair of contacts that are closed if the device is actuated, resulting in a signal at the control unit. The control unit can only display the circuit or zone that is actuated, and it does not know what device has operated. This type of system makes response more difficult.

Fire alarm systems may also actuate features to make the building safer in the event of a fire. These features are called emergency control functions and include the following:
- Door unlocking
- Elevator recall
- Elevator shutdown (shunt trip)
- HVAC shutdown
- Smoke control
- Stairway pressurization
- Door releasing service

Fire alarm controls may be a single, stand-alone panel or multiple panels interfaced together operating as a single system. The controls may also be interfaced with other systems, such as HVAC, burglary, or access control. Some manufacturers produce fully integrated fire alarm/access control/burglary systems, but the majority of fire alarms are of the stand-alone type. No matter what technology is used, a fire alarm system in a building must be designed and installed to operate as a single system. The designer will determine the type and capacity of the control unit, based upon many variables such as the number of field devices, cost, availability, and owner preference. Addressable technology has become very affordable, and it is often cost effective to install addressable systems even in small properties.

Initiating Devices

Input devices, called initiating devices, are connected to the control unit by field wiring extending throughout the building or protected space. Wireless systems use radio frequencies rather than conductors. Initiating devices can produce both alarms and supervisory signals, depending on the type of device and how the controls are programmed or wired.

Alarm-Initiating Devices. Alarm-initiating devices consist of heat detectors, smoke detectors (open area and duct-type), manual fire alarm boxes, radiant energy (flame and spark/ember) fire detectors, pressure-type sensing switches, and paddle-type water flow switches. **See Figure 1-7.** Alarm-initiating devices are wired or programmed to cause alarm signals, which are indicative of a fire condition.

Supervisory Signal-Initiating Devices. Fire alarm systems cannot extinguish fires, but they often supervise the conditions of suppression systems, which are essential for proper operation of a suppression system. Suppression systems can consist of wet pipe sprinklers, pre-action sprinklers, deluge systems, dry-pipe sprinklers, clean agent systems, dry and wet chemical systems, and foam systems.

Supervisory signal-initiating devices may consist of pressure switches, valve tamper switches, water level switches, and temperature switches. Supervisory signal-initiating devices are wired or programmed to cause supervisory signals, which indicate an off-normal condition on a suppression system. For example, a valve supervisory switch senses valve position and causes a supervisory signal when the valve is in a position other than fully open. **See Figure 1-8.**

Conventional Initiating Devices. Conventional initiating devices are generally simple contact closure devices. They are designed to cause an alarm by closing contacts on an initiating device circuit (IDC) operating at 12 or 24 VDC. Once actuated, the device causes a

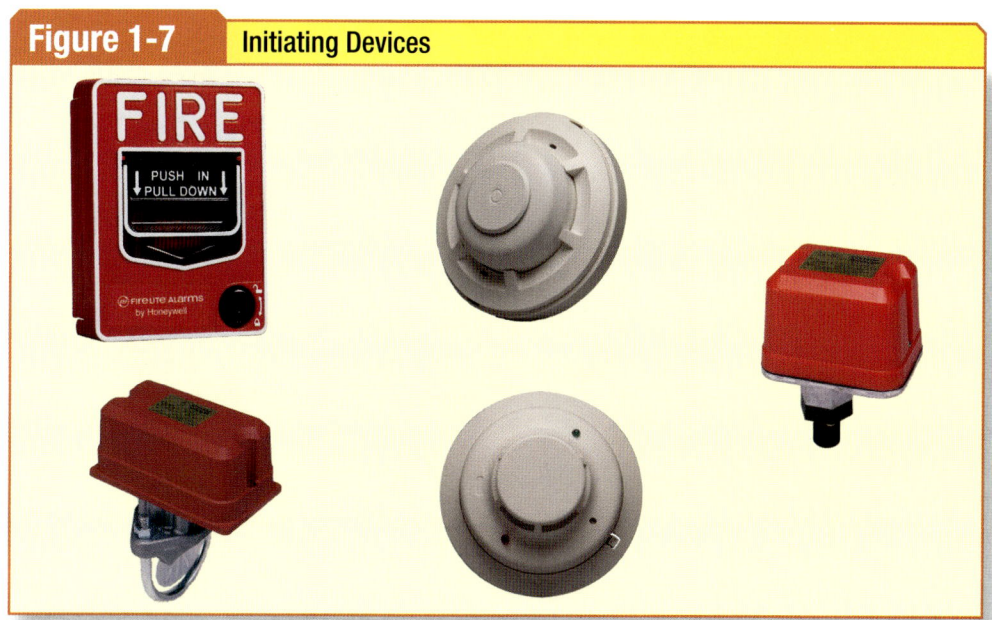

Figure 1-7. Smoke detectors and other initiating devices come in many forms.

voltage change on the IDC, which in turn is sensed as an alarm by the controls. The controls report the alarm by zone only and do not report by individual device.

Addressable Initiating Devices. Addressable initiating devices operate much differently from conventional devices. The device is connected to a signaling line circuit (SLC) and communicates with the controls using a proprietary language. Each device on the SLC has a unique address. Addressable initiating devices send information to the controls, and the decision to alarm is made by the controls, not the device. The controls report exactly which device is at alarm threshold and the physical location of the device in the building.

Signals

There are three types of signals produced by a fire alarm system: trouble signals, supervisory signals, and alarms. Trouble signals are indicative of a fault in the wiring or power supply, or a failure of a component. Supervisory signals indicate an off-normal condition of a suppression system, but may be used for duct smoke detector actuation in some cases. Alarm signals indicate a fire condition or other life-threatening condition. Alarm signals always have priority over all other signals, even

burglary hold-up alarms. In fact, hold-up alarms may only have priority over supervisory signals with permission of the authority having jurisdiction (AHJ).

Notification Appliances

Notification appliances are used to alert occupants of a fire or other life-threatening event. This is generally referred to as

For additional information, visit qr.njatcdb.org
Item #1004

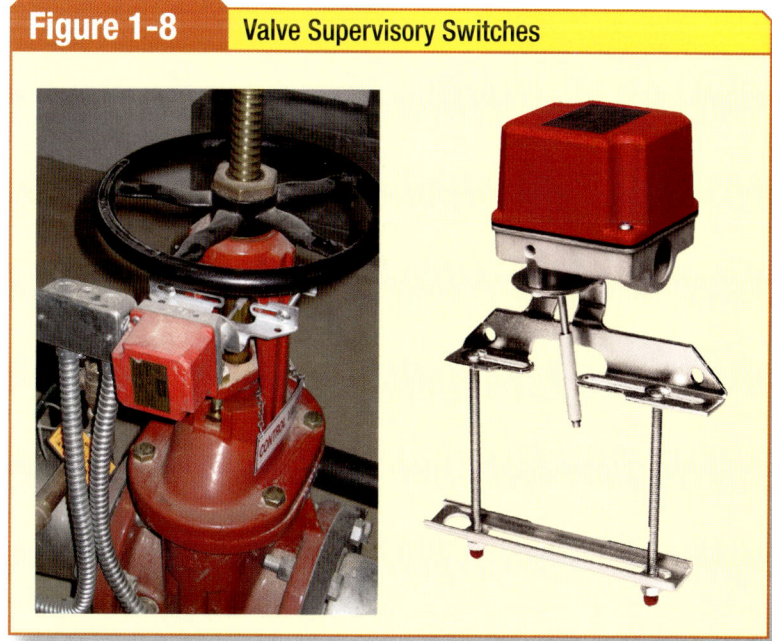

Figure 1-8. The valve on a supervisory switch sends a signal when it is not fully open.

Figure 1-9 Notification Appliances

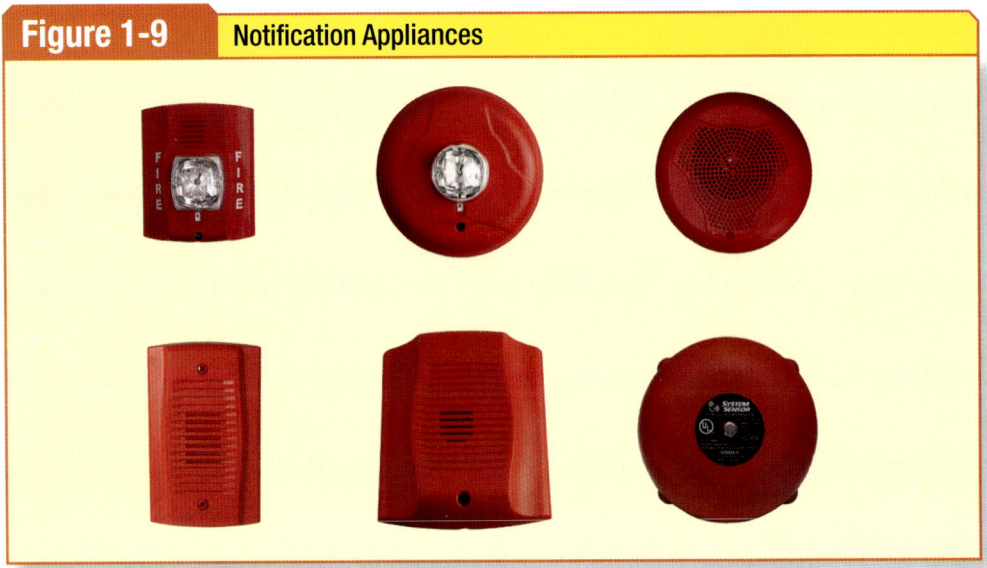

Figure 1-9. *Some notification appliances combine audible and visible signals.*

public mode notification. Typical notification appliances include horns, bells, chimes, sirens, speakers, and strobes. **See Figure 1-9.**

Some notification appliances are integral with the controls. They are used to alert operators of an alarm signal or maintenance-related signals such as trouble and supervisory signals. Public mode occupant-notification appliances are installed throughout an occupied area, based upon their coverage capability. Operators and responders will be notified of supervisory signals by audible and visible signals at the control unit or remote annunciator. Trouble signals are usually annunciated at the control unit or remote annunciator. **See Figure 1-10.**

Figure 1-10 Control Unit and Remote Annunciator

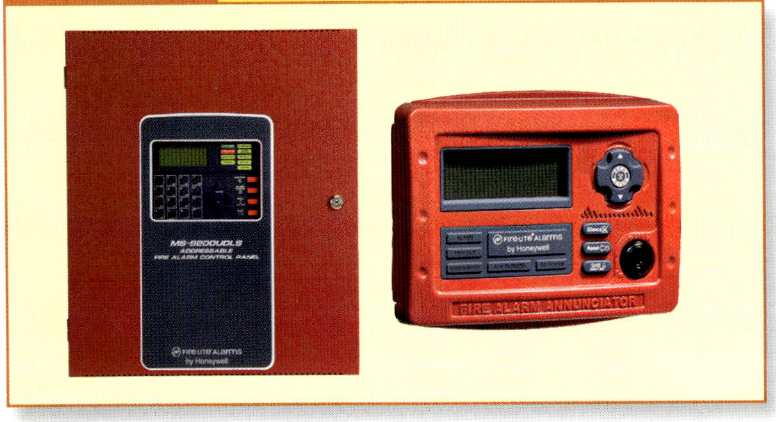

Figure 1-10. *Typical operator controls may include an LCD, among other operator interfaces.*

Graphic annunciators are a form of notification appliance and are frequently used to show the building or campus in a graphical, floor plan, or site plan format. These are especially useful in large or complex buildings. They generally have a floor plan of the building or campus showing fire alarm zones and the types of devices that are actuated. Many local jurisdictions require graphic annunciators for all buildings exceeding a certain area or height. **See Figure 1-11.**

CIRCUIT TYPES

Fire alarm circuits are those circuits that connect field devices to the system controls. They are powered, monitored, and controlled by the fire alarm system control unit. Most fire alarm controls utilize three basic types of circuits: initiating device circuits (IDCs), signaling line circuits (SLCs), and notification appliance circuits (NACs).

Initiating Device Circuits

Initiating device circuits are conventional zone-type circuits to which only conventional initiating devices are con-

nected. Manufacturers produce modules or zone cards, which are connected to the controls in a modular fashion. Each initiating device zone requires a separate card or module. Actuation of an initiating device on an IDC results in a "zone" indication, where the responders must search an area for the actuated device. **See Figure 1-12.**

On an IDC, there is usually an "end-of-line" (EOL) device, which sets up a voltage on the loop. Most manufacturers use resistors as EOL devices, while at least one manufacturer uses capacitors. The value of resistance or capacitance depends on the product. Most initiating device circuits on the market today operate at either 12 or 24 VDC. They use a simple contact closure on the circuit to

Figure 1-11 Example of a Graphic Annunciator

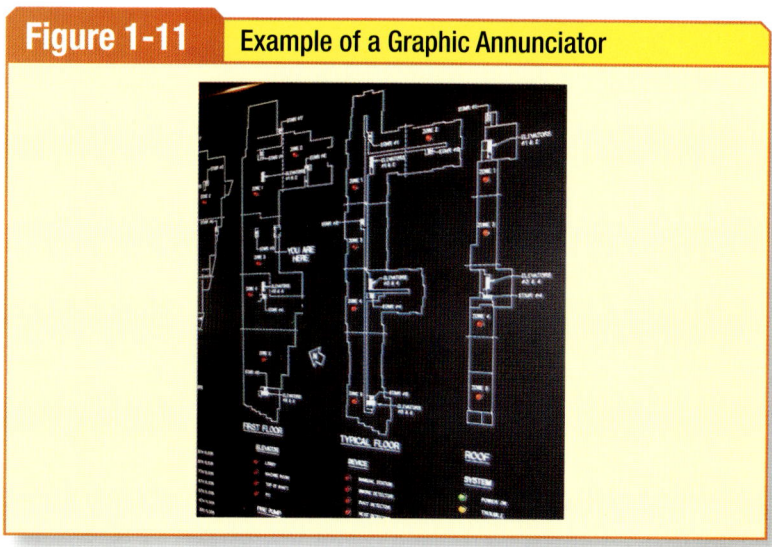

Figure 1-11. *Graphic annunciators provide a means of locating the origin of signals in large or complex buildings or compounds.*

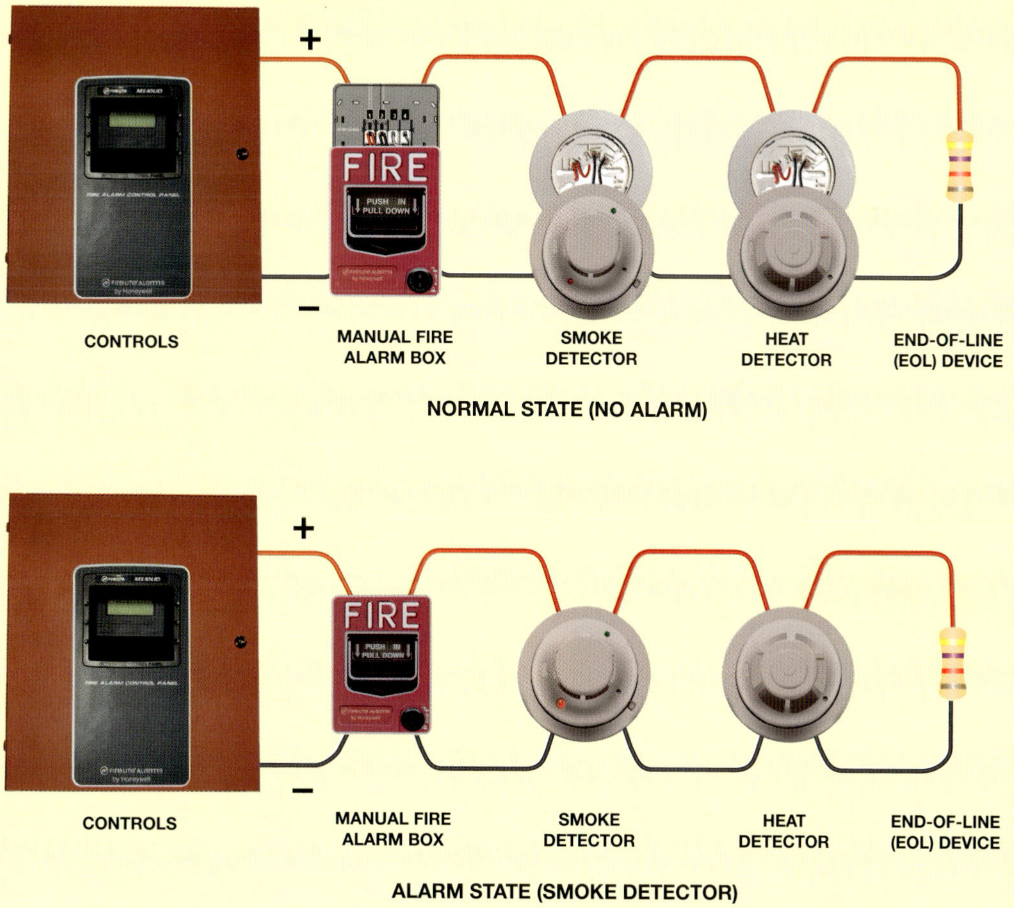

Figure 1-12 Initiating Device Circuits

Figure 1-12. *Initiating device circuits are conventional zone-type circuits.*

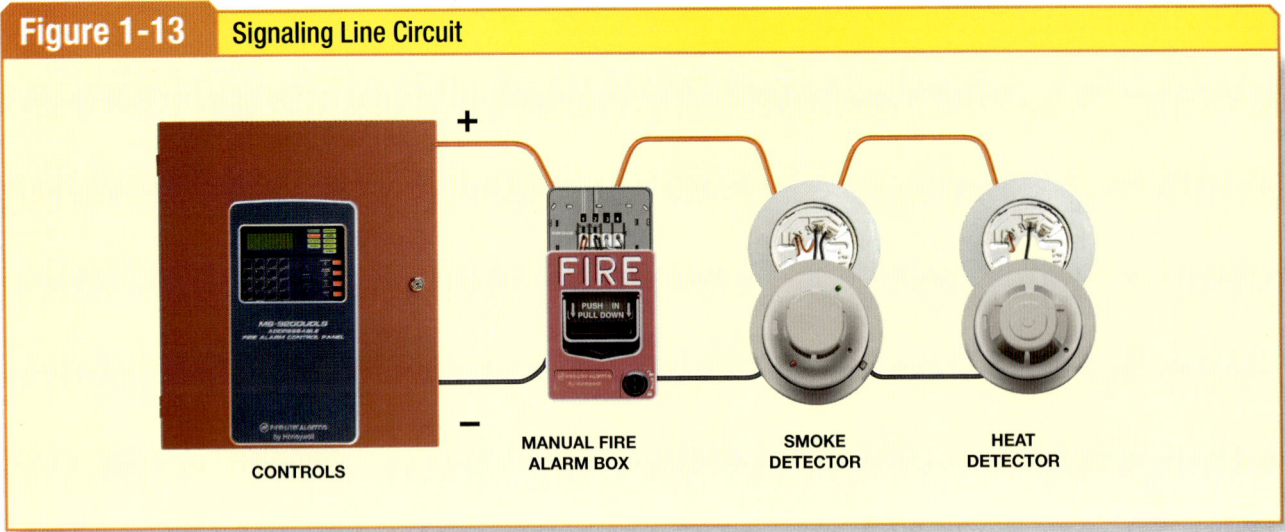

Figure 1-13. SLCs communicate with connected devices in a two-way manner.

create an alarm. The voltage at the controls drops suddenly when contacts are shorted, which is sensed as an alarm.

Signaling Line Circuits

Signaling line circuits are used to connect addressable initiating devices and other addressable devices and equipment with control units. An SLC can also be connected to addressable control relays, other compatible control units, and other addressable devices used on the system. Each device or component on the SLC has a unique address and is programmed with a specific physical location or function. When a detector on an SLC reaches threshold limits set by the software, the system software creates an alarm.

Signaling line circuits communicate with the devices on the circuit in a two-way or "multiplex" manner. Most SLCs operate between 12 and 35 VDC, depending on the product. **See Figure 1-13.** Each manufacturer uses a proprietary "language" or method of communication. Because the means of communication is proprietary, all

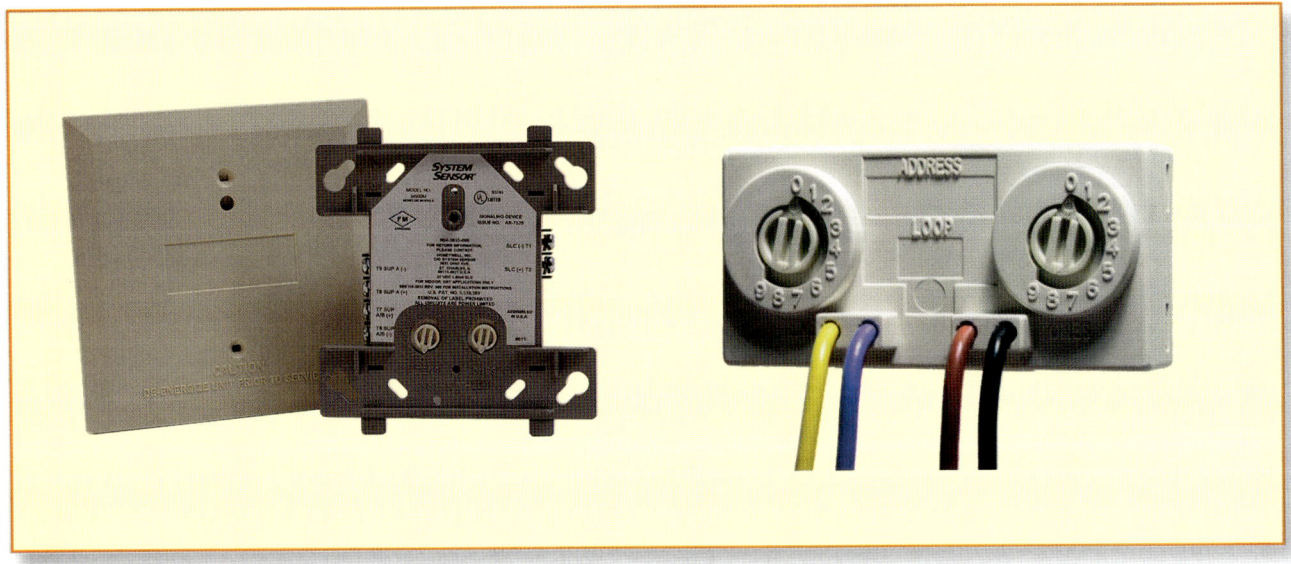

Addressable devices must be listed as compatible with the fire alarm system that they are connected to.

devices and equipment used on an SLC must be listed as compatible.

Notification Appliance Circuits

Notification appliance circuits are output circuits that are used to connect notification appliances to the system controls. NACs are either in a normal state (OFF) or an alarm state (ON). They operate at either 12 or 24 VDC, depending on the product. Notification appliances are connected to an NAC in a parallel fashion and contain an internal blocking diode to prevent monitoring current from passing through the appliances. In an alarm state, NACs switch polarity and allow current to pass through the appliances. **See Figure 1-14.**

SUPPRESSION SYSTEMS

Most suppression systems utilize flow detection devices and valve supervisory (tamper) devices installed on control valves. Some suppression systems, like pre-action systems, use detection devices and controls to operate main charging valves, which can prevent accidental discharge of the suppression system. Other supervisory initiating devices can supervise low or high pressure, low room temperature, water level, and fire pump conditions. Fire alarm systems often supervise these and other critical control valves and other conditions with supervisory signal–initiating devices. Wet sprinklers are the most common type

Fact

Suppression systems, such as sprinklers, are designed to protect the building, rather than the occupants and contents. However, 98% of all fires are controlled or extinguished with two or fewer sprinkler actuations.

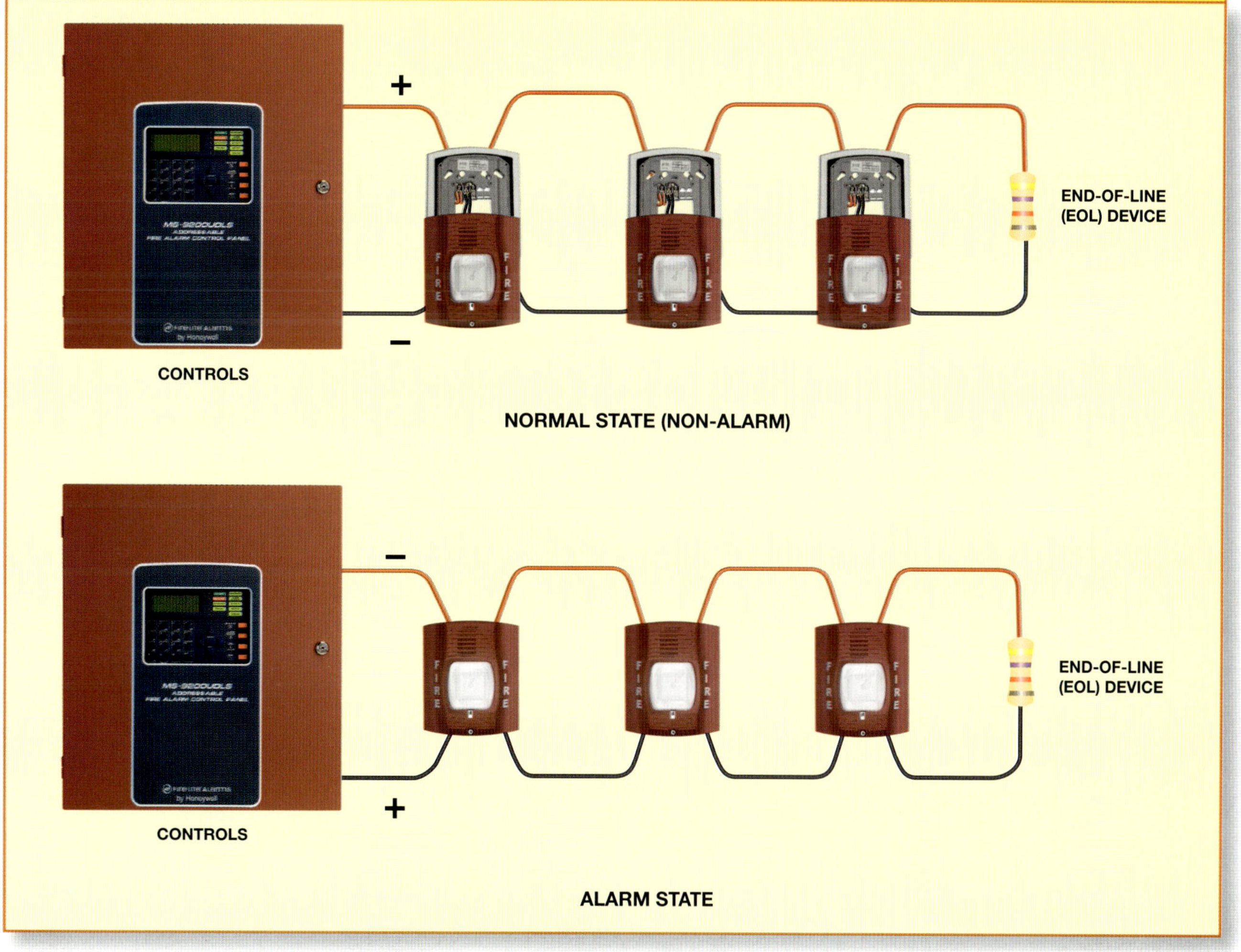

Figure 1-14. NACs connect notification appliances to system controls.

Figure 1-15: Suppression System Types

Type	Description	Typical Uses	Features	Conditions Supervised
Wet Pipe Sprinklers	A system that has piping extending throughout the protected area. Sprinklers actuate to allow water to be applied to the fire.	Ordinary buildings not subject to freezing, such as warehouses, office spaces, restaurants, etc.	Relatively low initial cost and maintenance	Water flow, Valve position, Fire pump trouble
Dry-Pipe Sprinklers	Sprinkler piping filled with pressurized air in protected space.	Areas subject to freezing, such as parking structures and freezers	Higher initial cost	High/low air pressure, Water flow, Valve position
Pre-Action Sprinkler System	System that requires actuation of detection devices to flood pipes in the protected space.	Computer rooms or places where nuisance operation would cause large loss or disruption	Expensive. Requires detection devices and control unit.	Water flow, Valve position, Control trouble
Deluge System	System that uses open sprinkler heads to flood space quickly.	High-density warehouses and areas with explosion hazards	Quick response. Very large water flow. Requires detection devices and control unit	Water flow, Valve position, Control trouble
Gaseous Suppression Systems (FM-200®, CO_2, Halon)	System with storage tank(s) of inert suppression agent connected to piping and discharge nozzles in protected space.	Vaults, computer mainframes and electronics, high-cost assets	Extinguishes small fires. Designed to protect valuable assets, not the building.	Control trouble, Valve position
Foam Systems	System designed to fill the protected space with foam suppression agent through generators, piping, and nozzles in protected area.	Tire warehouses, aircraft hangars, and other large spaces with special hazards	Fast response. Expensive. Foam is corrosive	Control trouble, Valve position, Alarm signals on flow
Wet/Dry Chemical Systems	System with storage tanks of chemical agent connected to piping and nozzles to protect the hazard.	Kitchen hoods, deep fat fryers, gasoline dispensing stations	May be automatically operated by detection or manual operation by cable pull. Relatively inexpensive.	Alarm contacts on trip head

Figure 1-15. There are many types of suppression systems and attachments.

of suppression system installed. **See Figure 1-15.** Other suppression systems include:
- Dry sprinkler systems
- Pre-action sprinkler systems
- Deluge sprinkler systems
- Wet/dry chemical suppression systems
- Gaseous suppression systems

Dry sprinkler systems have no water in the sprinkler piping. Instead, the piping is filled with compressed air to allow the piping to be extended into areas subject to freezing. A dry valve holds back the water. The air pressure must be monitored by the fire alarm system to ensure the air pressure is adequate to hold back water. When a sprinkler actuates, the air pressure drops and allows the dry valve to trip. Water can then enter the piping. When a sprinkler fuses (opens) because of heating by a fire, water will flow onto the affected area.

Pre-action systems are designed to protect areas where wet sprinklers are undesirable. Pre-action sprinklers most often use an interlock that requires the actuation of two automatic detectors in order to operate a charging valve. Once the charging valve is opened, water may flow into the piping. However, water will only flow once a sprinkler is actuated. This arrangement generally uses engineered systems, and the monitored equipment includes detection devices, isolation valves, and releasing control units. Pre-action systems are often used in data storage centers or computer server rooms.

Deluge sprinkler systems operate similarly to pre-action systems, but the installed sprinklers are open and will flow water immediately upon being pressurized. Monitored equipment includes initiating devices, isolation valves, and releasing control units. Deluge systems are used in high-hazard facilities, like munitions manufacturing.

Wet and dry chemical suppression systems include kitchen hoods and are usually manually operated, but they may also operate automatically by fusible link. A small switch operates when the system is actuated. This switch can be installed to cause an alarm signal when the system operates. The system has no other devices that are monitored.

Gaseous suppression systems are frequently used in computer rooms because they use a suppression agent that will not damage the sensitive electronic components. They contain a tank of compressed gas, such as FM-200® or Inergen®, held back by a control valve. The valve opens when two or more detection devices actuate in the protected space. Monitored equipment includes detection devices and releasing control equipment. Gaseous suppression systems are often used in computer rooms or data centers.

CODES AND STANDARDS

Codes and standards are an essential part of the electrical and life-safety industry in the United States. Codes and standards in the United States were developed as a direct result of large fires in New England near the end of the nineteenth century. Codes and standards establish minimum levels of performance and safety for the industry. There are many codes and standards in use today. These include product standards for nearly every product manufactured including chain saws, refrigerators, and fire alarm system components. Qualified testing laboratories, such as Underwriters Laboratories, develop and use these product standards to ensure product safety. Standards also include installation-related documents, such as the *National Electrical Code*. Without codes and standards, there would certainly be more fires and deaths from fires.

Codes and Standards Development in the U.S.

In the late nineteenth century, the northeastern United States was the center of the U.S. Industrial Revolution. Rapid growth of manufacturing facilities at this time was fueled by readily available hydroelectric power. Unfortunately, poor workmanship was the cause of many electrical fires in the burgeoning electrical industry. Additionally, many fire sprinkler systems installed to protect these facilities failed to operate properly, also because of shoddy installations and poor designs. The combination of poor electrical work and poor fire alarm installation often resulted in large losses due to fires.

Creation of NFPA and Underwriters Laboratories (UL). The National Fire Protection Association was founded in 1896 to address the losses suffered in industrial facilities. It developed standards related to sprinklers and fire alarms. The Underwriters Association of the Middle Department developed the first edition of the document that was to become the *National Electrical Code* (*NEC*) in 1897. NFPA became the steward of the *NEC* in 1911 and has remained the developer since that time.

In addition to NFPA and the Underwriters Association of the Middle Department, Underwriters Laboratories (UL) was founded in 1894 to address product safety. It developed product standards and tested products to these standards. To this day, UL continues to develop product standards for sprinklers, fire alarm equipment, electrical equipment, toasters, chainsaws, floatation devices, and most everything in between. Many of the standards developed by UL and NFPA over a hundred years ago still exist today, though they have been revised many times to keep up with new technology. These standards provide the very basic minimum installation requirements for proper operation of a system.

For additional information, visit qr.njatcdb.org Item #1533

Key Fires Affecting Code Development. Several large disastrous fires, such as the Iroquois Theater fire in Chicago (1903), the Cocoanut Grove fire in Boston (1942), the MGM Grand fire in Las Vegas (1980), and more recently the Station Nightclub fire in Warwick, Rhode Island (2003), have significantly contributed to code development. Through lessons learned, these and other events have resulted in many changes to building and safety codes. Devices like panic hardware, fire separations, automatic sprinklers, and automatic door unlocking controls have become standard building features as a result of these tragedies. Fires are still prevalent in the developing world, primarily because codes and standards are not widely used there. This was evidenced in November 2012 in Dhaka, Bangladesh, where more than 200 workers were killed in a factory fire.

Building Codes. In the early twentieth century, there was little code enforcement except in large cities and industrial establishments. Many jurisdictions, such as Chicago and New York, developed their own building codes to address safety concerns. Many cities followed their examples, but eventually there would be only four major building code developers in the U.S. Other standards and codes were also developed following the turn of the century as America grew. Safety codes for many other building systems, such as elevators, were also developed at this time.

Today, almost every jurisdiction in the U.S. adopts some sort of building code. Most jurisdictions in the U.S. adopt the *International Building Code* (*IBC*), while some adopt *NFPA 101, Life Safety Code*, or *NFPA 5000, Building Construction and Safety Code*. These codes reference other applicable codes and standards such as *NFPA 72, National Fire Alarm and Signaling Code*, and *NFPA 13, Standard for the Installation of Sprinkler Systems*. Additionally, *NFPA 70, National Electrical Code*, is adopted in all 50 states and is used in 16 foreign nations. Product standards are, in turn, referenced by codes and standards that regulate the installation of the applicable products.

The *International Building Code* was developed in the late 1990s. Many of the requirements in the *IBC* were taken from three other documents used in three different geographical regions. The three codes that formed the basis of the *IBC* were the *Standard Building Code*, *BOCA Building Code*, and the *Uniform Building Code*. The International Building Code Council consolidated these documents to form the 2000 *IBC*.

Shortly after the development of the *IBC*, the National Fire Protection Association developed *NFPA 5000, Building Construction and Safety Code*, to compete with the *IBC*. *NFPA 101, Life Safety Code*, is not a building code, but it contains many similar requirements to building codes, particularly in the area of egress, fire separation, and fire protection systems.

Modern codes and standards relating to fire protection generally focus on limiting risk in buildings. They accomplish this in several ways:
- By limiting height and area and requiring fire/smoke separation barriers
- By requiring a minimum number of exits and maximum occupant loading
- By requiring automatic sprinklers in certain occupancy types or use groups
- By requiring detection and alarm systems for early warning and occupant notification
- By mitigating hazards associated with equipment or processes, such as electricity or natural gas

The safety system in the United States comprises codes and standards, product standards, and enforcement. Without these three aspects, buildings would not be as safe as they are. All three elements of this system are needed to ensure that listed/approved products are installed to minimum performance requirements.

Standards Developing Organizations. Most codes and standards are revised every three years. This allows changes to be

Fact

High-profile fires in the United States have shaped the codes we use today. Without our safety system, fire losses would be significantly higher.

introduced into the *Code* in a timely fashion. Changes to codes often involve new technology, new methods, technical changes based upon field problems, and administrative (non-technical) changes. All standards developing organizations allow public input during their revision cycles.

Examples of American SDOs include, but are not limited to the following:
- National Fire Protection Association (NFPA)
- International Code Congress (ICC)
- Underwriters Laboratories (UL)
- FM Global (FM)
- National Electrical Manufacturers Association (NEMA)
- American Society of Mechanical Engineers (ASME)
- Institute of Electronics and Electrical Engineers (IEEE)
- National Electrical Contractors Association (NECA)
- American National Standards Institute (ANSI)

The American National Standards Institute is the overseer of codes and standards development in the United States. Standards developing organizations, such as the NFPA, must conform to ANSI procedures in order to become accredited. Accreditation helps ensure fair processes which result in good documents. Standards developers that become ANSI accredited must undergo frequent audits to ensure compliance with ANSI regulations. ANSI regulations require all ANSI-approved documents to be revised at intervals not exceeding five-year periods. Many committees place their documents in three-year cycles to keep pace with technical advances.

Most standards developers, such as the NFPA, use volunteer technical committees to develop codes and standards. The committees must approve all changes in a consensus-based process. However, others, such as Underwriters Laboratories, use a canvass method, where the volunteers are solicited for input to a particular code or standard, but staff liaisons are primarily responsible for the content of the documents. In most cases, expert volunteers donate considerable time and expense to assist in the development of standards.

Some SDOs practice a consensus-based approach where committees are balanced. In this case, a single interest group, such as manufacturers, cannot hold more than one-third of the seats on the committee. This prevents any group from having a controlling interest on the committee. Other standards developers do not take the same approach and allow majority control of their committees. The American National Standards Institute accredits many SDOs that participate in a consensus approach.

Not everybody can participate in committee activities because of limits on committee size, cost, or schedules. For individuals who cannot participate in committee activities, there is still a way that will allow them to help develop new documents. These individuals can submit proposals or comments to the committee. Submitting proposals and comments is often a rewarding experience, especially if the committee accepts the proposed change. Committees do not always agree with the submitter, but they will provide reasons for the disagreement.

Using Codes and Standards

Codes and standards are different. A simplistic, but easy way of remembering the difference between codes and standards is that codes specify "when and where" and standards specify "how." Codes usually contain text that requires a minimum level of coverage or type of protection in a particular occupancy type or use group. Code requirements are usually based upon the occupancy type, occupancy loading, and the hazards (risks) involved. Most codes have chapters or sections that apply to specific use group or occupancy classifications. Some codes may also have a chapter that provides separate or additional requirements for fire protection systems, such as fire alarm and sprinkler systems. This arrangement exists because not all occupancies require the same level of protection. For example, a single-story mercantile occupancy, such as a convenience store, will not pose the same level of risk as a high-rise hospital.

Adoption of Codes and Standards. Codes are generally suited for outright adoption by a local jurisdiction or government agency. The two basic methods of adoption

Fact

One of the best ways to effect changes to the codes and standards which affect our work is to become a member of the technical committee or code-making panel. This requires a commitment of time and expense. The next best alternative is to submit proposals and comments through the process.

Fact

NFPA 72, National Fire Alarm and Signaling Code formerly consisted of eleven different standards. These standards were consolidated into a single document with the title *National Fire Alarm Code* in 1993.

are by transcript and by reference. Adoption by transcript means the local authorities or legislature prepare and pass legislation that adopts a certain document as law. Adoption by reference means that another code provides a reference to the document. For example, the *International Building Code* references *NFPA 72, National Fire Alarm and Signaling Code*, and requires fire alarm systems in certain occupancy use groups. Where the *IBC* is adopted by transcript, *NFPA 72* is also adopted and has the same weight as if adopted by transcript. Many jurisdictions develop changes, or amendments, to adopted codes and standards to suit local needs.

Standards generally specify installation methods, equipment requirements, system performance, or maintenance. Standards generally do not specify requirements based upon occupancy. Rather, they provide the detailed how-to requirements for a specific technology. This category also includes product standards used to verify that products are safe. An example of a product standard is UL 268, *Smoke Detectors for Fire Alarm Signaling Systems*. Many standards are referenced by codes or other standards, thereby enabling them to be enforced.

NFPA 72, National Fire Alarm and Signaling Code, primarily contains standards-like requirements, such as detector spacing or secondary power requirements. However, Chapter 29 of *NFPA 72* contains requirements for smoke alarms and fire alarm systems in dwelling units. These requirements also exist in *NFPA 101* and *NFPA 5000* but are extracted in *NFPA 72* for ease of use. There are other code-like requirements in *NFPA 72*, particularly in the area of visible signaling in high-noise ambient environments. However, *NFPA 72* does not generally indicate where fire alarm systems are required. *NFPA 72* is called a code because it contains some code requirements and covers a broad range of subjects. Because *NFPA 72* primarily contains installation and performance requirements, applicable building codes must first be consulted to determine the fire alarm system requirements based upon occupancy or use.

In addition to codes and standards, some standards developers also publish recommended practices and guides. These documents are not enforceable because they contain no mandatory text. Recommended practices and guides often reflect industry standards or good practice. In order to develop these documents into standards or codes, technical committees are required to have technical substantiation supporting the changes to more restrictive requirements. Technical substantiation may not be available because of funding or a lack of published studies.

The national codes presented in this chapter, including *NFPA 72*, are not always adopted by local jurisdictions as they are written. Many local jurisdictions amend codes at a local level to suit local needs. The District of Columbia, for example, has a local amendment to Article 760 of the *NEC* requiring metallic cable or conduit on horizontal runs of fire alarm circuits in high-rise buildings. The designer and installer must be aware of all local amendments. Most jurisdictions publish local amendments in hard copy or on their websites.

Some jurisdictions do not adopt nationally developed codes. Instead, they develop and adopt local codes. For example, the city of Chicago has locally developed building and electrical codes. Locally developed codes and standards are often based upon national codes and contain many requirements to suit local needs or interest groups. One should always be familiar with codes in his or her area before designing or installing any system.

Enforcement of codes and standards is often conducted by both public and private authorities having jurisdiction (AHJ). Public authorities having jurisdiction may include the fire marshal, electrical inspector, and a building code official. In some cases, the AHJ may be a third party that acts on behalf of the jurisdiction. Other authorities having jurisdiction may include insurance interests or the building owner. However, the public AHJ may only enforce the requirements contained in the applicable codes and standards. Other authorities having jurisdiction (e.g., insurance interests) may enforce the additional requirements they impose. In some cases, a third party may be contracted to ensure compliance with these additional requirements. A typical third party might be a consulting firm specializing in the type of system being installed.

Codes and standards contain minimum requirements. Good designers generally develop additional requirements, based upon fire protection goals. The goals are developed in concert with the owner, insurance interest, authority having jurisdiction, and other stakeholders. The additional requirements are usually contained in project specifications that supplement the applicable codes.

Example of Code Requirements

To see how the codes and standards get applied in real life, consider the following example of the applicable fire alarm requirements for a 10-story high-rise office building, using the *IBC* and *NFPA 101*. Assume the building has elevators, and a large (over 20,000 CFM) rooftop HVAC unit serving the entire building. Both *NFPA 101* and the *IBC* define high-rise as having occupiable levels more than 75 feet above the lowest level of fire department access.

IBC, Chapter 9, "Fire Protection Systems," Section 907 requires an alarm and detection system with smoke detectors in all electrical and mechanical rooms. The *IBC* requires the building to have a sprinkler system throughout, so water flow and tamper switches will be required. Since there are elevators, the *IBC* references ASME A17.1, *Safety Code for Elevators and Escalators*, which requires smoke detectors in the elevator lobbies and elevator machine room. Because the building is a high-rise, an in-building fire emergency voice/alarm communications system is required, which requires speakers and amplifiers to be installed. Visible signaling in all common areas is required, using strobes. In addition, the *IBC* references the *International Mechanical Code* (IMC), which requires HVAC shutdown for HVAC systems over 2,000 CFM. **See Table 1-1.**

Occupancy/Use Group	Use Group B and High Rise
Applicable Sections	907
High-Rise?	Yes
Alarm/Detection System Required?	Yes
Sprinklers Required?	Yes
Elevator Recall?	Yes, per ASME A17.1
Automatic Detection Required	Sprinkler W/F and tamper; electrical and mechanical rooms, elevator lobbies, elevator machine room
HVAC Shutdown?	Yes, per *NFPA 90A* for systems over 2,000 CFM
Occupant Notification?	Audible/Visible
EVAC System Required?	Yes
Firefighters' Telephones?	Yes
Annunciator?	Yes
Emergency Forces Notification?	No

Table 1-1. *The IBC has many fire alarm requirements, especially in Section 907.*

Example of Code Requirements (continued)

Under *NFPA 101*, the building is classified as both new business occupancy (Chapter 38) and high-rise (Chapter 11). Section 38.3.4 requires an alarm and detection system in all business occupancies over two stories or occupant loads greater than 300. Therefore, a fire alarm system is required. Section 38.3.4 permits initiation by manual, full coverage of automatic detection, or by sprinkler waterflow in accordance with Section 9.6. Section 9.6 in turn references *NFPA 72, National Fire Alarm and Signaling Code*, for the installation requirements. Section 11.8 requires sprinkler protection in all high-rise occupancies; therefore, detection is not required throughout the building. *NFPA 101* references ASME A17.1, *Safety Code for Elevators and Escalators*, which requires smoke detectors in the elevator lobbies and elevator machine room. Section 11.8 also requires an emergency voice/alarm communications system, which requires speakers and amplifiers to be installed. Visible signaling in all common areas is required, using strobes. *NFPA 101* also references *NFPA 90A, Standard for Ventilating Systems*, which requires HVAC shutdown for HVAC systems over 2,000 CFM. **See Table 1-2.**

Occupancy/Use Group	Business and High Rise
Applicable Sections	Chapters 9, 11, and 28
High-Rise?	Yes
Alarm/Detection System Required?	Yes
Sprinklers Required?	Yes
Elevator Recall?	Yes, per ASME A17.1
Automatic Detection Required	Sprinkler W/F and tamper; electrical and mechanical rooms, elevator lobbies, elevator machine room
HVAC Shutdown?	Yes, per *NFPA 90A* for systems over 2,000 CFM
Occupant Notification?	Audible/Visible
EVAC System Required?	Yes
Firefighter's Telephones?	Yes
Annunciator?	Yes
Emergency Forces Notification?	No

Table 1-2. NFPA 101 covers many fire alarm requirements for life safety.

NFPA 72 Organization and Revision

NFPA 72, National Fire Alarm and Signaling Code, is organized into 29 chapters and nine annexes. Fifteen of these chapters are reserved for future use. **See Figure 1-16.**

Chapters 1 through 29 are mandatory text and are enforceable. The annexes are not enforceable as code. Changes to the *Code* are identified by a vertical rule in the left margin. This vertical rule only indicates there was a change and does not identify the nature of the change. A dot or "bullet" in the left margin indicates that part of the *Code* was deleted in the last revision.

An asterisk next to a *Code* section indicates a reference to Annex A, which provides explanatory material. The annex section has the same number as the section but is preceded by an upper case "A." For example, if 12.1.3* appeared in the *Code*, the Annex A section would be A.12.1.3.

With the exception of the NFPA "stock" definitions in Section 3.2, a parenthetical reference following definitions indicates the responsible committee for the definition. In the following example, the Technical Committee on Fundamentals of Fire Alarm System is responsible for the definition of "Alarm." This reference is for committee use and has no bearing on the use of the *Code*.

3.3.11 Alarm. A warning of danger. (SIG-FUN)

Figure 1-16 NFPA 72 Organization

Chapter	Title	Chapter	Title
Chapter 1:	Administration	Chapter 24:	Emergency Communications Systems (ECS)
Chapter 2:	Referenced Publications	Chapter 25:	Reserved
Chapter 3:	Definitions	Chapter 26:	Supervising Station Alarm Systems
Chapters 4-9:	Reserved	Chapter 27:	Public Emergency Alarm Reporting Systems
Chapter 10:	Fundamentals	Chapter 28:	Reserved
Chapter 11:	Reserved	Chapter 29:	Single- and Multiple-Station Alarms and Household Fire Alarm Systems
Chapter 12:	Circuits and Pathways	Annex A:	Explanatory Material
Chapter 13:	Reserved	Annex B:	Engineering Guide for Automatic Detector Spacing
Chapter 14:	Inspection, Testing, and Maintenance	Annex C:	System Performance and Design Guide
Chapters 15 and 16:	Reserved	Annex D:	Speech Intelligibility
Chapter 17:	Initiating Devices	Annex E:	NEMA SB 30, Fire Service Annunciator and Interface
Chapter 18:	Notification Appliances	Annex F:	Sample Ordinance Adopting *NFPA 72*
Chapters 19 and 20:	Reserved	Annex G:	Wiring Diagrams and Guide for testing Fire Alarm Circuits
Chapter 21:	Emergency Control Functions and Interfaces	Annex H:	Informational References
Chapter 22:	Reserved	Annex I:	Cross Reference Table
Chapter 23:	Protected Premises Fire Alarm Systems		

Figure 1-16. NFPA 72 *is organized into 29 Chapters, 15 of which are reserved for future use. There are also nine Annexes.*

Summary

Fire alarm systems have been in existence for more than 150 years. They provide life safety, property protection, mission continuity, heritage protection, and environmental protection. Manufacturers continually introduce new technology to develop better controls, advanced detection, and faster response times. The advent of the microprocessor has greatly increased the capability of fire alarm systems.

New technology helps to provide systems that are more user friendly, versatile, and stable, while providing early detection and improved communications. Codes and standards are necessary to ensure a minimum level of safety. As new technologies are developed, the *Code* will change to take advantage of them.

As technology improves, however, it grows more and more complex. Designers, installers, and codes and standards must all keep up with the changing technology in order to provide functional and reliable systems. Training and participation in code development help keep individuals abreast of changes in the field.

Review Questions

1. For the purposes of fire protection, items found in a protected space are often referred to as __?__.
   a. commodities
   b. contents
   c. fuels
   d. furnishings

2. According to NFPA data, structure fires occur approximately once every __?__ in the United States.
   a. 2 minutes
   b. 65 seconds
   c. 45 seconds
   d. 20 seconds

3. Which of the following contributes to fire alarm system reliability?
   a. Codes
   b. Installation
   c. Price of the project
   d. Property protection features

4. The first heat detector used this technology.
   a. Bimetallic strip
   b. Fusible link
   c. Glass bulb
   d. Rate-of-rise

5. The Americans with Disabilities Act (ADA) became law in __?__.
   a. 1991
   b. 1992
   c. 1993
   d. 1994

Review Questions

6. Which of the following types of systems are most likely to be used in a high-rise building (using partial evacuation)?
   a. Emergency voice alarm communications
   b. Protected premises
   c. Public fire reporting system
   d. Auxiliary fire alarm system

7. Supervisory signals indicate this condition.
   a. Loss of primary power
   b. Loss of secondary power
   c. Off-normal suppression system
   d. Open in system wiring

8. Door unlocking is an example of __?__.
   a. an emergency control function
   b. initial safety
   c. occupant alerting
   d. phase I operation

9. Inergen is an example of a __?__.
   a. deluge system
   b. dry suppression system
   c. gaseous suppression system
   d. pre-action system

10. Underwriters Laboratories uses __?__ to develop its standards.
    a. ANSI accredited volunteers
    b. committees
    c. staff engineers
    d. the canvass method

11. Codes specify __?__.
    a. how
    b. when and how
    c. when and where
    d. none of the above

12. Local jurisdictions may change national codes and standards through the use of __?__.
    a. adoption
    b. amendments
    c. references
    d. transcripts

13. A high-rise is generally a building that is more than __?__ from lowest level of fire department access to the highest level of occupancy.
    a. 50'
    b. 75'
    c. 100'
    d. 125'

Fundamentals and System Requirements

Chapter 10 of *NFPA 72* contains fundamental requirements that apply to all types of systems. These requirements also apply to qualifications of system designers and installation personnel, power supplies, system functions, performance and limitations, protection of controls, monitoring for integrity, documentation, and system impairments. Chapter 12 of *NFPA 72* contains *Code* requirements for circuit classes and wiring.

Objectives

- » List general requirements for personnel, equipment, wiring, control equipment, and zoning
- » Explain the various signal types
- » Describe the methods for monitoring system integrity
- » Describe the various circuit classes
- » Describe requirements for primary and secondary power supplies

Chapter 2

Table of Contents

General Requirements.................................32
- System Designer Qualification Requirements 32
- Installer Qualification Requirements 32
- Inspection and Testing Personnel Qualification Requirements 32
- Supervising Station Operator Qualification Requirements 33
- Listed Equipment.................................33
- Compatibility 34
- Alarm Signal Deactivation 34
- Operating Conditions 34
- Wiring 35
- Protection of Control Equipment..........35
- Zoning and Annunciation......................35
- Acceptance and Completion Documentation 35

Signals and Signal Types 37
- Alarm Signals 37
- Supervisory Signals..............................37
- Trouble Signals 38
- Emergency Communications Systems 38
- Annunciation of Signals 38

Monitoring for Integrity 38
- Basic Requirements 38
- Circuit Types...39
- T-Tapping..44
- Circuit Class 45

Power Supplies..50
- Primary Supply 50
- Secondary Supply................................51

Summary 54

Review Questions..54

GENERAL REQUIREMENTS

General requirements for personnel and equipment are located in Chapter 10 of *NFPA 72*. Some of the requirements were already located there, some were relocated, and some were added. Section 10.5 now contains all requirements for personnel qualifications.

System Designer Qualification Requirements

Section 10.5.1.1 requires fire alarm system and emergency communications system plans and specifications to be developed by persons who are experienced in the proper design, installation, and testing of these systems. Section 10.5.1.2 also requires local licensing laws to be followed. The *Code* provides examples of certification that may be acceptable, but does not limit an individual to these. Among the examples in the *Code* are the following:

- Certification by state or local authority
- Factory training/certification in the specific model and brand of system by persons who are acceptable to the authority having jurisdiction
- Certification by a nationally recognized fire alarm certification organization acceptable to the AHJ, such as the National Institute for Certification in Engineering Technologies (NICET) or the International Municipal Signal Association (IMSA)

NICET and IMSA both offer certifications applicable to interior fire alarm systems design. Similar to the requirements for installers, Section 10.5.1.3 requires the designer to provide evidence of qualifications when requested by the AHJ.

Installer Qualification Requirements

The requirements in Section 10.5.2.1 call for the system installer to be qualified or supervised by persons who are qualified in the installation, inspection, and testing of fire alarm and emergency communications systems. 10.5.2.2 requires all local licensing laws to be followed for determination of qualified personnel. The *Code* provides examples of acceptable certifications, but does not limit individuals to any specific type of certification. Among the examples in the *Code* are the following:

- Certification by state or local authority
- Factory training/certification in the specific model and brand of system by persons who are acceptable to the authority having jurisdiction
- Certification by a nationally recognized fire alarm certification organization acceptable to the AHJ, such as NICET or IMSA

NICET and IMSA both offer certifications applicable to interior fire alarm systems installation. These programs will meet the intent of the *Code* for qualification. Section 10.5.1.4 gives the authority having jurisdiction the right to ask for the proper credentials to validate the required qualifications.

Inspection and Testing Personnel Qualification Requirements

Requirements for inspection and testing personnel are found in Section 10.5.2. They are very similar to the qualification requirements for the installer and designer. Service personnel must be qualified and experienced in the inspection and testing of fire alarm systems. Evidence of qualifications may include the following:

- Certification by state or local authority
- Factory training/certification in the specific model and brand of system by persons who are acceptable to the authority having jurisdiction
- Certification by a nationally recognized fire alarm certification organization acceptable to the AHJ
- Personnel who are employed and qualified by a nationally recognized testing laboratory for the servicing of systems covered by the *Code*

Certification by a nationally recognized fire alarm certification program includes NICET and IMSA certification programs. Although not stated in the *Code*, a NICET

Fact

Many local jurisdictions have local licensing requirements. Be sure to know and follow all local licensing laws before starting any job.

Level II or IMSA Level I certification is intended by these requirements. Personnel employed and qualified by listed alarm service companies are generally considered qualified for purposes of the *Code*. These organizations are listed by testing laboratories, such as Underwriters Laboratories, for the servicing of fire alarm systems. There are approximately 100 such organizations listed in the United States. Section 10.5.3.3 requires service personnel to provide evidence of qualifications when requested by the AHJ.

Supervising Station Operator Qualification Requirements

Requirements for supervising station operating personnel are found in Section 10.5.4. They are very similar to the requirements for the installer, designer, and service personnel. Supervising station operators must demonstrate competence in all tasks required by Chapter 26 of *NFPA 72*. Evidence of competence may include the following:

- Certification by the manufacturer of the receiving system or equipment, or the alarm-monitoring automation equipment
- Certification by an organization acceptable to the AHJ
- License issued by state or local authority
- Other training or certification acceptable to the AHJ

Certification programs administered by an independent organization, such as the training program sponsored by the Central Station Alarm Association (CSAA), will generally be acceptable to the AHJ. Section 10.5.4.2 requires all operators covered by this section to present evidence of qualifications when requested by the AHJ. Finally, 10.4.4.3 requires operator trainees to be under direct supervision by a qualified operator at all times.

Listed Equipment

Section 10.3.1 of *NFPA 72* requires all equipment used in conformity with the *Code* to be listed for the purpose for which it is used. Qualified testing laboratories evaluate products and services against known standards. If the products pass all required tests, the product is added to a listing (directory) of products and services published by the laboratory. The manufacturer is then permitted to apply a listing mark to the listed product. The listing can apply to products ranging from smoke detectors to chainsaws. It is important to note that testing laboratories do not test every single piece of equipment produced by manufacturers under their listing programs. Only a small statistical sample is tested. Furthermore, listing does not indicate the quality of the product. It only indicates that the product complies with defined product standards.

Periodic evaluation usually requires the laboratory to evaluate another statistical sampling of production line equipment on a periodic basis. As long as the product continues to meet the standards, the laboratory permits the manufacturer to apply the listing mark. Many listings are now available online at laboratory internet websites.

Unlike the *National Electrical Code* (*NEC*), *NFPA 72* requires products to be listed by a qualified testing laboratory, such as Underwriters Laboratories or Factory Mutual (FM). There are scores of other qualified testing laboratories in the United States, some of which test fire alarm products. Approval of the testing laboratory is the decision of the AHJ, but most authorities having jurisdiction accept both UL and FM. Some local jurisdictions have their own listing and approvals agencies. For example, in California, products must also have the State Fire Marshal's approval. The manufacturer must obtain this approval in addition to an approved listing mark in order to sell products there.

Fact

Many jurisdictions have additional requirements beyond the listing. California, for example, requires all equipment to be California State Fire Marshal (CSFM) approved. Be sure the listing and equipment you choose is "approved" by the authority having jurisdiction.

Section 10.3.2 requires all listed equipment to be installed, tested, and maintained in accordance with the instructions and the requirements of the *Code*. This essentially comprises the installation instructions that accompany the equipment part of the *Code*.

Compatibility

Section 10.3.3 requires all alarm initiating devices that receive power from the system controls to be listed as compatible for use with the controls. This is especially important because addressable devices generally interface with controls using proprietary communications protocols. This is similar to two people trying to communicate with each other when neither speaks the same language. Therefore, detectors made by one manufacturer are almost always incompatible with any other manufacturer's equipment. The same is often true of conventional detectors that must be voltage and current matched. Most manufacturers provide compatibility information as a service to their customers. Underwriters Laboratories also provides a cross reference of compatible devices and controls.

Alarm Signal Deactivation

Alarm signals can be disruptive to fire service communications during a response. For this reason, Section 10.13 permits alarm signal deactivation. A means for turning off all notification appliances is permitted, provided that it is key operated or equivalently protected from unauthorized use. For instance, a physical deactivation device could be kept in a locked cabinet, while a computerized device might be protected by a password.

Subsequent actuation of other initiating devices must cause any silenced signals to re-actuate. This requirement is intended to ensure that a spreading fire is properly annunciated, even when the signals have been silenced. Additionally, any alarm silencing switch that is left in the "silence" position must cause a signal to sound when the alarm signals are cleared. This feature is intended to ensure the switch is left in the "normal" position when the controls are clear of alarms.

It should be noted that this section formerly permitted the individual deactivation of both audible and visible notification appliances, but now requires simultaneous deactivation of all appliances, in order to comply with the Americans with Disabilities Act (ADA). Similar requirements for the deactivation of supervisory and trouble signals exist in Sections 10.14.7 and 10.15.10, respectively.

Operating Conditions

Fire alarm controls, initiating devices, and notification appliances are sensitive to their environments. Equipment should never be placed where the environment will cause degradation or failure or cause nuisance alarms. For example, most controls are listed for a dry location. This implies an indoor, air-conditioned space. However, some manufacturers do produce equipment and devices that are listed for more extreme ambient conditions, exceeding *Code* requirements. In this case, the equipment would be permitted to be installed in ambient conditions that satisfy the listing.

Section 10.3.5 requires equipment to be designed to operate within 85% - 110% of nameplate voltage, in temperatures between 32°F (0°C) and 120°F (48°C), and a relative humidity of 85% at 86°F (30°C). These requirements apply to all equipment, but Chapters 17 and 18 of the *Code* contain additional requirements for initiating devices and notification appliances, respectively. It is imperative that fire alarm equipment be installed in conditions that will not degrade the equipment or cause malfunctions. Temperature and humidity extremes are one of the leading causes of nuisance alarms and fire alarm equipment failure. Fire alarm equipment is usually listed for dry locations (as defined by Article 100 of the *NFPA 70, National Electrical Code*).

Section 10.14.4.3 further reinforces the requirements to follow manufacturer's instructions for all equipment. Additionally, 10.4.5.1 requires the installation of initiating devices to be selected and installed so as to minimize nuisance

Fact

Most fire alarm equipment will prematurely fail when subjected to temperature and humidity extremes. Equipment which is not listed or approved for use in the expected ambient must not be used in those conditions.

alarms. This generally means that detectors should not be installed where there is sufficient environmental stimulus to operate the device. This would include, for example, placing a smoke detector in a dusty or humid environment. Equipment should always be installed within its listing and be suited to its environment.

Wiring

NFPA 72 provides requirements for fire alarm system installation and performance. Section 12.2.4 specifically provides requirements for installation of fire alarm system wiring. Since the scope of *NFPA 72* does not actually include wiring requirements, all wiring must be installed to the requirements of the *National Electrical Code*. Articles 760, 770, and 800 are specifically referenced by Section 12.2.4, and of particular importance is a requirement to provide transient protection in accordance with Article 760 of the *NEC*. Transient protection generally requires protection of circuits as they leave or enter a building.

Protection of Control Equipment

Section 10.4.4 requires protection of control equipment where the equipment is installed in a location that is not constantly attended. A smoke detector is usually installed to provide early warning of a fire in the space containing the control equipment. The requirement is intended to detect and annunciate an alarm before a fire in that space could destroy the equipment. Where conditions prohibit the installation of smoke detection, an exception permits automatic heat detection to be used. However, if ambient conditions do not permit smoke detection, then the ambient environment is probably not suitable for the controls.

Zoning and Annunciation

Zoning and annunciation is intended to help responders quickly determine where the signals originate in order to minimize damage to the premises. Section 10.18 provides requirements for zoning and annunciation. Of particular importance is the requirement in

> **10.4.4* Protection of Fire Alarm System.** In areas that are not continuously occupied, automatic smoke detection shall be provided at the location of each fire alarm control unit(s), notification appliance circuit power extenders, and supervising station transmitting equipment to provide notification of fire at that location.
>
> *Exception: Where ambient conditions prohibit installation of automatic smoke detection, automatic heat detection shall be permitted.*
>
> (Excerpt from NFPA 72.)

10.18.3.2 to locate the annunciator where required by the AHJ, to facilitate an efficient response. The AHJ determines the location because he or she represents the responders who use the equipment during the emergency. At a minimum, Section 10.18.5.1 requires each floor of a building and each separate building to constitute a separate zone.

Many local and national building codes, such as the *International Building Code*, will contain additional annunciation and zoning requirements. Some jurisdictions have local requirements that exceed national codes like the *IBC*. For example, some jurisdictions require a graphic annunciator for all buildings over four stories that exceed 25,000 square feet. *NFPA 72* by itself contains very few zoning and annunciating requirements. It is important to know and understand local annunciating requirements before any work begins.

Acceptance and Completion Documentation

A new Chapter 7, Documentation, was added in the 2013 edition of *NFPA 72*. The primary purpose of this new chapter is to strengthen all documentation requirements and locate them in a single location. New requirements were added for drawing content, risk analysis, and other documentation.

Section 10.20.2 requires notification of the AHJ before the installation or

Figure 2-1 Record of Completion

SYSTEM RECORD OF COMPLETION

This form is to be completed by the system installation contractor at the time of system acceptance and approval. It should be permitted to modify this form as needed to provide a more complete and/or clear record.
Insert N/A to all unused lines.
Attach additional sheets, data, or calculations as necessary to provide a complete record.

Form Completion Date: _____ Supplemental Pages Attached: _____

1. PROPERTY INFORMATION
Name of property: _____
Address: _____
Description of property: _____
Name of property representative: _____
Address: _____
Phone: _____ Fax: _____ E-mail: _____

2. INSTALLATION, SERVICE, TESTING, AND MONITORING INFORMATION
Installation contractor: _____
Address: _____
Phone: _____ Fax: _____ E-mail: _____
Service organization: _____
Address: _____
Phone: _____ Fax: _____ E-mail: _____
Testing organization: _____
Address: _____
Phone: _____ Fax: _____ E-mail: _____
Effective date for test and inspection contract: _____
Monitoring organization: _____
Address: _____
Phone: _____ Fax: _____ E-mail: _____
Account number: _____ Phone line 1: _____ Phone line 2: _____
Means of transmission: _____
Entity to which alarms are retransmitted: _____ Phone: _____

3. DOCUMENTATION
On-site location of the required record documents and site-specific software: _____

4. DESCRIPTION OF SYSTEM OR SERVICE
This is a: ☐ New system ☐ Modification to existing system   Permit number: _____
NFPA T2 edition: _____

4.1 Control Unit
Manufacturer: _____ Model number: _____

4.2 Software and Firmware
Firmware revision number: _____

4.3 Alarm Verification   ☐ This system does not incorporate alarm verification.
Number of devices subject to alarm verification: _____   Alarm verification set for _____ seconds

© 2012 National Fire Protection Association   NFPM 72 (p. 1 of 3)

Figure 2-1. A Record of Completion provides information on how to maintain the system after installation.

alteration of any wiring or equipment. In most jurisdictions, a permit will be required before starting any work other than repairs. The permitting process ensures that designs are reviewed by the AHJ, and that all parties are qualified to conduct the work at hand. The review process requires the designer/installer to submit all shop drawings, calculations, floor plans, product submittal sheets, and other pertinent information for approval.

Once the installation is completed, Section 7.5.8 allows the AHJ to require a written statement from the installing contractor stating that all equipment was installed according to manufacturer's requirements, approved plans, and the requirements of the *Code*. This option is usually exercised in jurisdictions that use third-party agencies to conduct acceptance on behalf of the AHJ.

Section 7.8.2 requires the installer to complete a "Record of Completion." The Record of Completion is the "birth certificate" of the system. It contains all necessary information to maintain the system after installation. This form is actually Figure 7.8.2(a) of *NFPA 72*. **See Figure 2-1.**

The Record of Completion must be given to the AHJ after successful completion of the installation. The owner must retain the Record of Completion for the life of the system. Section 7.5.6.4 requires the Record of Completion to be kept current at all times. It must be updated every time the system is changed, including software changes.

Section 7.5.6.5 requires the Record of Completion to be stored in a documentation cabinet. However, when not stored at the location of the fire alarm control unit, the system documentation must be stored at a location identified at the system controls. The storage location must be labeled as "FIRE ALARM DOCUMENTS."

SIGNALS AND SIGNAL TYPES

Fire alarm systems must produce at least three different signals: alarm, supervisory, and trouble. Each signal type has a distinct meaning and must be responded to differently. Section 10.10.1 requires all fire alarms, supervisory signals, and trouble signals to be descriptively and distinctively annunciated. In other words, they must be accurate and convey the required information regarding the system.

Sections 10.10.1, 10.10.2, 10.10.5, and 10.10.6 of *NFPA 72* require alarm, supervisory, and trouble signals to be distinctive or different. In addition, alarm signals cannot be used for any other purpose, such as burglar alarms. The *Code* permits supervisory and trouble to share the same audible signal provided that there is a visible indication at the control unit to distinguish them from each other. Fire alarm, supervisory, and trouble signals also have priority over all other signals, except that Mass Notification Systems (MNS) and Emergency Communications Systems (ECS) signals may be deemed to have higher priority by risk analysis. The AHJ may permit other life-threatening signals to have priority over supervisory and trouble signals.

Alarm Signals

Alarm signals are reserved for fire or other life-threatening emergencies. Alarm signals may be used to notify occupants and responders of a fire. With the introduction of emergency communications systems into Chapter 24 of the *Code*, alarm signal content takes a new twist. Alarm signals might be used to signal a terror alert, gas leak, Amber alert, tornado warning, or some other emergency. Emergency communications systems can also be used to communicate non-emergency information, like traffic delays. Alarm signals on a fire alarm system may be initiated by smoke detectors, heat detectors, manual fire alarm boxes, radiant energy sensing fire detectors, or waterflow switches.

Coded alarm signals are found on older systems that use spring-wound manual fire alarm stations that transmit a specific (coded) signal when actuated. Where they are used, 10.9.3 requires all coded alarm signals to consist of not less than three complete rounds of the number transmitted. 10.9.4 requires each round to consist of not less than three impulses.

Supervisory Signals

Supervisory signals generally indicate an off-normal condition of a suppression

Fact

One of the primary reasons the *Code* was reorganized was to allow for the addition of mass notification systems (MNS) and emergency communications systems (ECS). New Chapter 24 was added to accommodate ECS and MNS.

system. This may include a closed sprinkler valve, a fire pump in an OFF condition, or even power loss on a suppression system control unit. However, duct smoke detectors are sometimes programmed to cause supervisory signals rather than alarm signals. This arrangement has limitations and will be thoroughly discussed in later chapters.

Trouble Signals

Trouble signals are used to indicate a loss of power, equipment failure, a communications fault, or a fault on system wiring. Open circuits and ground faults are always annunciated as a trouble. Detectors that are no longer in their listed sensitivity range may also initiate a trouble signal. Trouble signals are annunciated at the controls and are responded to by maintenance personnel.

Emergency Communications Systems

Requirements for emergency communications systems are found in Chapter 24 of the *National Fire Alarm and Signaling Code*. ECSs are used to provide live or recorded information to large numbers of people during an event such as a tornado or terrorist attack. This technology is new to *NFPA 72* and reflects an ever-changing world.

Requirements for in-building fire emergency voice/alarm communications systems (EVACS) that use voice signals are located in Chapter 24. ECS signals may be permitted to have priority over fire alarm signals in some cases. Specifically, Section 24.4.2.7.2 permits ECS signals to have priority where risk analysis indicates other non-fire events may be more immediately life threatening. ECS will be discussed more in later chapters.

Annunciation of Signals

Signals must be heard, so equipment such as annunciators and remote controls may be required in some cases. Section 10.18.3.2 requires annunciation to be located where required to facilitate an efficient response to the emergency condition. Additional annunciation requirements are covered by Section 10.18.5 of *NFPA 72*. Annunciation may be required by a local building code, the *IBC*, or *NFPA 101, Life Safety Code*.

The primary purpose of annunciation is to provide information to first responders in the event of an emergency. As a minimum, Section 10.18.5.1 requires each floor and building to be separately zoned and annunciated.

MONITORING FOR INTEGRITY

Since fire alarm systems provide life safety, Section 12.6.1 of *NFPA 72* requires most circuits to be monitored for integrity. This section was relocated to Chapter 12 in the 2013 edition because it directly relates to circuits and pathways. Monitoring circuits for integrity provides an assurance that the circuits have continuity and are free of ground faults. In some cases, system controls monitor for short circuits. Monitoring for integrity does not guarantee correct circuit or system operation. It only provides an indication that the conductors are neither opened nor grounded.

Basic Requirements

The *Code* requires fire alarm system controls to indicate a trouble signal when a circuit is opened or grounded. Loss of a carrier signal or loss of communications on wireless systems and optical fiber cables is also monitored and annunciated as a trouble. Power supplies are monitored for integrity, and loss of either primary or secondary power supplies must result in a trouble signal. Some devices such as smoke detectors are also monitored so that their failure causes a trouble signal. Trouble signals must appear within 200 seconds of the fault condition and must self-restore within 200 seconds after the fault is cleared. A trouble signal is indicated by an audible and visible signal at the system controls.

Many requirements in the *Code* have exceptions to permit special cases or conditions. There are several key exceptions to these requirements. For example, Section 12.6.8 permits exclusion of monitoring the integrity for circuits within a common enclosure. This exception exists because most wiring inside enclosures is factory-installed, limited in length, and access to the wiring is restricted to authorized persons. The wiring is not subjected to the same potential damage as field wiring, so monitoring these circuits is not necessary.

Section 12.6.6 does not require monitoring of circuits providing power to notification appliances located in the same room as the main control equipment, provided that the circuit conductors are protected against mechanical damage. Protection of the conductors supplying the notification appliance(s) is generally achieved through the use of conduit or raceway or fishing the cables inside a wall rather than installation of unprotected cables on surfaces. This exception exists because the main controls will indicate alarm signals. Notification appliances in the same room would be redundant.

Another exception exists in Section 12.6.9 as it relates to field wiring. This section does not require wiring between enclosures to be monitored for integrity if the enclosures are less than 20 feet (6 m) apart and the wiring is installed in conduit or equivalently protected against mechanical damage. Again, it is assumed that the wiring in the enclosures is accessible to authorized personnel, and the runs between equipment are limited to 20 feet (6 m) or less.

Other exceptions exist for wiring between central controls and peripherals, such as a keyboard, monitor, or mouse, provided that the connecting conductors are less than eight feet in length and the equipment is listed for fire alarm use. Wiring within a central station is also exempt from the requirements for monitoring for integrity. In these cases, it is assumed that problems will be discovered and reported quickly by operators. Additionally, access to these types of environments is controlled, resulting in a lower assumed risk.

Power supplies are also monitored for integrity, as required by Sections 10.6.9, 10.6.9, and 10.6.10.6. Loss of electrical power or a battery charger failure must result in a trouble signal at the control unit. However, it is impossible to monitor the output of an emergency generator unless it is running. Therefore, Section 10.6.9.1.6 permits an exception to this arrangement. Section 10.6.9.1.5 pertains to central station power supplies, which are exempt because the station is staffed 24/7 and outages would not go unnoticed. Other supervising stations are not exempt and must monitor their power supplies. Section 10.6.9.3 requires supervising station systems to monitor the power supplies at the protected premises. However, the transmission of trouble signals to supervising station receivers must be delayed by 60 to 180 minutes so that transmission of trouble signals caused by widespread power outages do not jam communication lines.

Circuit Types

There are three basic types of fire alarm circuits: initiating device circuits, signaling line circuits, and notification appliance circuits. These circuits connect the field devices that detect fires or abnormal conditions and alert occupants and responders to the conditions.

Initiating devices are field-installed devices that cause alarms or supervisory signals. Examples of initiating devices are smoke detectors, heat detectors, manual fire alarm boxes, and valve supervisory switches. Initiating devices are installed on either initiating device circuits or on signaling line circuits. Initiating devices that are installed on signaling line circuits (SLCs) are addressable. Even conventional devices can be connected to an SLC by using an interface, usually referred to as a "monitor module" or "zone address module." Notification appliances are installed to alert occupants and operators that there is an alarm, supervisory, or trouble condition requiring attention. Notification appliances are installed on notification appliance circuits.

Initiating Device Circuits. Initiating device circuits (IDCs) are often called "conventional" or "non-addressable" circuits. The control unit cannot distinguish which initiating device is actuated. In other words, the circuit and all of its devices represent a single zone. Devices on an IDC usually cause signals by closing contacts and shorting the two line conductors together. This

 Fact

End-of-line (EOL) devices are used on initiating device circuits (IDC) and notification appliance circuits (NAC). Many EOL devices are resistors, but some may be capacitors, depending on the manufacturer. Signaling line circuits (SLC) generally do not use EOL.

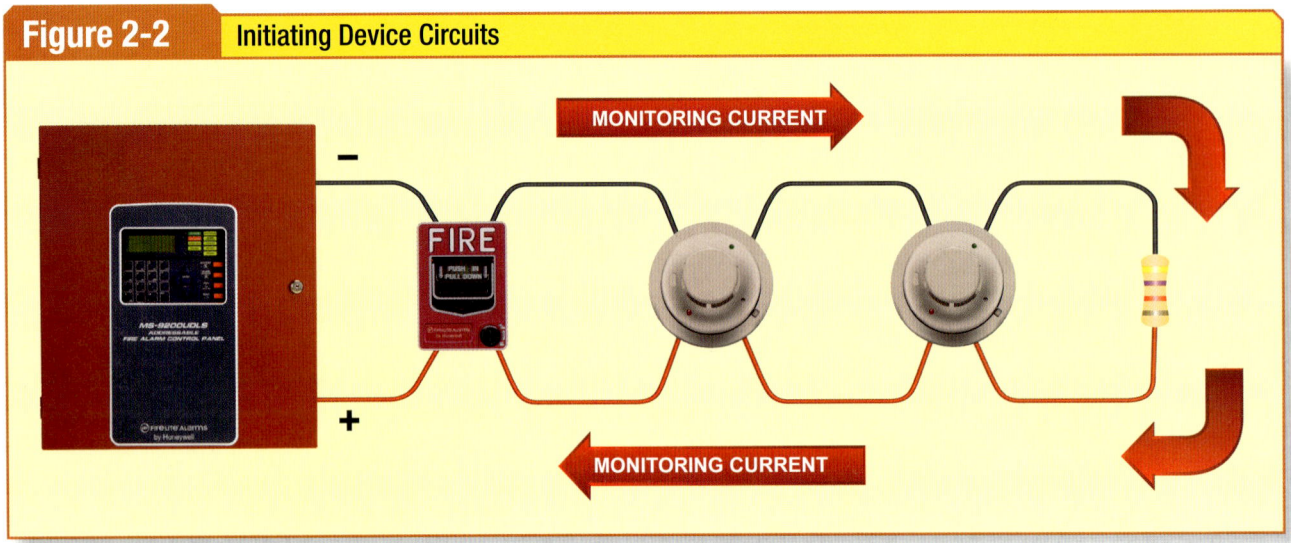

Figure 2-2. In normal IDC operation, the monitoring current flows uninterrupted around the circuit.

causes a drop in voltage at the controls, which is sensed as an alarm signal by the control unit. The controls only indicate the zone in alarm, usually through the use of an indicator lamp or LED.

A small current is circulated through the circuit conductors and end-of-line device to provide a means of monitoring for integrity. Opens and ground faults will interrupt this current and cause a trouble signal. **See Figure 2-2.**

In the event an IDC becomes open, the open circuit will prevent the monitoring current from passing through the circuit, thereby causing a trouble signal at the controls. Devices between the controls and the open circuit may operate, but devices located electrically "downstream" from the open circuit will not operate. **See Figure 2-3.**

Devices that require power to operate, such as smoke detectors, sometimes receive power from the same two conductors that transmit the alarm signal. Some manufacturers require the use of an extra pair of conductors to supply power, which results in four conductors to the device. In this case, the power supply conductors must also be monitored for integrity. An end-of-line power supervision relay usually accomplishes this required monitoring.

It should be noted that conventional (zone-type) initiating device circuits could be wired to cause either a supervisory signal or an alarm signal. In this case, the manufacturer usually has two

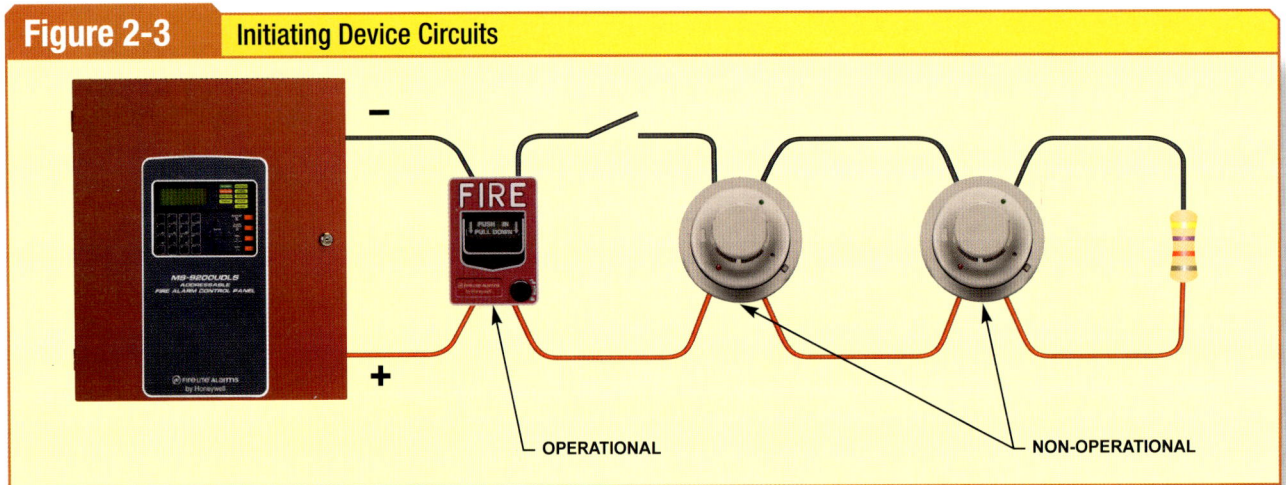

Figure 2-3. In an open IDC circuit, the monitoring current cannot complete its circuit and a trouble signal occurs at the controls.

zone module types: one for alarm signals and one for supervisory signals. Many installers have put alarm and supervisory initiating devices on the same IDC, which usually results in the wrong signals being annunciated. An example of this mistake is a waterflow switch and a valve supervisory device on the same alarm initiating device circuit. Valve closure should result in a supervisory signal, but in this case will cause an alarm signal instead. **See Figure 2-4.**

The correct method of installing a waterflow and valve supervisory (tamper) switch is to install them on separate zones. **See Figure 2-5.**

In this case, two separate zone cards or modules are needed to create two separate signals. However, some manufacturers produce equipment that can detect the difference between supervisory and alarm initiating devices on the same circuit. These devices and circuit controls must be listed for this purpose.

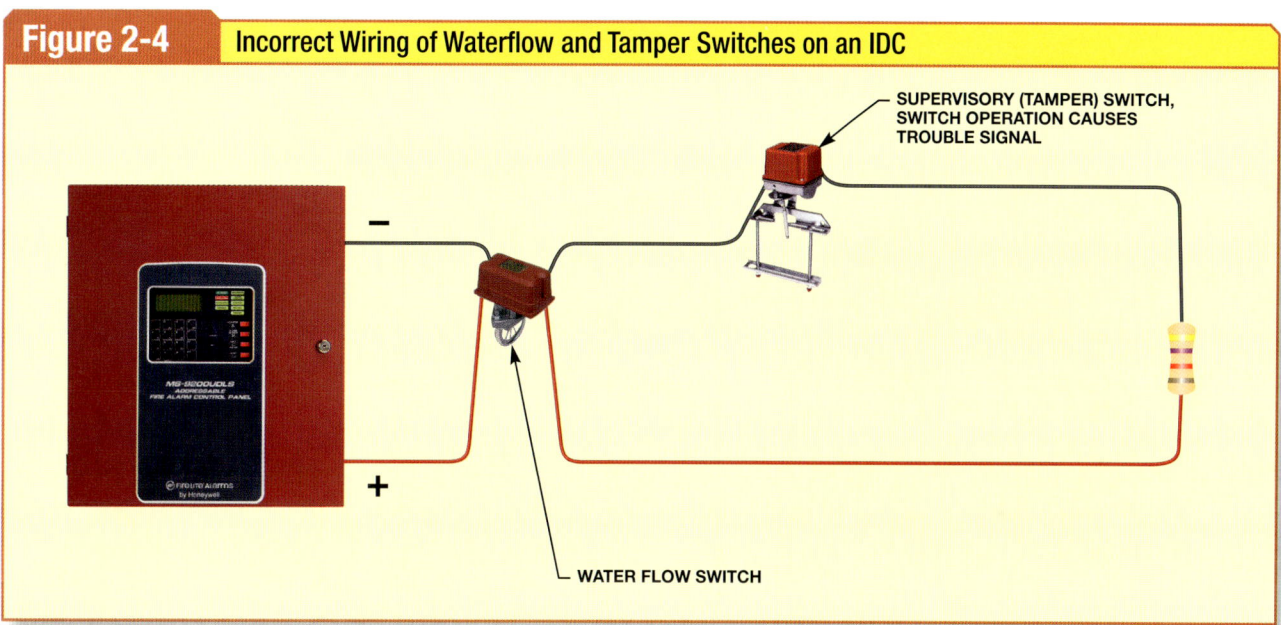

Figure 2-4. Putting alarm and supervisory devices on the same IDC usually results in the wrong signal.

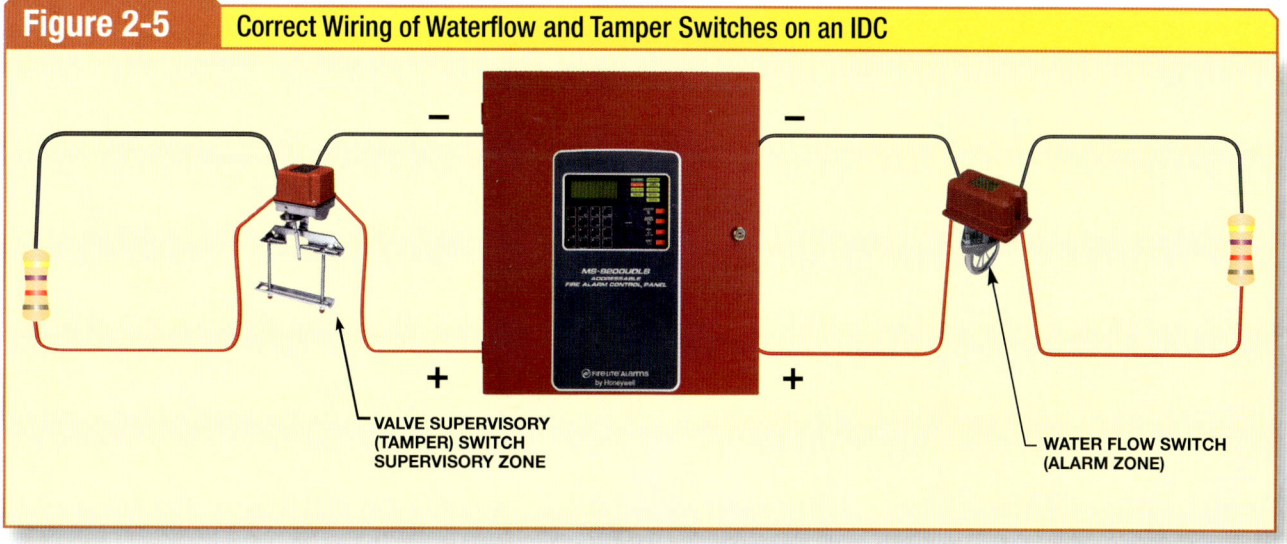

Figure 2-5. Alarm and supervisory devices should be on separate circuits to ensure correct signaling.

Signaling Line Circuits. Signaling line circuits (SLCs) allow two-way or multiplexed communication between the controls and field devices through the use of digital technology. Field devices include addressable smoke detectors, heat detectors, control relays, and other equipment. Even though addressable devices, such as smoke and heat detectors, are connected on a signaling line circuit, they are still referred to as initiating devices. Each device on the circuit has a unique address, which must be programmed into the system software.

When an initiating device actuates, the control unit knows what device is actuated and generally displays the location and address of the device on a screen or display. An example of this arrangement might be:

SMOKE DETECTOR 058, ELECTRICAL CLOSET 12TH FLOOR

The system software will be programmed to indicate the type of signal, whether it is supervisory or alarm. Signaling line circuits usually require a single pair of conductors for both power and signals.

Controls that operate signaling line circuits are software driven. The controls communicate with all devices on the

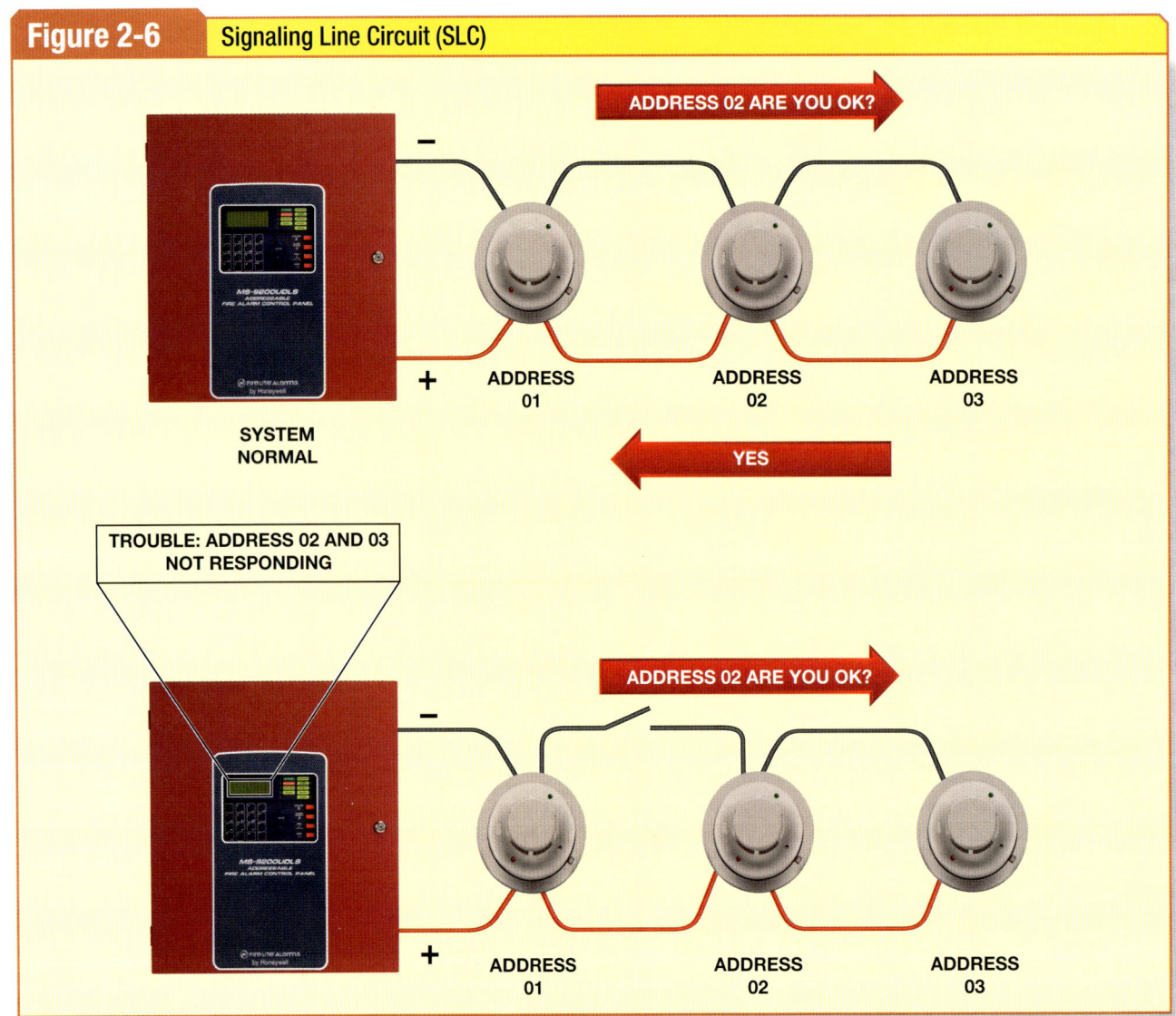

Figure 2-6. SLC devices are polled for their status. When an SLC has an open circuit, the polling fails.

circuit through the use of software protocols much the same way computers and printers communicate. The controls periodically interrogate or poll each address to determine the status of the device or equipment. Failure of the device to respond will result in a trouble signal. For this reason, monitoring for integrity does not employ the use of monitoring current. Additionally, end-of-line devices may or may not be used, depending on the manufacturer. **See Figure 2-6.**

Notification Appliance Circuits. Notification appliance circuits (NACs) operate similarly to initiating device circuits. Notification appliance circuits have two states: ON or OFF. The circuit state is actuated by the fire alarm system controls, which reverse the polarity of the circuit. Most manufacturers make equipment so the fire alarm control unit does not have control over individual appliances on an NAC. However, one manufacturer recently developed signaling line circuit technology that allows for addressable notification appliances.

Monitoring notification appliance circuits for integrity is accomplished in the same manner as for initiating device circuits, by using a small monitoring current. An open or ground fault will interrupt the current flow and cause a trouble signal. An internal blocking diode, which is reverse biased under normal (monitoring) conditions prevents monitoring current from passing through the appliance circuitry. However, when the circuit is placed in the alarm state, the NAC polarity is reversed causing the blocking diode to become forward biased. Alarm current is then allowed to flow through the devices, thereby producing an output. **See Figure 2-7.**

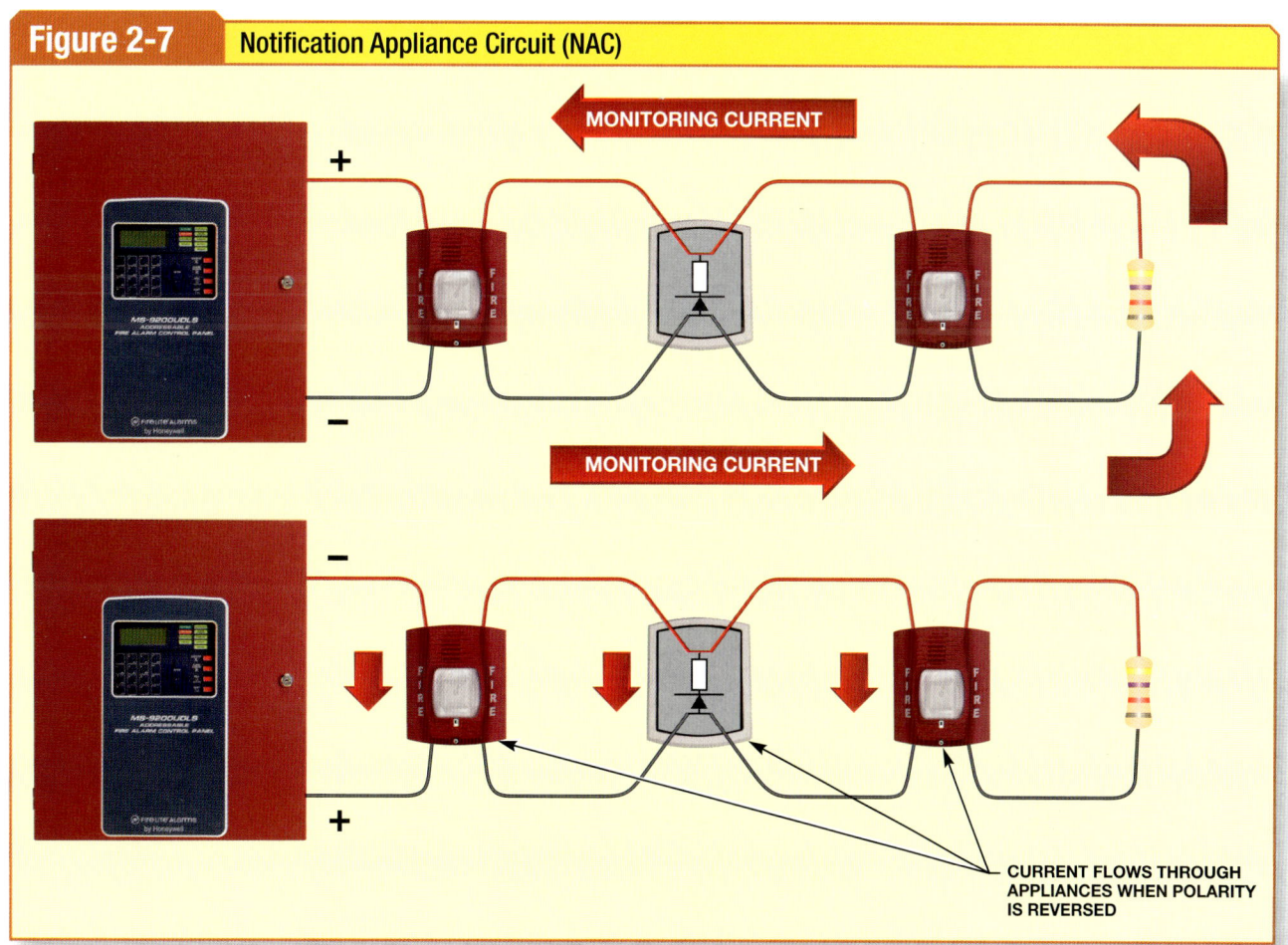

Figure 2-7. An NAC has two states: ON and OFF. An NAC is in alarm when the polarity is reversed.

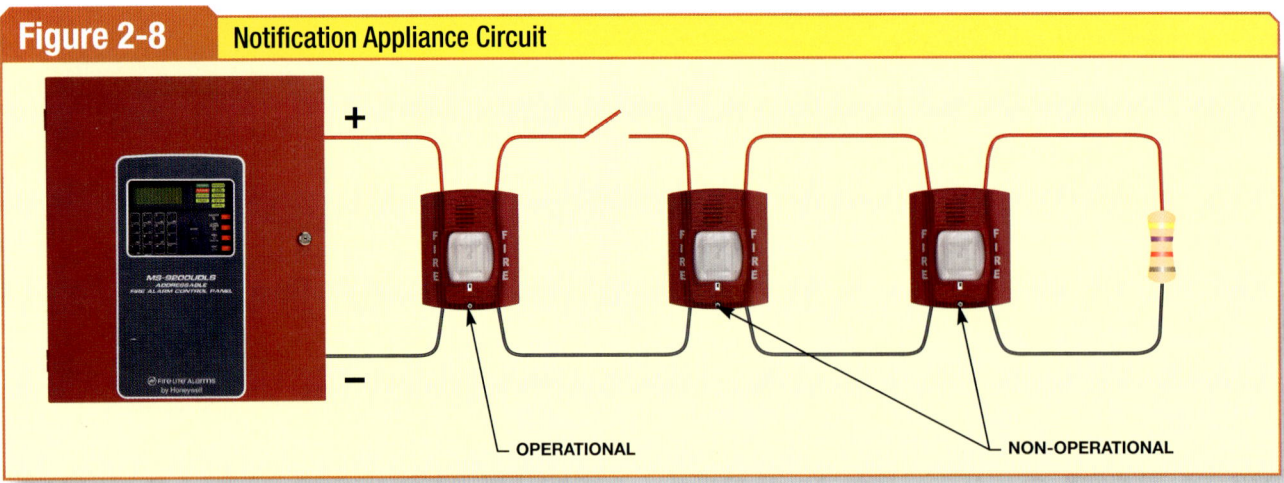

Figure 2-8. When the NAC current is interrupted, only some of its appliances will operate and a trouble signal will be generated.

When the circuit is opened, the monitoring current is interrupted; only those appliances between the controls and the open circuit will operate and a trouble signal will be generated. **See Figure 2-8.**

T-Tapping

Most field devices have duplicate terminals, or pigtails, resulting in four connection points. Where provided, all terminals or leads must be used. Duplicate terminals and leads are provided to ensure a loose connection is annunciated as a trouble. For this reason, multiple field conductors cannot be installed under a single terminal. This arrangement is called T-tapping, which could result in disconnected devices without indication of a trouble signal. Another common technique, which consists of looping a conductor under a terminal, is also unacceptable because the loop could become disconnected from the device without causing a trouble signal. The only circuits that are permitted to be T-tapped are Class B signaling line circuits. No other circuits may be T-tapped. **See Figure 2-9.**

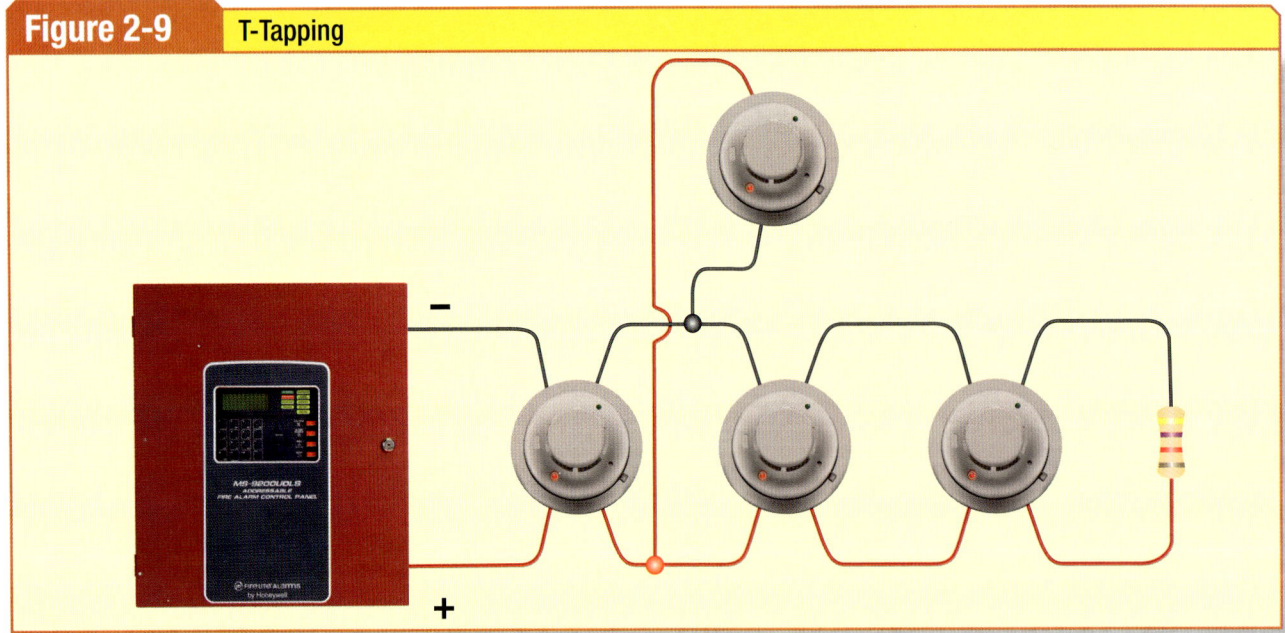

Figure 2-9. T-tapping is only allowed on Class B signaling line circuits.

Circuit Class

Until the 2007 edition of *NFPA 72*, there were several classes and styles for each type of fire alarm circuit. However, the committee responsible for this material in *NFPA 72* simplified the *Code* considerably by reducing the tables that identified their requirements. The Committee removed the tables and placed them in Annex A for reference on older systems. These tables are now found in A.12.3 of *NFPA 72–2013* edition.

Circuit classes were significantly changed for the 2010 edition of *NFPA 72*. Circuit "styles," once used in *NFPA 72*, are no longer used. There are also several new circuit classes now in use. Chapter 12 w covers circuits and pathways. In the *Code*, the words "circuits" and "pathways" are used interchangeably. The class of a circuit or pathway is defined by its ability to operate during fault conditions.

Class A Circuits. Class A circuits remain unchanged from the 2007 edition of the *Code*. Class A circuits are defined by return conductors that extend from the last device on the circuit to the fire alarm system controls. A fault, such as a single open circuit, will condition the circuit so it operates from both ends.

Section 12.3.1 of *NFPA 72* requires a Class A circuit to have a redundant path. The redundant path simply means the two conductors from the last device return to the fire alarm controls. **See Figure 2-10.**

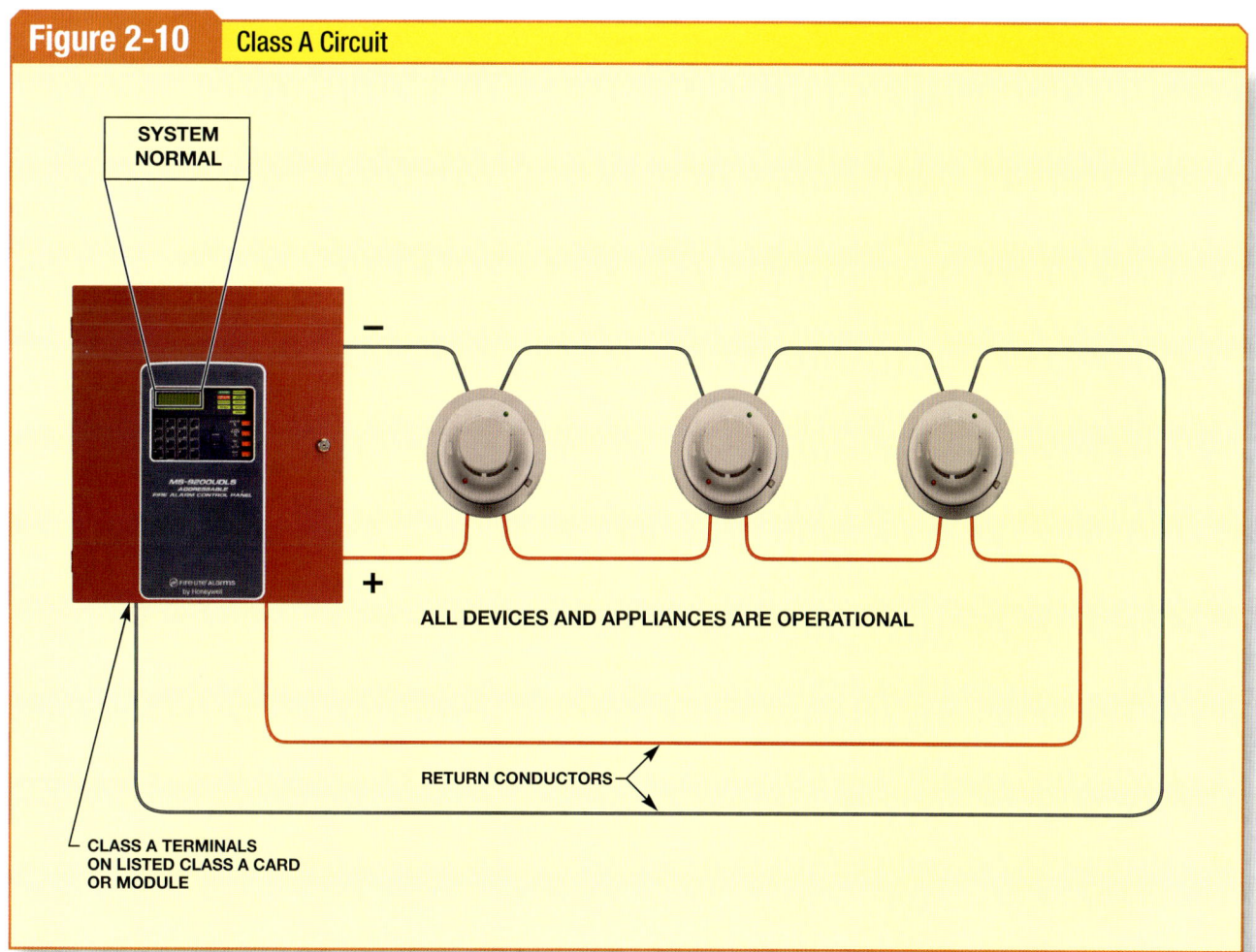

Figure 2-10. A Class A circuit must have redundant paths.

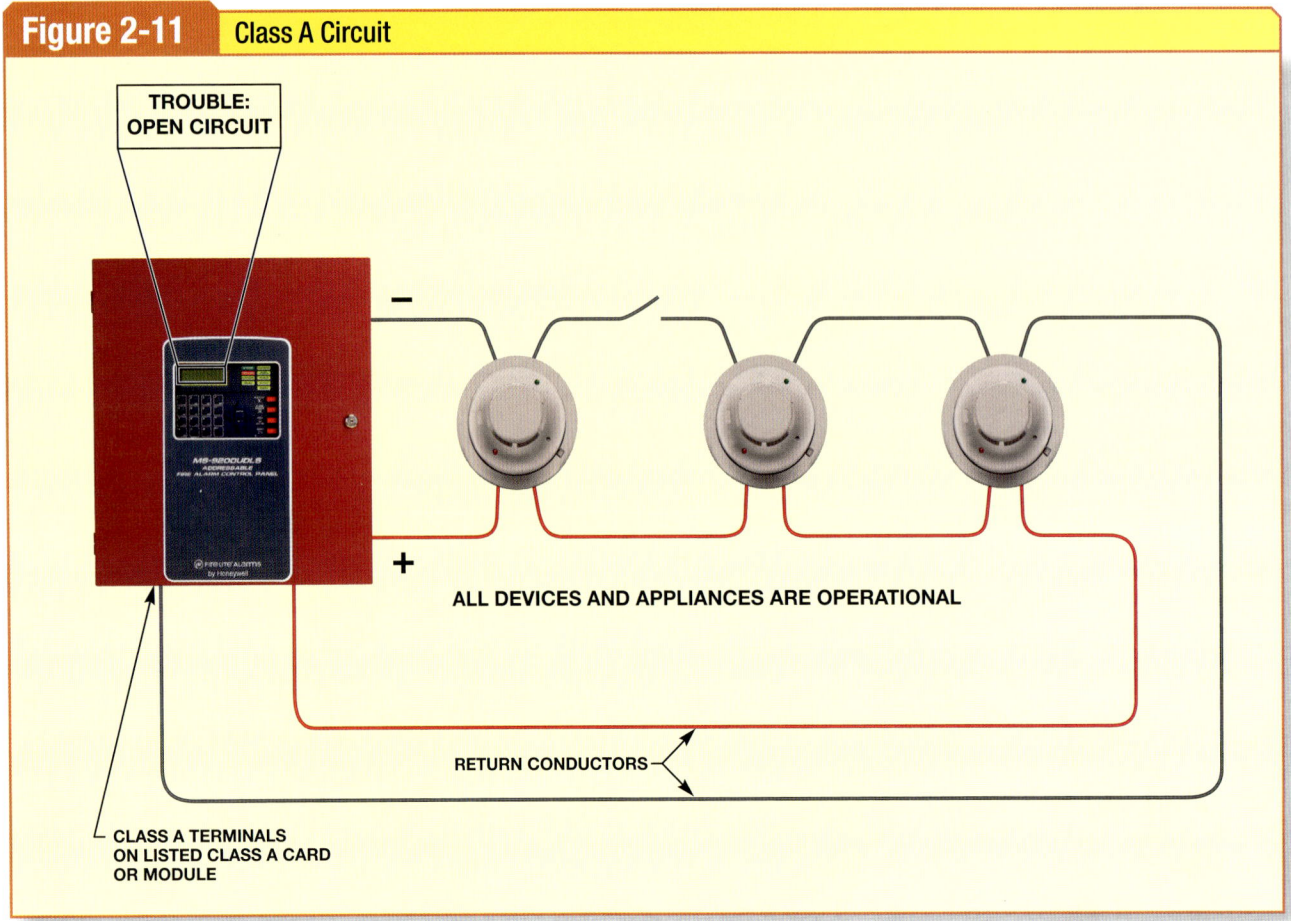

Figure 2-11. Class A circuits must operate beyond a single open circuit.

Section 12.3.1 also requires Class A circuits to operate beyond a single open circuit. Class A circuits generally offer more reliability because all devices or appliances are required to operate for a single open or single ground fault. Class A circuits are required to be monitored for integrity according to Section 12.6 of *NFPA 72*. **See Figure 2-11.**

Class A circuits require a pair of return conductors from the last device to the control unit in order to provide power and signaling for the entire circuit under a single fault condition. However, extra conductors alone do not enable the circuit to operate under fault conditions. The circuit must also contain special controls that sense the fault conditions and provide power and communications in two directions. Class A circuits are more reliable and hence are used where there are hazards that warrant their use. For example, a hotel will have guests sleeping on the premises and more reliability may be required to ensure the system works as intended.

Class B Circuits. Class B circuits are covered by Section 12.3.2 of *NFPA 72*. Class B circuits do not have redundant pathways and are not required to operate beyond a single open. Class B circuits are, therefore, usually considered less reliable than Class A circuits. **See Figure 2-12.** Class B circuits are required to be monitored for integrity according to Section 12.6 of *NFPA 72*.

Class B circuits have a single pathway from the controls to the devices. Devices are connected in parallel along the circuit. Class B circuits end with the last device or appliance on the circuit, but an end-of-line (EOL) device is usually installed at the last device or appliance on the circuit. The EOL is designed to provide a path for monitoring current so the occurrence of an open or ground will be detected and

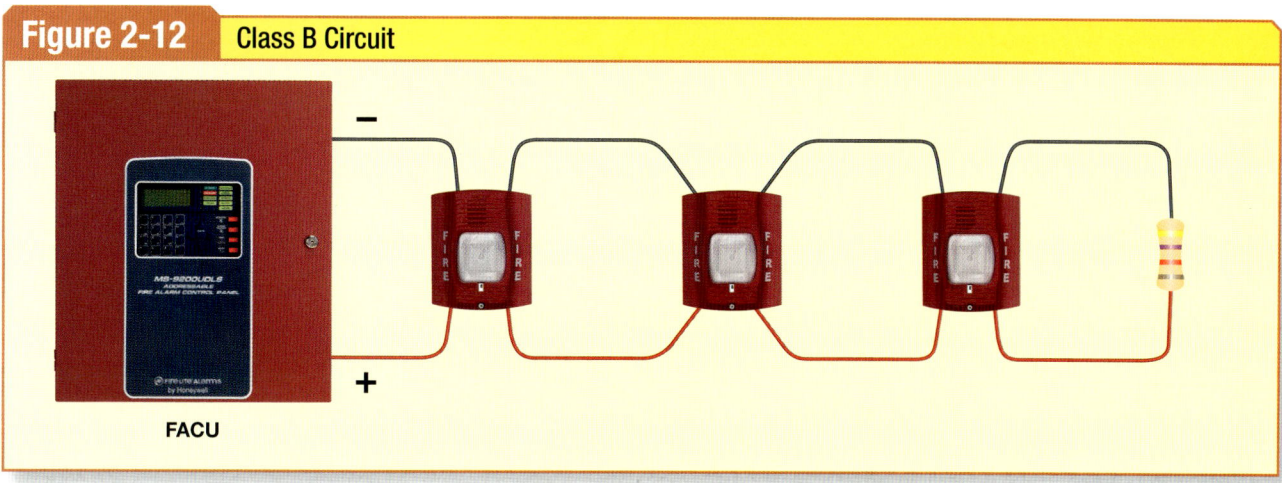

Figure 2-12. Class B circuits may be less reliable than Class A circuits because they do not have redundant pathways.

reported. Most signaling line circuit technologies do not require the use of an EOL since they monitor by polling the devices rather than using monitoring current. An open will cause a loss of all devices or appliances electrically downstream of the fault. A ground fault may cause a loss of all devices or appliances electrically downstream of the fault. Class B circuits are often used in lower risk environments, such as a small mercantile occupancy or a small office building.

Class C Circuits. Class C circuits are covered by 12.3.3 and have a redundant path, where end-to-end communications are verified. A loss of end-to-end communications is annunciated at the system controls. However, monitoring for integrity is not required for Class C circuits. **See Figure 2-13.** Class C circuits are new and are intended to include a local area network (LAN) connection between control units, or an Internet connection to a supervising station. The public switched telephone network (PSTN) is an example of a Class C circuit. These circuits are sometimes referred to as managed facilities-based voice networks (MFVN). It should be noted that *NFPA*

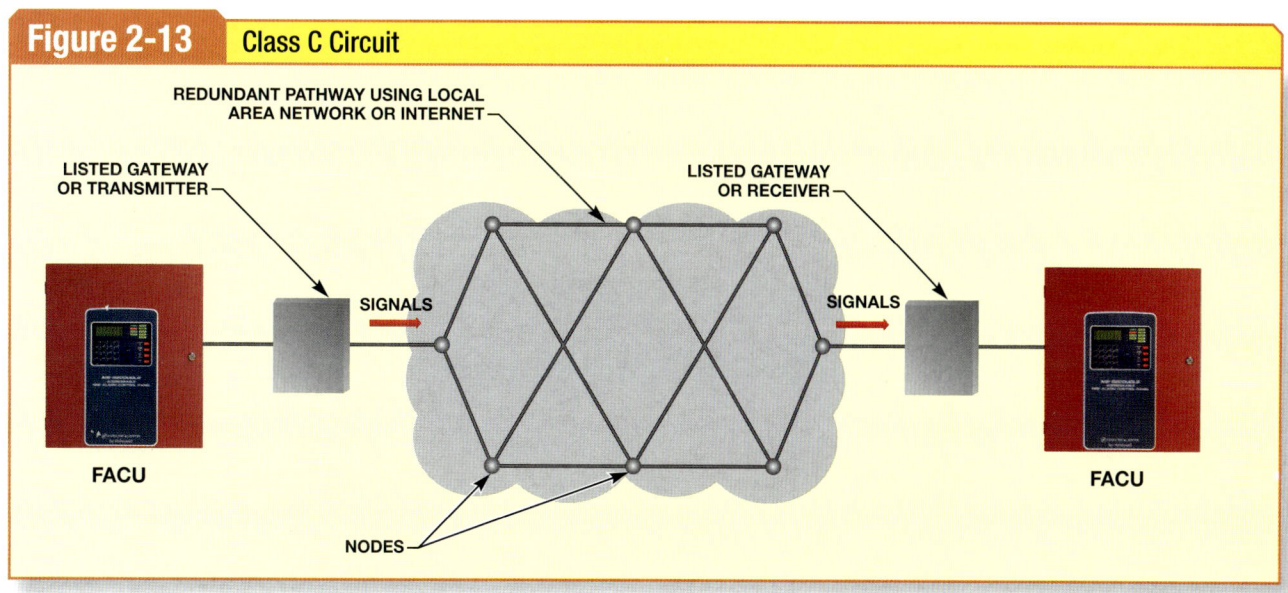

Figure 2-13. One kind of Class C circuit is the LAN.

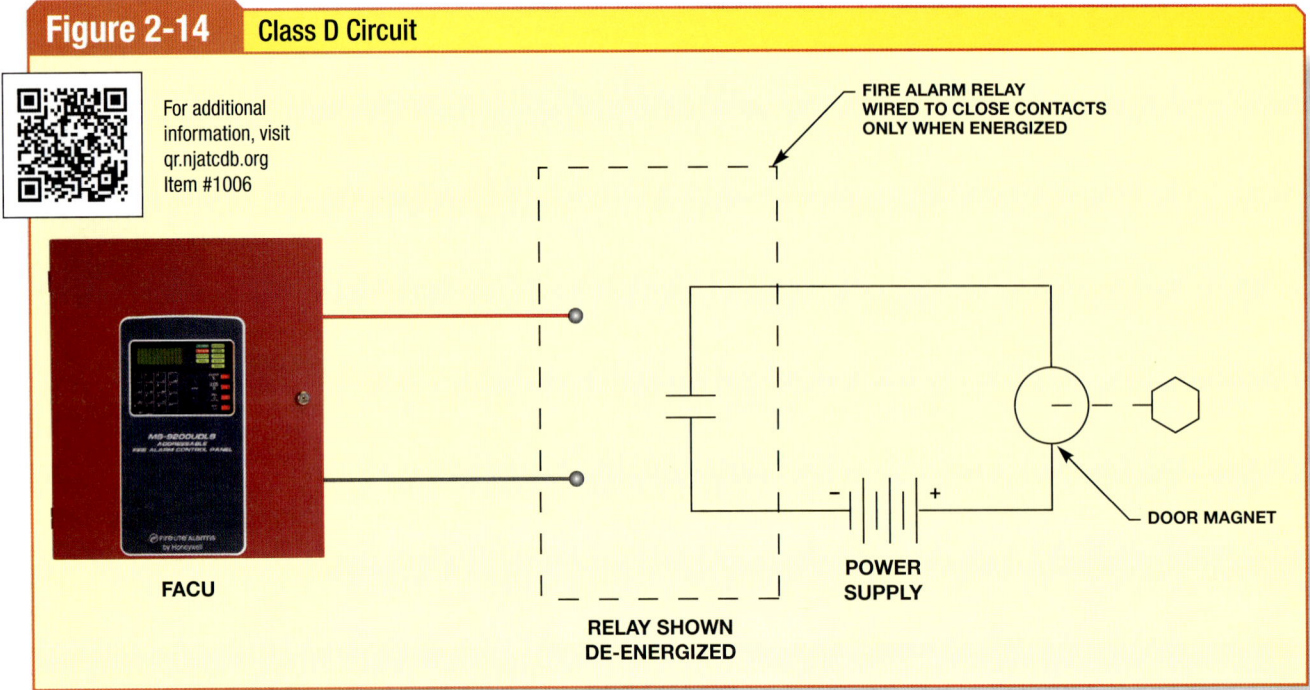

Figure 2-14. When a Class D circuit fails, its devices assume a safe mode of operation. For instance, if a fire door is supposed to close in the event of a fire, it will close if its circuit loses connection.

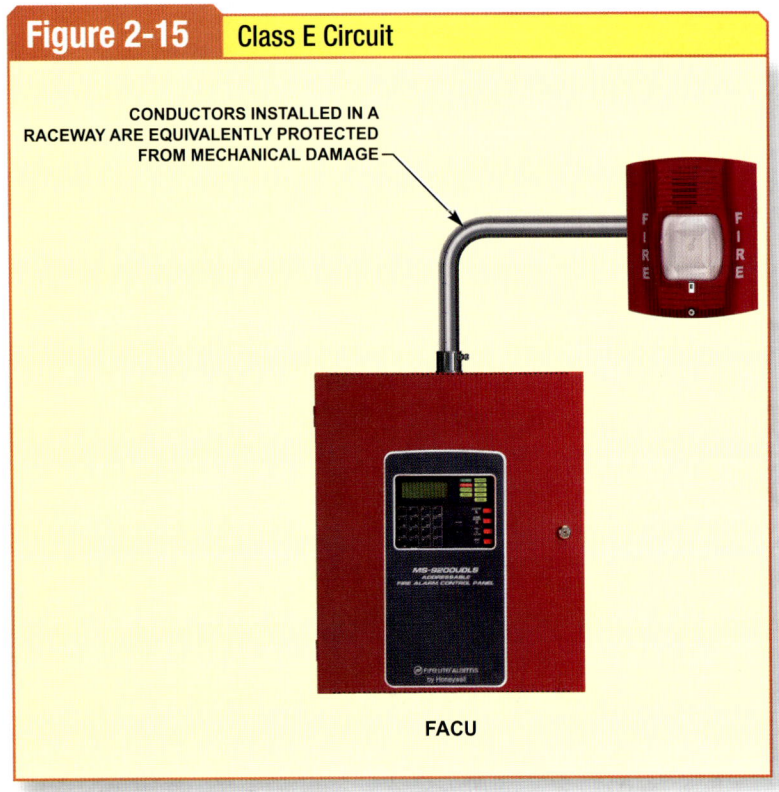

Figure 2-15. Class E circuits are not monitored for integrity because their failure will be obvious.

72 has no scope over MFVNs, and the requirements of the *Code* cannot be enforced on service provider's pathways.

Class D Circuits. Class D circuits are covered by Section 12.3.4 and are intended to be used for safety control functions, such as door unlocking or smoke door closure. Because Class D circuits are considered fail-safe, they are not monitored for integrity. Failure of the circuit, such as an open or short, will cause the control function to fail in the safe mode. Again, fail-safe circuits have been used for many years, but the class of circuit was created to address them in the *Code*. **See Figure 2-14.**

Class E Circuits. Class E circuits are covered by Section 12.3.5 of *NFPA 72*. Class E circuits are not required to be monitored for integrity by Section 12.6 and do not annunciate failure at any control unit. An example of a Class E circuit is a notification appliance circuit in the same room as the controls where conductors are protected from mechanical damage. **See Figure 2-15.**

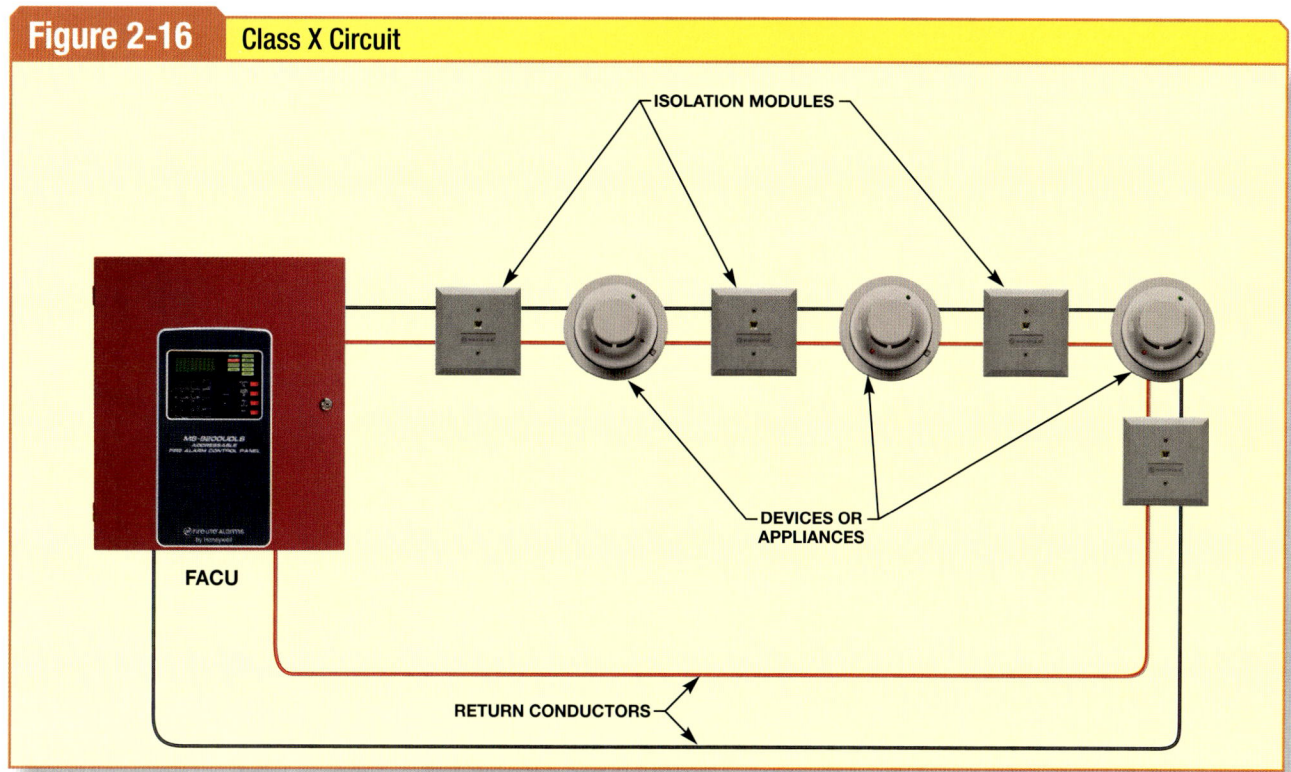

Figure 2-16. Class X circuits are enhanced versions of Class A circuits.

Class X Circuits. Class X circuits are essentially the same as Class A, Style 7 circuits in previous editions of the *Code*. Section 12.3.6 of *NFPA 72* provides requirements for Class X circuits. All Class X circuits must have a redundant pathway and must operate beyond a single open, short, or ground fault. The way this is accomplished is to start with a basic Class A circuit and add an isolation module between each device or appliance on the loop. This isolation module allows the same operation as a Class A circuit but offers additional short-circuit protection. **See Figure 2-16.**

Class X circuits are used where extremely high reliability is desired. An example of where this type of application is used is at a nuclear power plant.

The designer is responsible for choosing the circuit class. He or she will select the class of circuit based upon fire protection goals and the hazards involved on the premises. Higher risks generally demand the use of more reliable Class A or Class X circuits. For example, hospital fire alarm systems are frequently designed with Class A circuits, but low-rise office buildings often use Class B circuits. Where higher reliability is needed, designers generally specify Class X circuits.

Where Class A or Class X circuits are used, Section 12.3.7 of *NFPA 72* contains special requirements for the separation of outgoing and return conductors. Where Class A or Class X circuits are installed, the outgoing and return conductors must be physically separated to prevent total failure of the circuit in the event of fire. Section A.12.3.7 recommends a separation for Class A or Class X circuits of one foot where run vertically and four feet where run horizontally. However, Annex A is non-mandatory but may be amended locally. For example, the Federal Aviation Administration (FAA) enforces A.12.3.7 as a requirement.

There are three exceptions to the Class A and Class X separation requirements. These "exceptions" are written in positive language, rather than actual exceptions. The first exception permits outgoing and return conductors to be run together within 10 feet of a device, appliance, or

For additional information, visit qr.njatcdb.org
Item #1005

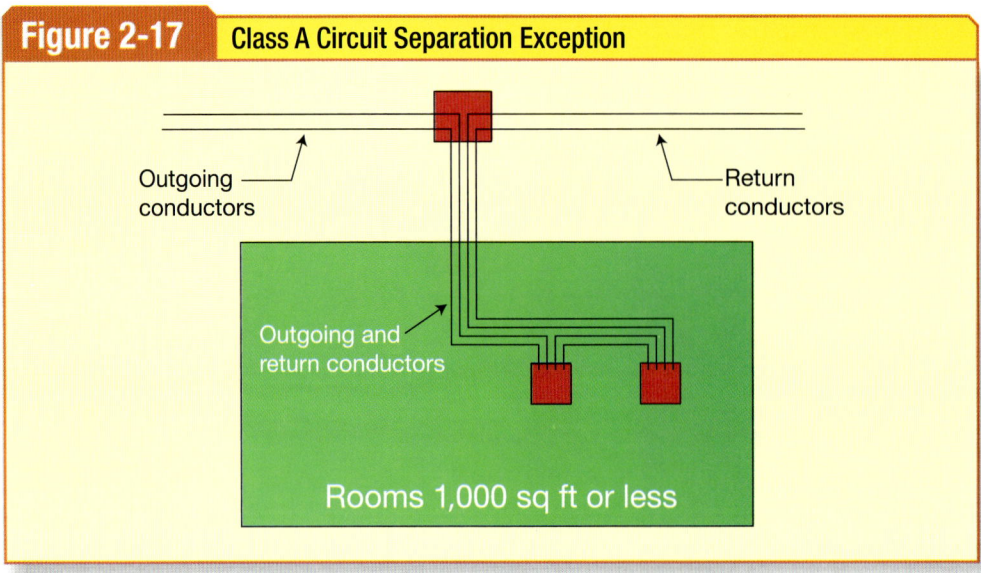

Figure 2-17. In rooms under 1,000 square feet, unlimited drops are allowed.

equipment. This exception applies to any wiring method (for example, cable or raceway). The second exception permits unlimited length conduit drops to single appliances or devices. The third exception allows unlimited drops to multiple devices in rooms not exceeding 1,000 square feet in area. **See Figure 2-17.** The last two exceptions assume that a failure will affect a limited number of appliances or devices.

POWER SUPPLIES

Section 10.6 provides requirements for all power supplies. Fire alarm system power supplies must be very reliable. For this reason, Section 10.6.3.2 requires fire alarm systems to have two independent power supplies: one primary and one secondary. Section 10.6.5.1 requires the primary supply to consist of a commercial utility source, an engine-driven generator, or an engine-driven generator arranged for co-generation where a trained operator is in constant attendance. The most common primary power supply is commercially available utility-provided power, but the other options exist where utilities are not available or reliable.

Primary Supply

Primary wiring must conform to Chapters 1 through 4 of *NFPA 70, National Electrical Code*. However, *NFPA 72* adds some requirements that are unique to fire alarm systems. Section 10.6.5.1 requires the fire alarm system to be powered by an individual (dedicated) branch circuit that carries no other loads. No other systems or outlets can be connected to this circuit, including receptacles or luminaires. Multiple fire alarm control units can be connected to this single branch circuit provided that it has adequate capacity.

Section 10.6.5.3 requires the primary power supply circuit to be mechanically protected. Generally, this can be accomplished through the use of conduit, raceway, or by fishing cable into a wall cavity. Section 10.6.5.2.2(1) requires the fire alarm circuit disconnecting means to be labeled, "FIRE ALARM CIRCUIT." Labeling for an emergency communications

---

10.6.5.1 Branch Circuit. A branch circuit supplying the fire alarm equipment(s) or emergency communication sustem(s) shall supply no other loads and shall be supplied by one of the following:
(1) Commercial light and power
(2) An engine-driven generator or equivalent in accordance with 10.6.11.2, where a person specifically trained in its operation is on duty at all times
(3) An engine-driven generator or equivalent arranged for cogeneration with commercial light and power in accordance with 10.6.11.2, where a person specifically trained in its operation is on duty at all times

(Excerpt from NFPA 72.)

system disconnecting means must be "EMERGENCY COMMUNICATIONS," and a combination fire alarm/emergency communications system disconnecting means must be labeled "FIRE ALARM/ECS." Additionally, this section requires the disconnecting means to be red in color and accessible only to authorized personnel. Accessibility may be limited through the use of listed circuit-breaker locks, a locked cabinet, or a locked room. Finally, Section 10.6.5.2.1 requires the location of the disconnecting means to be identified at the control unit, for the purposes of servicing the equipment.

Secondary Supply

According to national statistics, commercial (utility) power supplies are about 99.7% reliable. However, this leaves 22 hours per year when a building is left without power. This figure represents an average, so some buildings will suffer more outages, some fewer. Therefore, the fire alarm system must have a back-up power supply. There are different requirements for secondary power supplies depending on whether they are used on supervising station fire alarm systems or protected premises (local) systems.

Section 10.6.7.3.1 requires a protected premises (local) fire alarm system's secondary power supply to consist of either a battery, or a dedicated branch circuit of an engine-driven generator plus four hours of battery capacity. The four hours of standby batteries are required to carry the system in case the generator fails to start. Additionally, Section 10.6.11.7 requires a separate battery and charger for the engine-driven generator, which cannot be used for any other loads.

Automatic engine-driven generators used as secondary power supplies on protected premises systems are required to meet the requirements of *NFPA 110, Standard for Emergency and Standby Power Systems*, Chapter 4, for a Type 10, Class 24, Level 1 power system. Manual-starting, engine-driven generators used to provide secondary power are not permitted. Of course, wiring for emergency generators must comply with Article 700 of *NFPA 70, National Electrical Code*.

> **10.6.7.3* Secondary Power Supply for Protected Premises Fire Alarm Systems and Emergency Communications Systems.**
>
> **10.6.7.3.1** The secondary power supply shall consist of one of the following:
> (1) A storage battery dedicated to the system arranged in accordance with 10.6.10
> (2) An automatic-starting, engine-driven generator serving the branch circuit specified in 10.6.5.1 and arranged in accordance with 10.6.11.3.1, and storage batteries dedicated to the system with 4 hours of capacity arranged in accordance with 10.6.10
>
> *(Excerpt from NFPA 72.)*

Section 10.6.11.3.2.1 contains requirements for secondary power supplies in supervising station systems. These requirements are different from other fire alarm systems.

Engine-driven generators used as secondary power supplies on supervising station systems are covered by the requirements of Section 10.6.11.3.2.1. Automatic-starting, engine-driven generators used to provide secondary power for a supervising station shall comply with *NFPA 110, Standard for Emergency and Standby Power Systems*, Chapter 4, requirements for a Type 60, Class 24, Level 2 system. Installation shall be in accordance with *NFPA 70, National Electrical Code*, Article 701. Manual-starting, engine-driven generators used to provide secondary power for a supervising station shall comply with *NFPA 110, Standard for Emergency and Standby Power Systems*, Chapter 4, requirements for a Type M, Class 24, Level 2 system. Installation must be in accordance with *NFPA 70, National Electrical Code*, Article 702.

Section 10.6.7.3.2 requires circuits that supply secondary power from outside the control unit to be mechanically protected. This could include a remote battery supply or an uninterruptible power supply (UPS).

Secondary Power Calculations. Section 10.6.7.1.1 requires the fire alarm system to operate normally while using the

Figure 2-18. Secondary Power Calculations

Component Description	Quantity	Standby Current	Total Standby Current	Alarm Current	Total Alarm Current
Control Unit	1	0.575 A	0.575 A	1.230 A	1.230 A
Manual Fire Alarm Boxes	35	0.005 A	0.175 A	0.010 A	0.350 A
Smoke Detectors	90	0.007 A	0.630 A	0.010 A	0.900 A
Horns	53	0.000 A	0.000 A	0.155 A	8.215 A
Control & Monitor Modules	15	0.010 A	0.150 A	0.015 A	0.225 A
Strobes	62	0.000 A	0.000 A	0.175 A	10.850 A
NAC Booster Panels	3	0.325 A	0.975 A	0.390 A	1.170 A
		Totals	2.505 A		22.940 A

Figure 2-18. A typical system's load requirements helps to calculate backup requirements.

secondary power supply, except that a trouble signal indicating a power loss must be annunciated.

The capacity of the secondary power supply must be no less than 24 hours in standby (quiescent) load, followed by five minutes of alarm. In-building fire emergency voice/alarm communications systems are required to have no less than 24 hours of standby (quiescent) load, followed by two hours of emergency operation. However, two hours of emergency operation can be simulated by 15 minutes of all connected load, commonly known as "all call." This vastly simplifies calculations, since two hours of emergency operation may be different for every building.

Secondary battery power supply capacity is calculated in several steps. The first step is to determine the standby and alarm current draw for all devices, appliances, and equipment. The current is multiplied by the time required to obtain ampere-hour capacity. Power supply calculations for generators require the connected alarm system load and generator fuel-use rates. Combination systems that include fire alarm system functions and other functions, such as access control, must also have adequate capacity to carry the entire system for the prescribed time. **See Figure 2-18.**

Multiplying the standby and alarm currents by the time they must operate and adding them together will provide the battery capacity. For example:

Standby Capacity = (2.505 A) × (24 hours)
= 60.12 Ampere Hours (AH)

Alarm Capacity = (22.940 A) × (0.083 hours)
= 1.904 Ampere Hours (AH)

The total capacity is the sum of the standby and alarm capacity:

60.120 AH + 1.904 AH = 62.024 AH

Section 10.6.7.2.1(1) requires a 20% safety factor for all batteries. Therefore:

62.024 X 1.20 = 74.429 AH

Fact

Section 10.6.10.1.1 now requires batteries to be marked with the month and year of manufacture. Where not marked by the manufacturer, the installer must obtain the information and mark the batteries.

The next higher standard size battery should be selected. It is not critical to calculate battery capacity to three decimal places. One decimal place is usually sufficient.

In the example, the alarm current was calculated with all initiating devices in alarm. This is a conservative approach. Some designers will use the number of devices in the largest area or zone for the purposes of calculating alarm current. Because one cannot determine how many devices will be actuated during the alarm, however, it is usually best to plan for the worst case and include all devices in the alarm current calculation.

Alarm current is almost always larger than standby current. The conductors connecting the battery to the fire alarm controls must be sized so that they can safely deliver the current necessary to operate the system in both alarm and standby. Additionally, the battery must be capable of delivering the required current at rated voltage. The battery specification sheets or discharge curves will show if this is the case. Some batteries cannot deliver the required current at nominal voltage. This condition will result in notification appliances that do not deliver the required output.

Section 10.6.6 provides requirements for continuity of power supplies. If signals are lost, interrupted, or delayed by more than 10 seconds, an uninterruptible power supply will be required. However, if batteries are provided, a UPS is usually not necessary.

Section 10.6.8.1 contains requirements for remotely located fire alarm equipment power supplies. Requirements for power supplies that supply remotely located fire alarm equipment are the same as for all other fire alarm equipment. This requirement includes power supplies for notification appliance booster power supply panels, remote control units, and transmitters.

Section 10.6.10.2 requires batteries to comply with Article 480 of *NFPA 70, National Electrical Code*.

It is imperative that fuel-use calculations for engine-driven generators supplying loads other than fire alarm systems include the other connected loads. Section 10.6.11.6 requires sufficient fuel capacity for engine-driven

Engine-driven generators can be used to provide both primary and secondary power for fire alarm systems. An automatic transfer switch and four hours of batteries are required when using a generator.

generators to provide six months of testing plus the required standby and alarm time specified for the system in Section 10.6. However, Section 10.6.11.6.2 permits fuel storage capacity for engine-driven generators to be 12 hours at full load if replenishment is available within two hours.

Engine performance is adversely affected by stale gasoline. For this reason, gasoline-powered engine-driven generators should be supplied from a frequently replenished tank. Where piped-in natural gas is used, on-site storage is not required by Section 10.6.11.6.3 unless the installation is located in a seismic risk zone 3 or greater. The risk zone is defined in ANSI A-58.1, *Building Code Requirements for Minimum Design Loads in Buildings and Other Structures.*

Summary

Monitoring for integrity is a basic tenet of fire alarm signaling. Life-safety systems are critical to the safety of occupants and, therefore, are required to be monitored for integrity.

Monitoring applies not only to the system as a whole, but also to its parts. Devices must be tested to ensure they operate. Circuits must be monitored to be sure they are functioning and that the devices on them can operate.

Different kinds of circuits require different kinds of monitoring. Some are monitored to determine if they are ON or OFF. Others use a polling technique to check their devices individually.

Finally, a safe and reliable fire alarm system depends on having dependable backup in the event of power failure. This backup may consist of batteries or generators or some combination of the two.

Review Questions

1. Loss of AC power is indicated by a(n) __?__ signal.
    a. alarm
    b. guard's tour
    c. supervisory
    d. trouble

2. Coded systems must transmit at least __?__ complete rounds.
    a. 2
    b. 3
    c. 4
    d. 5

3. Transmission of trouble signals by a supervising station system must occur between __?__.
    a. 30 and 60 minutes
    b. 60 and 90 minutes
    c. 60 and 120 minutes
    d. 60 and 180 minutes

Chapter 2 Fundamentals and System Requirements

Review Questions

4. Monitoring for integrity is intended to detect __?__.
    a. grounds
    b. loss of operating power
    c. opens
    d. all of the above

5. Non-addressable (conventional) devices can be connected to an addressable system by using a __?__.
    a. control module
    b. monitor module
    c. monitor relay
    d. relay module

6. Monitoring for integrity of IDC and NAC is accomplished through the use of a(n) __?__.
    a. ELO
    b. EOL
    c. LEO
    d. OEL

7. Addressable circuits monitor for integrity by __?__.
    a. a small current through the circuit
    b. interrogation and response
    c. monitoring current
    d. none of the above

8. Class __?__ circuits have return conductors between the last device and the controls.
    a. A
    b. B
    c. C
    d. D

9. The __?__ is responsible for selection of the circuit class used on a system.
    a. designer
    b. electrical inspector
    c. fire marshal
    d. insurance interest

10. Secondary power supplies must operate for a period of at least __?__.
    a. 12 hours
    b. 24 hours
    c. 48 hours
    d. 60 hours

11. *NFPA 72* requires fire alarm equipment to operate between __?__.
    a. 40°F and 100°F
    b. 40°F and 120°F
    c. 32°F and 100°F
    d. 32°F and 120°F

Initiating Devices

Initiating devices are the eyes and ears of the fire alarm system. The primary purpose of initiating devices is to contribute to life safety, property protection, and fire protection by providing a reliable means of initiating and sending signals to the fire alarm controls. They are placed throughout the protected area and are connected to and monitored by the system control unit. Although *NFPA 72* does not specify where detectors are required, it does specify installation requirements. *NFPA 101,* Life Safety Code, and building codes specify the location and coverage requirements for initiating devices. Depending on the type of device used, the signal generated may be an alarm or a supervisory signal. Initiating devices can be addressable devices on signaling line circuits or conventional devices on initiating device circuits.

Objectives

- » Distinguish alarm from supervisory devices
- » Explain operating theory for heat detectors, smoke detectors, and multi-sensor detectors
- » Describe the installation requirements for:
    - › General coverage
    - › Mechanical protection
    - › Heat detectors
    - › Smoke detectors
    - › Waterflow initiating devices
    - › Manual fire alarm boxes
    - › Supervisory signal–initiating devices

Chapter 3

Table of Contents

Types of Initiating Devices 58
    Alarm Signal Initiating Devices 58
    Supervisory Signal–Initiating Devices ..59
Initiating Device Operating Theory 59
    Heat Detectors 59
    Smoke Detectors 63
Installation and Spacing Requirements for Initiating Devices 65
    Detection Coverage Requirements.......66
    Heat Detectors 67
    Smoke Detectors..................................73
    Sprinkler Waterflow Alarm-Initiating Devices 78
    Manually Actuated Alarm-Initiating Devices 79
    Supervisory Initiating Devices 80
Summary 82
Review Questions..82

TYPES OF INITIATING DEVICES

Initiating devices may be alarm or supervisory signaling devices. Automatic alarm-initiating devices are designed to alert system operators, responders, and occupants of a life-threatening condition, such as a fire. There are many types of alarm-initiating devices on the market today. These include smoke detectors, heat detectors, radiant energy sensing (flame or spark/ember) detectors, and even video flame image detectors. Smoke detectors are frequently used in fairly clean environments, while heat detectors are generally used in more harsh environments. Radiant energy-sensing fire detectors and video image flame detectors are most often used in industrial applications; they may not be the best choice for all environments.

Manual fire alarm boxes are designed to allow occupants to manually create an alarm signal. These devices are usually located near exits from the building. Manual fire alarm boxes are the oldest form of initiating device on the market today and are widely used in nearly all occupancy types.

Supervisory signal–initiating devices are designed to alert system operators that a suppression system is in an off-normal condition. These devices include valve supervisory (tamper) switches, pressure switches, water level switches, and temperature switches.

It is essential that installers and designers understand the basic operation of initiating devices so that nuisance alarms are avoided while providing early warning. Early warning buys time to react, evacuate, and suppress or extinguish the fire. Since many fires grow exponentially, time is the most important element in the detection process. However, poorly designed systems will create nuisance alarms because the devices are improperly selected or installed. Nuisance alarms create apathy among occupants and must be avoided. On the other hand, poorly designed and installed systems may not create an alarm signal until it is too late to react. Both of these conditions must be avoided. Proper design and installation is critical to the success of the system. The line between early actuation and nuisance alarms is one that must be carefully assessed.

Alarm Signal Initiating Devices

Typical alarm signal initiating devices include the following:
- Heat detectors
- Smoke detectors
- Flame detectors (radiant energy-sensing detector)
- Spark/ember detectors (radiant energy-sensing detector)
- Manual fire alarm boxes
- Waterflow detection devices (may include vane-type flow switches and pressure switches)

Another technology is video image flame detection. The requirements for these systems are found in Section 17.7.7 of *NFPA 72, National Fire Alarm and Signaling Code*. This technology uses raster scan software to detect movement of smoke and flames on an image from a video camera. Video image flame detection systems are very complex and are engineered for each application. Video image flame detection has been used successfully in tunnel projects, such as the Ted Williams Tunnel in Boston.

> **Fact**
> Always select the alarm initiating device to provide the fastest response, without causing nuisance alarms. Many smoke detectors are inappropriately placed in environments that present a false stimuli, which results in a nuisance alarm.

One of the oldest forms of initiating devices is the manual fire alarm box.

Alarm signal initiating devices create alarm signals. As defined, a fire alarm signal is a signal initiated by a fire alarm initiating device, such as a manual fire alarm box, automatic fire detector, waterflow switch, or other device in which activation is indicative of the presence of a fire or fire signature. Alarm initiating devices are wired or programmed to cause an alarm signal, depending on how they are used. Conventional controls generally use zone cards, which create alarm signals when a device on the zone shorts its contacts. However, addressable controls use software to cause an alarm signal when alarm thresholds are reached by the initiating device.

Supervisory Signal–Initiating Devices
Typical supervisory signal–initiating devices include the following:
- Valve supervisory (tamper) switches
- Water level switches
- Air pressure switches
- Room temperature switches
- Water temperature switches
- Carbon monoxide (CO) and other gas detectors

Supervisory signal–initiating devices are intended to alert operators that a suppression system is off normal and requires attention. However, gas detection may be used to create a supervisory signal that alerts operators of a gas leak or other hazardous condition. Conventional controls generally use zone cards, which create supervisory signals when a device on the zone shorts its contacts. But, addressable controls use software to cause a supervisory signal when thresholds are reached.

Within each basic detector type, there may be one or more operating principles. For example, there are several different types of smoke detectors. The type of detection is based upon the environmental conditions in the space being protected. Designers attempt to minimize nuisance alarms while providing the fastest detection possible. For example, smoke detectors are not used in parking garages or kitchens because exhaust and cooking fumes may cause nuisance alarms. In these cases, heat detection may be a better choice. Section 17.7.1.9 of *NFPA 72* requires the

A valve supervisory (tamper) switch is a typical supervisory signaling device.

ambient conditions to be evaluated for the presence of conditions that will cause nuisance alarms. Failure to properly evaluate and apply good judgment will result in nuisance alarms.

INITIATING DEVICE OPERATING THEORY
There are several kinds of initiating devices, each operating on differing principles.

Heat Detectors
Heat detectors are the oldest form of automatic detection available on the market today. Heat detectors respond to the heat produced by a fire and generally operate when the temperature of the element in the detector reaches a certain temperature limit or rate of temperature increase, depending on the type of detector. There are two different operating principles: fixed temperature and rate-of-rise. Fixed temperature heat detectors operate using

several methods: fusible link (non-restorable), bimetallic strip (restorable), rate compensation, and heat sensitive cable (line-type heat detection). New electronic heat detectors are also available for addressable systems.

Non-Restorable Fixed Temperature Heat Detectors. Fixed temperature heat detectors actuate when the sensing element reaches a predetermined temperature. By contrast, rate-of-rise heat detectors actuate on an increasing temperature change, usually 12°F to 15°F (6.6°C to 8.3°C) per minute. Alternately, electronic-type heat detectors use thermistors (electronic elements) to sense the temperature in a space and can report the actual temperature of the sensor to the control unit. This type of heat detector is used with addressable systems and the software in the controls will determine the output.

Fixed-temperature, non-restorable heat detectors generally use a fusible link type of element. A pair of contacts is held open by a spring-loaded plunger, which is in turn held in place by a eutectic solder. Eutectic materials melt very quickly when the materials reach their set point. A heat collector around the eutectic material helps to decrease the time required for response. The eutectic solder melts very quickly and releases the plunger once it absorbs sufficient heat.

As the plunger is released, the contacts close across the initiating device circuit, creating an alarm. Once the eutectic material has melted, the device must be replaced. **See Figure 3-1.**

Restorable Fixed-Temperature Heat Detectors. Fixed-temperature restorable heat detectors are usually of the bimetallic strip type. The bimetallic strip is made of two metals having different coefficients of thermal expansion. The bimetallic strip will bend in one direction when cool, and the opposite direction when heated. The contacts are mounted so that they are open when the device is cool and closed when heated to the set point. This design results in a restorable device, provided that the element is not overheated. **See Figure 3-2.**

One problem associated with heat detectors is the phenomenon known as thermal lag. Thermal elements will not immediately actuate when the surrounding ambient room temperature reaches the set point of the sensor. This is because the sensor must also absorb sufficient heat to bring it to the designed set point. Therefore, thermal lag is the time required to heat the detector element to its operating point. For example, a 165°F (73.8 °C) fixed-temperature heat detector might operate when the ceiling temperature is much higher than the set point of the

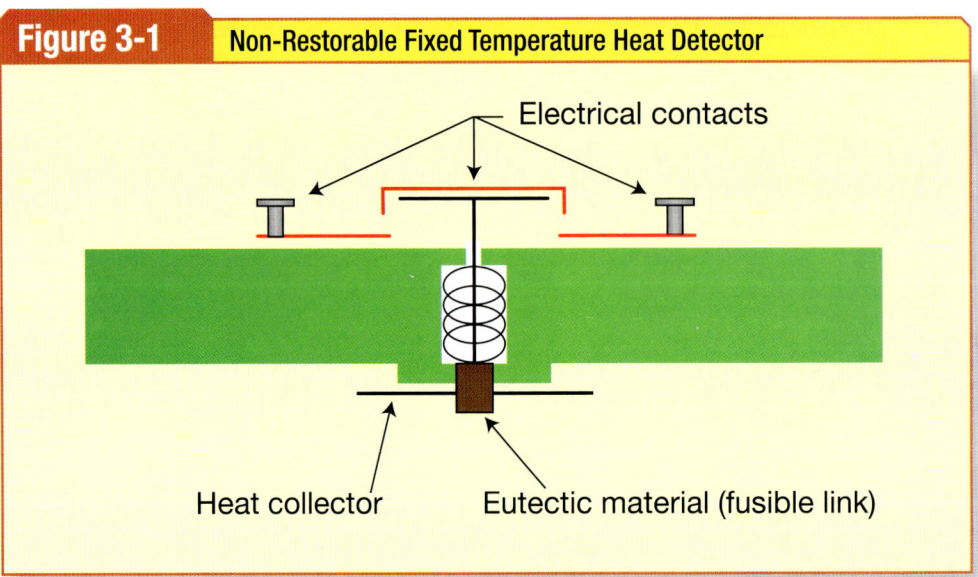

Figure 3-1. When the fusible link melts, the contacts close and an alarm is initiated.

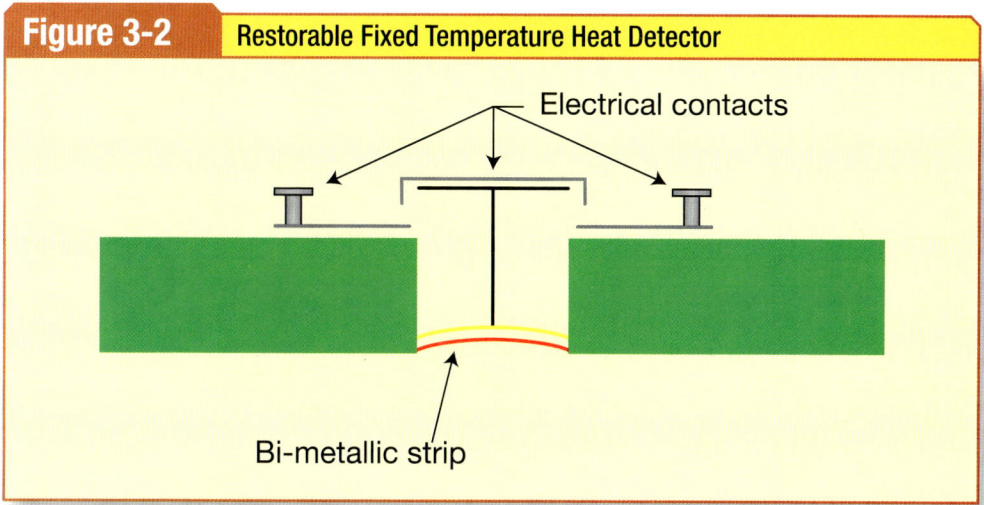

Figure 3-2. The bimetallic strip provides a reusable heat detector.

detector. Of course, the actual ceiling temperature at actuation will depend greatly on the fire growth rate and ambient conditions. For this reason, heat detectors are not well suited to life-safety protection.

Typical applications for fixed-temperature heat detectors are listed below:
- Unheated areas
- High-temperature ambient areas
- High-humidity areas
- Dirty or dusty areas
- Locations where flashing fires can be tolerated before detection
- Other areas as required by the AHJ or local codes

Rate-Compensated Heat Detectors. Rate-compensated, or rate-anticipated, heat detectors are designed to reduce thermal lag. Rate-compensated heat detectors are very sensitive and generally actuate when the ambient temperature in the protected space is much closer to the actual design set point. They are designed with an overall metal alloy tubular casing that quickly absorbs heat from the fire. Inside, two insulated struts hold electrical contacts apart until there is sufficient elongation (from heat) of the casing to close the contacts and create an alarm. When the device cools, the contacts open and the device is restored to normal. **See Figure 3-3.**

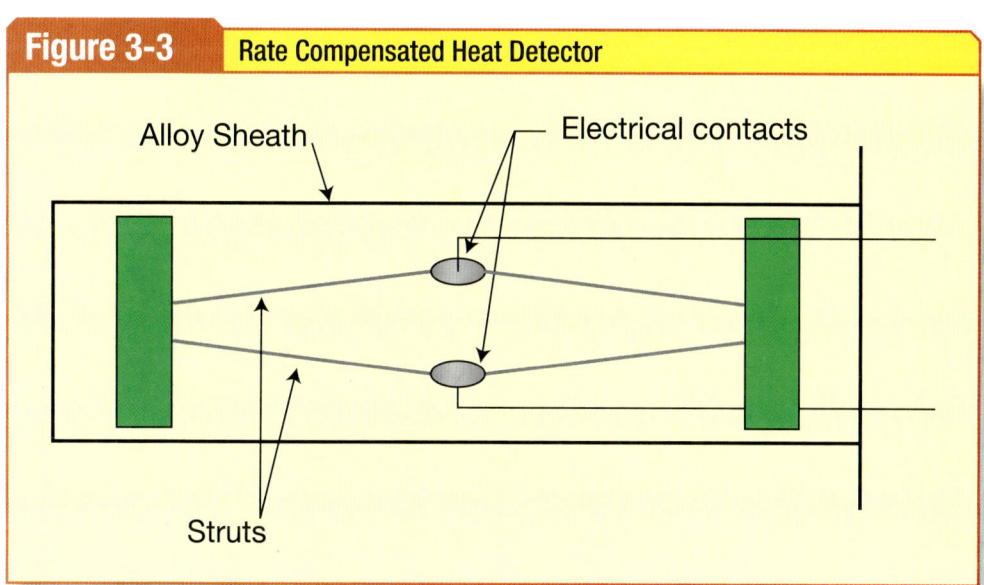

Figure 3-3. Heat causes the casing to expand, closing the contacts and issuing an alarm.

Line-Type Heat Detectors. Heat-sensitive cable is usually referred to as a line-type heat detector. This type of detector utilizes a pair of twisted spring-steel conductors that are insulated from each other by a heat-sensitive thermoset plastic. As the heat from a fire impinges on the cable, the plastic melts, allowing the conductors to short and cause an alarm. This type of detector is not restorable, but patch kits are available to repair damaged areas. Line-type heat detectors are generally used to protect coal conveyors, cable trays, tunnels, or other elongated hazards. They are also used in large areas requiring heat detection, such as cold storage warehouses.

Rate-of-Rise Heat Detectors. Rate-of-rise (ROR) heat detectors respond to an increasing temperature change. Most rate-of-rise heat detectors operate on a 12°F to 15°F (6.6°C to 8.3°C) change per minute. These detectors have a diaphragm inside that reacts to changes in air pressure caused by a fire. As the air is heated in a fire, it expands, and pressure is exerted on the diaphragm, resulting in a contact closure. If the rate of temperature rise is insufficient to cause an alarm, rate-of-rise detectors are also outfitted with non-restorable fixed temperature contacts. ROR heat detectors are provided with a small vent on the back of the unit. This vent is factory calibrated and sealed; it must not be blocked in order to ensure proper operation. **See Figure 3-4.**

Rate-of-rise heat detectors should not be installed in the following areas in order to avoid nuisance alarms:
- Kitchens
- Areas affected by sunlight where a rapid temperature increase may occur
- Areas subjected to air conditioning/heating being turned ON and OFF
- Excessively dusty areas or areas where debris may plug the vent, which prevents operation

Electronic Heat Detectors. Electronic-type heat detectors are manufactured primarily to operate on signaling line circuits. These detectors have a sensing element and sensing circuit, which report the actual temperature in the protected space. These units are generally available only in a 135°F (57.2°C) fixed-temperature rating because the electronic components cannot operate in sustained ambient temperatures above that point. In

Fact

Rate-of-rise heat detectors offer a faster response than fixed temperature heat detectors. However, they may cause a nuisance alarm in areas subject to a rapid temperature rise, such as an area where heating equipment is frequently energized.

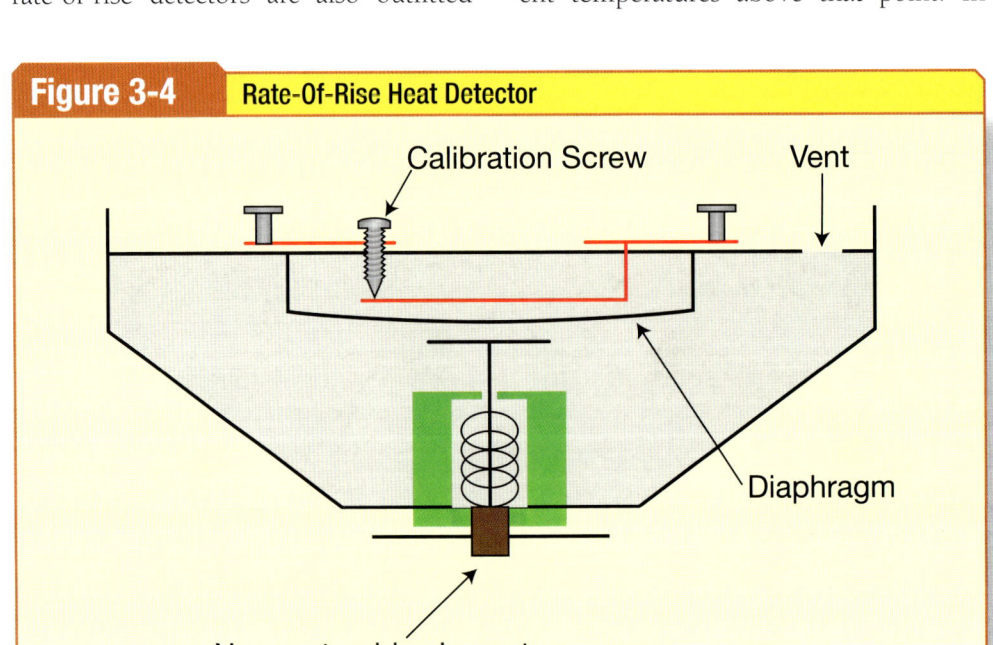

Figure 3-4. ROR heat detectors use a diaphragm and a fixed temperature contact mechanism.

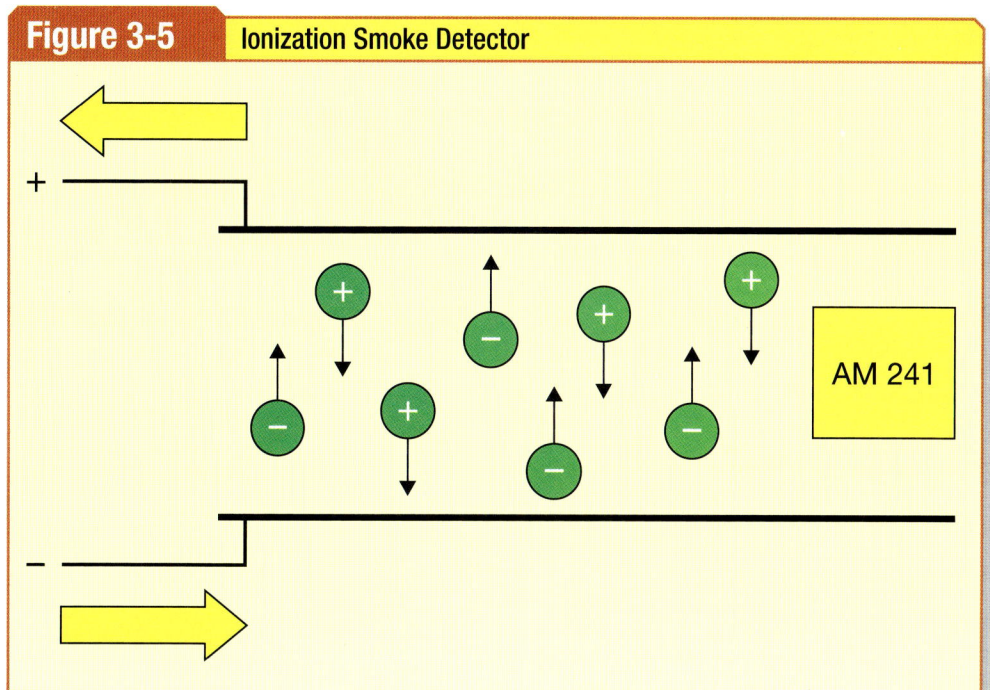

Figure 3-5. Ionization smoke detectors use very low levels of radioactivity to achieve results.

fact, many electronic heat detectors are not listed for ambient temperatures above 100°F (38°C). However, electronic heat detectors are also usable as rate-of-rise detectors because they are software controlled and can be programmed to respond to rapid temperature rises. These units are self-restoring, provided that they are not heated to the point at which the electronic components are damaged.

Smoke Detectors

Smoke detectors respond to particles of combustion produced by fires. There are five basic types of smoke detection:
1. Ionization
2. Photoelectric light scattering
3. Photoelectric light obscuration
4. Particle counting
5. Air sampling

Spot-type smoke detectors are available in either ionization or photoelectric light-scattering types. Photoelectric light obscuration is primarily available in projected beam smoke detection. Air sampling smoke detection is a very sensitive type of detection and uses lasers (particle counting) or light obscuration to detect very small quantities of smoke.

Ionization Smoke Detectors
Ionization smoke detection uses a small radioactive source to charge, or ionize, the air between two plates. The ionized air between the plates conducts a very small electrical sensing current across the gap between the plates. As smoke particles enter the chamber, they attach themselves to the air molecules and decrease the sensing current flow between the plates. This reduction in current is sensed as an alarm. Ionization detection uses a very low-level Americium 241 radioactive material as the ionizing source. This material is a very low-level source, which is relatively harmless to humans. Americium 241 will not oxidize and will not release more particles in a fire. Sitting within one foot (0.3 m) from an ionization smoke detector for an entire year would result in a dose of less than one millirem of radiation. The Nuclear Regulatory Commission has determined that 5,000 millirem per year is a safe limit. Ionization detectors are well-suited to detecting small particles of less than one micron in diameter. Flaming fires tend to produce particles in the one micron size range, which makes ionization-type smoke detectors well suited to detecting flaming fires. **See Figure 3-5.**

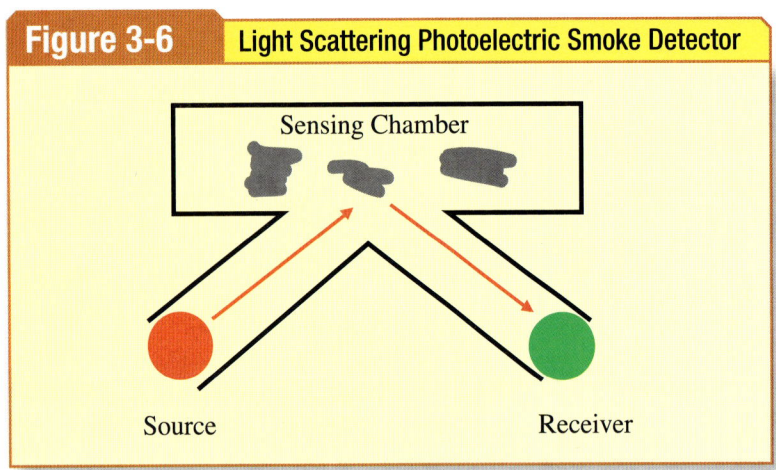

Figure 3-6. As more light is "scattered" onto the receiver by smoke particles, the photoelectric light scattering smoke detector issues an alarm.

For additional information, visit qr.njatcdb.org Item #1007

Photoelectric Light-Scattering Smoke Detectors. Photoelectric light-scattering smoke detection uses a light source to detect smoke particles. This type of detection typically has a sensing chamber with an infrared light-emitting diode as a source. The source and receiver are not directly aimed at each other. However, the light source is monitored for integrity by a receiver that is directly aimed at the source. As particles enter the sensing chamber, light is reflected, or scattered, onto a receiver. As more light is reflected onto the receiver, the sensing circuit perceives this as an alarm. Light-scattering smoke detection is most sensitive to large, light-colored particles (greater than one micron in diameter) because they reflect more light than small or dark particles. Smoldering fires tend to produce particles that are greater than one micron in size and photoelectric-type smoke detectors are well suited to detecting these products of combustion. **See Figure 3-6.**

Light Obscuration/Projected Beam Smoke Detectors. Light obscuration smoke detection also uses an infrared light source and receiver to detect smoke. However, the source and receiver are aimed directly at each other, across a large distance. Some models contain both the transmitter and receiver in the same housing and use a mirror to reflect the signal from the opposite wall. As smoke obscures the beam, there is less light received from the transmitter, which is sensed as an alarm. The transmitter and receiver may be placed hundreds of feet from each other, which make them well suited to protecting large open spaces, such as an atrium or gymnasium. The manufacturer's installation instructions and listing provide the area of coverage, which may vary from one product to another. Light

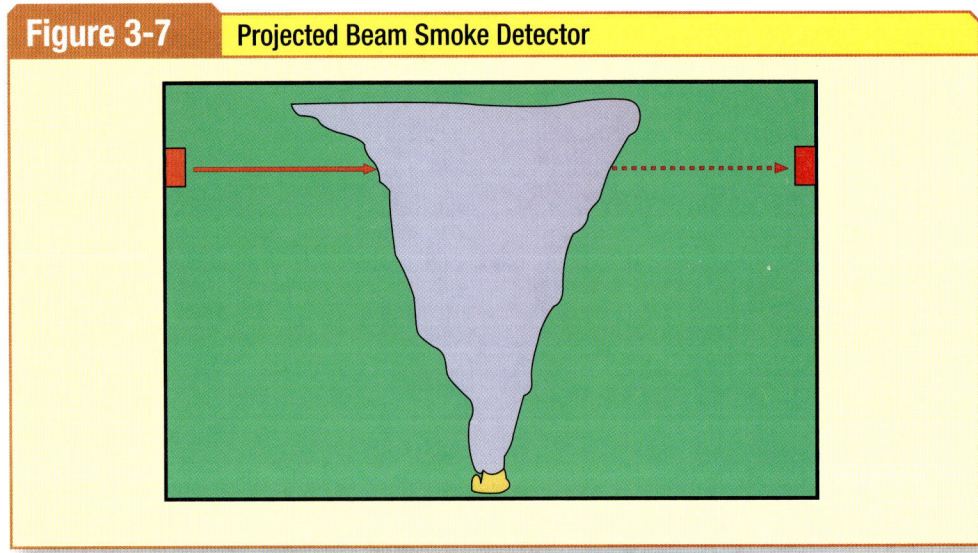

Figure 3-7. As smoke obscures the light beam, the receiver issues an alarm.

obscuration smoke detection is usually referred to as projected beam or linear smoke detection. **See Figure 3-7.**

Particle Counting (Laser) Smoke Detectors. Newer types of spot-type smoke detection have recently become available. These detectors use on-board lasers to count particles and are very sensitive. They are primarily designed for clean-room applications and are generally not well suited to ordinary open area protection. The spacing of these detectors is usually the same as for other smoke detectors, unless ambient conditions require an adjustment in the spacing.

Air Sampling Smoke Detectors. Air sampling smoke detectors can be more than a thousand times as sensitive as spot-type smoke detectors. They use a pipe network and a fan to sample the air in a protected space.

Multi-Sensor and Multi-Criteria Detectors. Multi-sensor detectors use two or more sensing technologies to create an alarm. The most common type of multi-sensor detector is a smoke and heat detector. Multi-sensor detectors will alarm if either sensor actuates (an "or" function) because there is no software algorithm to control the output. Multi-criteria detectors use multiple sensors and software to look for a condition that resembles a real fire, rather than false stimuli like cigarette smoke or dust.

The most commonly applied multi-criteria detectors also use both smoke and heat sensors. Carbon monoxide (CO) and carbon dioxide (CO_2) sensors are becoming more common. The difference between these detectors and multi-sensor technologies is the multi-criteria detector software will create an alarm when the combination of sensor inputs closely approximates a fire scenario or "signature." This output is controlled by a software algorithm in the programming. Multi-criteria detectors are becoming increasingly commonplace as manufacturers seek new ways to prevent nuisance alarms.

INSTALLATION AND SPACING REQUIREMENTS FOR INITIATING DEVICES

Section 10.3.3 of *NFPA 72, National Fire Alarm and Signaling Code,* requires all detectors to be compatible with the system controls. Conventional detectors are generally contact closure devices that might or might not receive power from the controls. Heat detectors and manual fire alarm boxes generally do not receive or require any operating power from the controls but are still required to be compatible because of contact voltage and current limitations.

Smoke detectors must be listed as compatible because they usually require operating power (12 or 24 volts direct current) from the controls. Most manufacturers prescribe limits on the number of smoke detectors on a single zone because of power limitations. Before any device is connected to an initiating device circuit, the device should be checked to ensure it is compatible with the zone card or controls. Addressable controls also require compatibility listings with the connected circuits. This is because they communicate in a proprietary "language" that is unique to each manufacturer or panel. With very few exceptions, initiating devices from one manufacturer can only be used with that manufacturer's controls/circuits.

Chapter 17 of *NFPA 72* covers the requirements for the selection, spacing, and installation of initiating devices. However, *NFPA 72* does not dictate where these

All detection devices must be compatible with their associated controls. The manufacturer will provide compatibility cross listings.

devices are required. Building codes, such as the *International Building Code* and *NFPA 5000, Building Construction and Safety Code*, address this issue. Although it is not a building code, *NFPA 101, Life Safety Code*, also determines what occupancies require detection.

Section 17.4 of *NFPA 72* contains general requirements for all initiating devices. These requirements are intended to ensure good installation practices and proper maintenance. Specifically, Section 17.4.3 requires detectors to be supported independently of their conductors. This is an important requirement since copper is a relatively soft material and will not indefinitely support the device without breaking. Section 17.4.4 also requires initiating devices to be installed in a manner to allow for periodic maintenance. Section 17.4.6 requires the use of duplicate terminals for the purpose of monitoring for integrity, unless another means of monitoring is provided. Looping conductors around device terminals is not permitted because a loose terminal may not result in a trouble signal.

There are several other requirements in Section 17.4 that pertain to all initiating devices. Specifically, Section 17.4.4 does not permit initiating devices to be installed in inaccessible areas. Installing initiating devices in inaccessible areas will prevent the maintenance and testing required by Chapter 14 of *NFPA 72*.

Section 17.4.2 requires all initiating devices subject to mechanical damage to be protected. For example, detectors in gymnasiums must be protected from damage by balls or other objects. Similarly, detectors installed in institutional occupancies, such as detention and correctional facilities, also require protection. A variety of mechanical guards are available, but must be listed for such use. In some cases, the initiating device may require a reduction in spacing because the mechanical guard reduces detector sensitivity. This is especially true for mesh-type guards, as opposed to open wire-type guards. The manufacturer's specification sheets will provide information related to derating of detectors when using guards.

One important requirement of Section 17.5.1 is that detectors cannot be recessed unless listed for such use. Recessing detectors will severely reduce their ability to detect fires because the smoke and heat from the fire will not always penetrate into a recessed space. Detectors should be mounted so they will be located in the flow of heat and smoke from the fire. There are currently no detectors listed for recessed mounting.

Detection Coverage Requirements

By itself, *NFPA 72* does not require detector coverage. Other codes, such as the *International Building Code*; *NFPA 5000, Building Construction and Safety Code*; and *NFPA 101, Life Safety Code*, provide requirements for where detection is required. Detection requirements in these documents are based upon specific use groups or occupancy types and reference *NFPA 72* for installation requirements.

Total (Complete) Coverage. Total (complete) coverage of detection devices is defined, but not required, by Section 17.5.3.1 of *NFPA 72*. Where required by another code, total coverage is intended to imply the use of automatic initiating devices in all rooms, corridors, storage spaces, basements, attics, lofts, spaces above suspended ceilings, and the inside of all closets, chutes, and hoistways. Inaccessible areas that do not contain combustibles do not require protection by detectors. However, inaccessible areas that contain combustibles must be made accessible and must be protected by detectors. This requirement does not require detection inside stud or joist cavities in walls or ceilings, but it does apply to pipe chases or other areas that contain combustibles or those that could be used for storage. There are

Fact

The requirements to have detection in a particular building, occupancy, or use group are found in other codes, such as the *International Building Code*; *NFPA 101, Life Safety Code*; or local codes. These codes reference *NFPA 72*, which provides the requirements for their installation.

other exceptions in Section 17.5.3.1 that apply to this option.

Total coverage is generally not an attractive option for building owners. High installation costs and, more importantly, maintenance costs will force most building owners to use other options. Enabling codes like the *IBC* or *NFPA 101* permit total coverage, but they also allow total sprinkler coverage or other means of initiation in place of total detection coverage.

Partial or Selective Coverage. Partial or selective coverage is anything less than total coverage. Section 17.5.3.2 provides requirements for partial or selective coverage. This is intended to permit coverage in certain areas of a building, such as corridors, means of egress, or other critical areas where required by *Code* or an AHJ. Where partial or selective coverage is employed in a design, all area coverage must meet the requirements of Chapter 17 of *NFPA 72* for detector spacing.

Non-Required Coverage. Non-required coverage is slightly different from partial coverage because the initiating devices are not required by any code or AHJ. Building owners often like to provide detector coverage in certain areas for early warning, but building owners usually do not want to install the correct quantity of detectors required by *NFPA 72* for the specific area. Where detectors are installed solely to protect against a specific hazard, Section 17.5.3.3.2 permits the deletion of any additional detectors that might normally be required to cover an entire area.

Non-required coverage is sometimes used to protect against a specific hazard, where area coverage is not otherwise required. An example might be where some detectors are installed over a critical piece of equipment in a large area without area protection. It should be noted that any detector coverage that is not required by another code must still meet the requirements of *NFPA 72*, unless the coverage is provided to protect against a specific hazard. In other words, detectors protecting against a specific hazard are not required to meet spacing requirements of Chapter 17 of *NFPA 72*. Section 17.5.3.3 of *NFPA 72* provides requirements for non-required coverage. AHJ approval should be acquired prior to assuming any non-required detection will be accepted.

> **10.4.4*** In areas that are not continuously occupied, automatic smoke detection shall be provided at the location of each fire alarm control unit(s), notification appliance circuit power extenders, and supervising station transmitting equipment to provide notification of fire at that location.
>
> *Exception: Where ambient conditions prohibit installation of automatic smoke detection, automatic heat detection shall be permitted.*
>
> (Excerpt from NFPA 72.)

NFPA 72 does not provide requirements for where area detection is required. However, Section 10.4.4 of *NFPA 72* does require all control equipment to be protected. All control units must be protected by a smoke detector unless they are located in an area that is continuously occupied (24/7).

This requirement exists to prevent a fire from destroying the controls before occupant notification or fire safety functions occur. Control equipment includes main and satellite controls, notification appliance circuit power boosters, and supervising station transmitters. An exception permits the use of heat detection, where conditions are not suitable for smoke detection. However, conditions required for smoke detection are similar to those for control units.

Detectors are spaced according to requirements found in Chapter 17 of *NFPA 72*. Generally, detector spacing is determined by the ceiling configuration and other ambient conditions. For example, ceilings with deep beams require detectors to be placed closer together to compensate for the delay in movement of fire gases and heat caused by the obstructions.

Heat Detectors

Heat detectors are permitted to be mounted on a wall or ceiling. Where the heat detector is ceiling-mounted, Section 17.6.3.1.3.1 requires heat detectors to be located at least four inches (100 mm) from

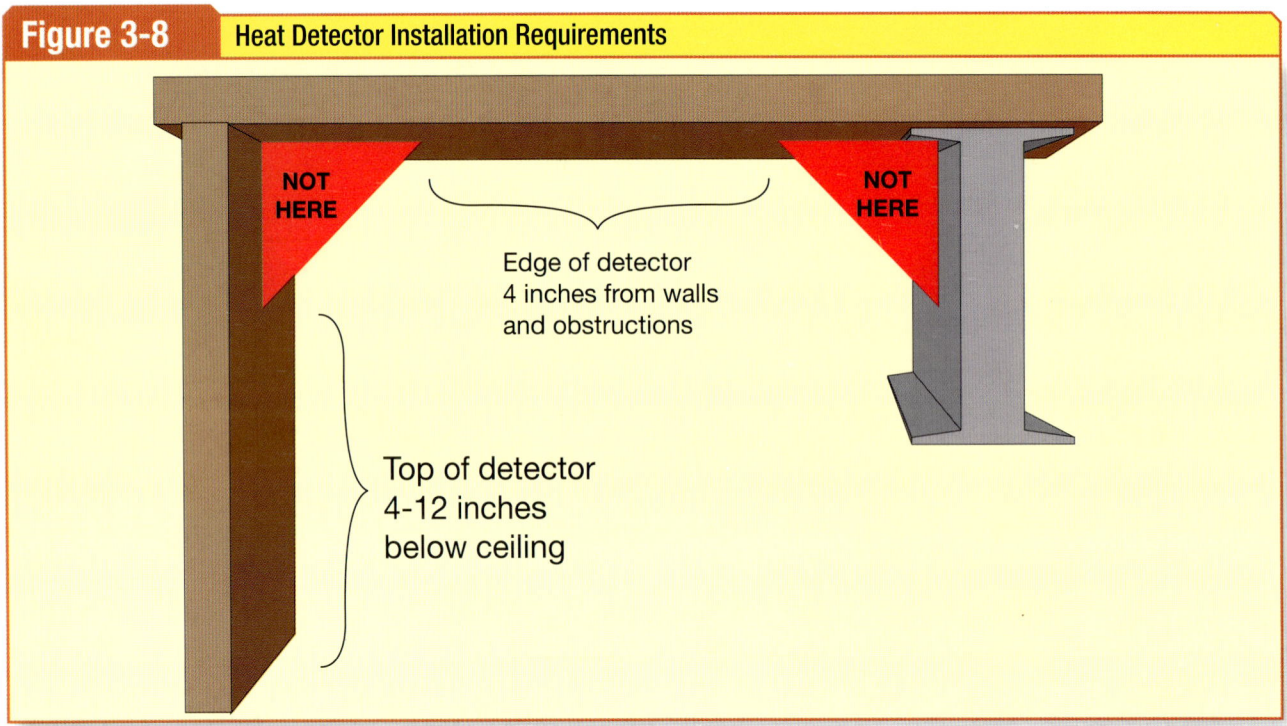

Figure 3-8. Heat detectors must be mounted far enough away from obstructions to allow them to sample the main air currents in a space.

a wall. If the detector is installed on a wall, the top of the heat detector must be between four inches (100 mm) and 12 inches (300 mm) from the ceiling. Line-type heat detectors are also permitted to be located on ceilings or walls. However, Section 17.6.3.1.3.2 permits line-type heat detectors to be located as much as 20 inches (510 mm) from the ceiling. These requirements exist because of the dead air pockets created by the ceiling-wall interface. This prevents heat from entering the pocket and delays the detector response. **See Figure 3-8.**

Listed Spacing. Heat detectors always have a listed spacing. Qualified testing laboratories test heat detectors to determine the maximum spacing of the devices through controlled conditions. Many heat detectors available on the market today have 50-foot (15.2 m) spacings. But, there are detectors available with other listed spacings from 25 feet (7.6 m) to 75 feet (22.8 m). Using a heat detector with the incorrect spacing may result in a system that will not operate as intended. The spacing found on the product cut sheets is referred to as the listed spacing. The product cut sheet should guide any installation.

To prevent nuisance alarms, Section 17.6.2.3 of *NFPA 72* requires that heat detectors having fixed temperature or rate-compensated elements must be selected to actuate at least 20°F (11°C) above the maximum ambient temperature expected at the ceiling in the space. For example, attic spaces in most of the United States commonly reach 140°F (60°C) in the summer months. A heat detector with a (standard) rating of 165°F (74°C) would be needed in this case to prevent nuisance alarms. One drawback to this is that a larger fire will be needed to actuate the detector during winter months. For most applications, heat detectors are available with standard ratings of 105°F (41°C), 135°F (57°C), 165°F (74°C), and 195°F (91°C). Ratings to 575°F (302°C) are available for special applications, such as steel mills.

Smooth Ceiling Spacing. Smooth ceilings are those that have obstructions less than four inches (100 mm) in depth. Heat detectors installed on smooth ceilings up to 10 feet (3 m) in height may use

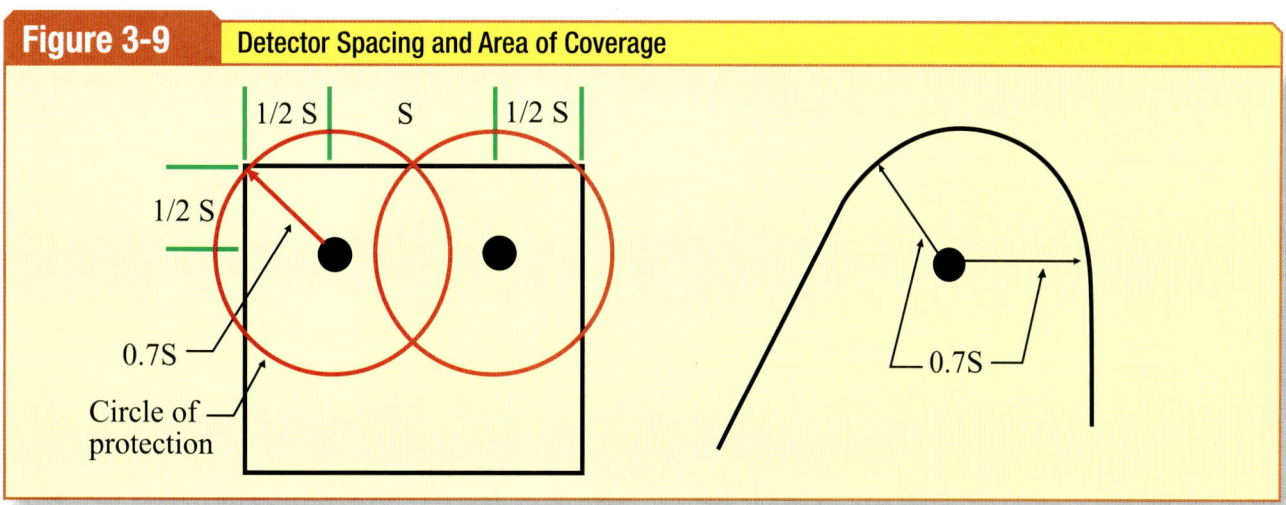

Figure 3-9. Detectors should be placed to ensure all areas are within their circle of coverage. Non-rectangular areas require special spacing.

the listed spacing. Non-smooth (beamed or joisted) ceilings and high ceilings over 10 feet (3 m) require correction of the spacing because these conditions may slow the response of the detector. Reduction of the spacing essentially makes the detection system more sensitive and will increase the response.

For smooth ceilings not over 10 feet (3 m) in height, the listed spacing is used between detectors. In this case, the detectors are placed so that they are one-half of the listed spacing from a wall, measured at right angles. The distance between rows of detectors is the listed spacing. This results in detectors within 0.7 times the listed spacing from any point on the ceiling.

Irregular (nonrectangular) areas must also have detectors within 0.7 times the listed spacing on all points of the ceiling. This method of spacing ensures that all points on the ceiling are within the circle of coverage of a detector. **See Figure 3-9**.

Walls and partitions that extend close to the ceiling height will obstruct the flow of heat. Therefore, Section 17.5.2 requires areas divided by partitions which extend within 15% of the ceiling height to be treated as separate areas with detection in each area.

The spacing of heat detectors on ceilings is a two-part process. The ceiling surface and ceiling height must be evaluated to determine if any compartment configurations will affect detector performance.

Smooth ceilings do not require any correction for obstructions. Obstructions, such as joists, impede the flow of smoke and heat in the direction perpendicular to the joists. The flow of heat and smoke parallel to the beams and joists is not affected. Smoke and heat will spill and fill into each cavity before rolling into the next cavity. Depending on the size, the cavity acts as a dead air space and prevents heat and smoke from penetrating the space. This phenomenon is similar to filling an ice cube tray. **See Figure 3-10**.

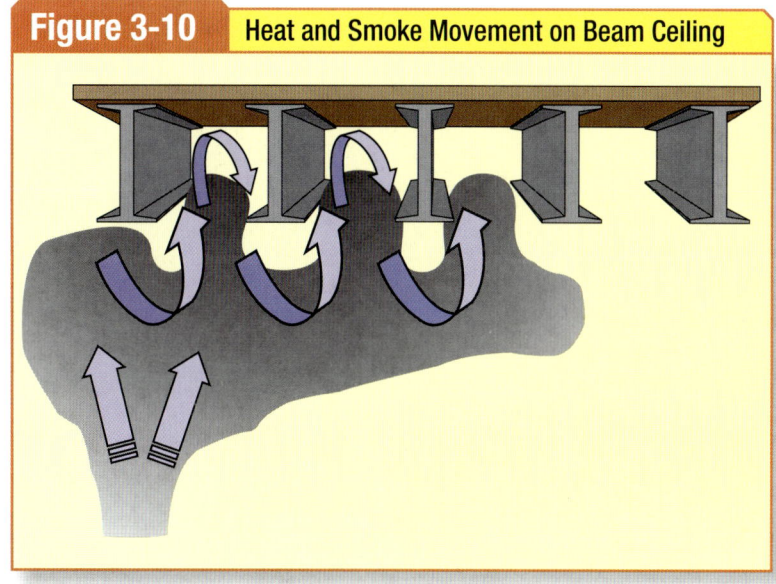

Figure 3-10. Smoke and heat "spills and fills" the empty spaces between joists, similar to how water fills an ice cube tray.

Joisted Ceilings. Non-smooth ceilings (beamed or joisted) can slow the spread of heat and smoke, which also increases the time required for detection of the fire. For this reason, the spacing of heat detectors must be reduced for non-smooth ceilings. For the purposes of the *Code*, non-smooth ceilings are those with obstructions (beams and joists) more than four inches (100 mm) in depth.

The impedance caused by beams and joists will result in a delayed response by the detector(s). For this reason, Section 17.6.3.2.2 of *NFPA 72* requires spot-type heat detectors to be mounted on the bottoms of solid joists, where heat will travel as it spills to the next pocket. Solid joists are defined as being obstructions more than four inches (100 mm) deep and 36 inches (910 mm) or less, center to center. This rule applies even if the obstructions are non-structural. Open web (bar) joists are not solid; therefore, this requirement does not apply unless the top chord is more than four inches (100 mm) deep.

Section 17.6.3.2.1 of *NFPA 72* requires the spacing of heat detectors to be reduced by one-half of the listed spacing, at right angles to joists. It is not necessary to reduce the spacing parallel to the joists because the channels created by the joists do not provide significant impedance to the flow of heat in that direction. When the spacing is reduced, the new spacing becomes the selected spacing, represented by the letter "S" by designers. **See Figure 3-11.**

Beam Ceilings. Section 17.6.3.3 provides requirements for heat detectors on

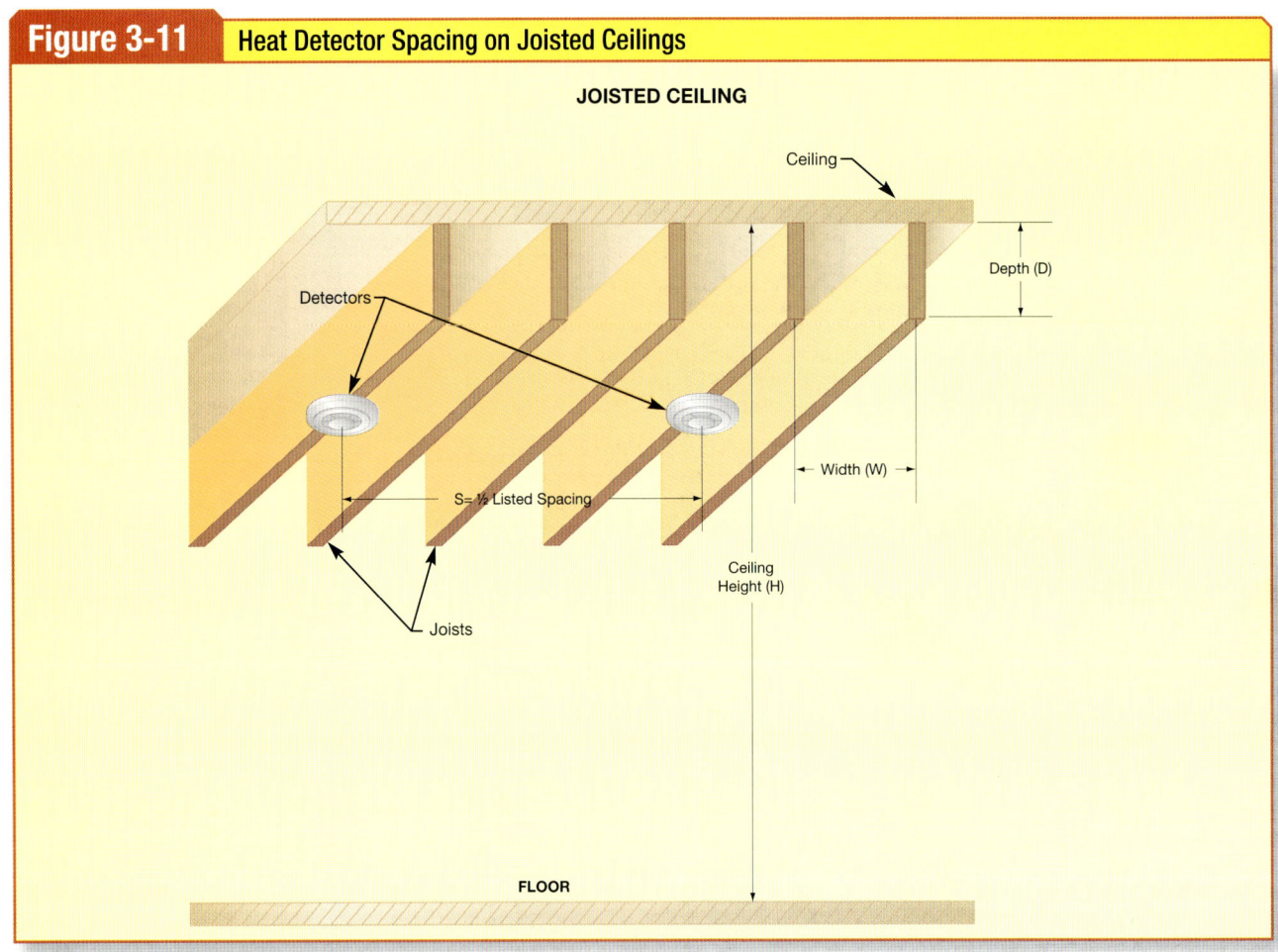

Figure 3-11. Joisted ceilings require special spacings for heat detectors.

Figure 3-12 Heat Detector Spacing on Beamed Ceilings

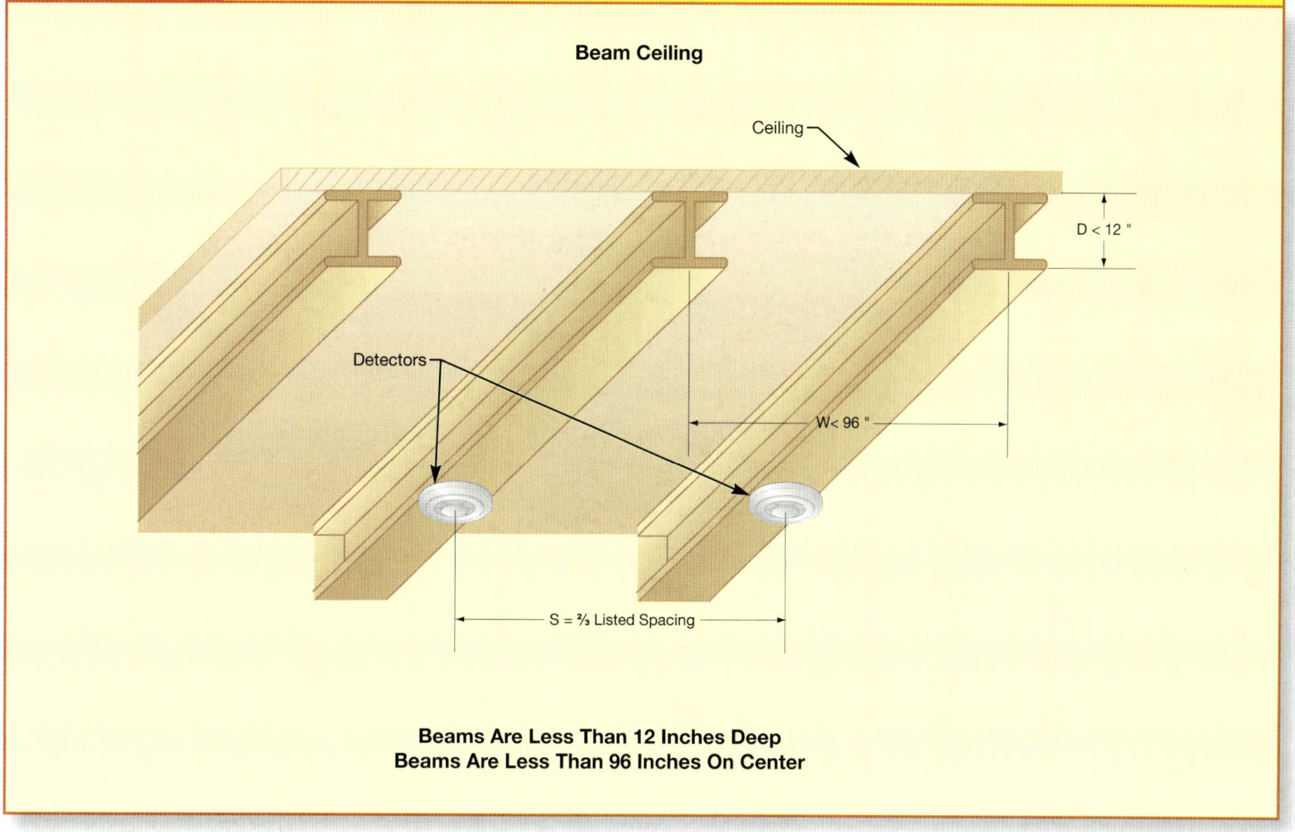

Beam Ceiling

Ceiling

D < 12"

Detectors

W < 96"

S = ⅔ Listed Spacing

Beams Are Less Than 12 Inches Deep
Beams Are Less Than 96 Inches On Center

Figure 3-12. Heat detectors must be specially spaced for beamed ceilings.

beamed ceilings. Where beams project less than four inches (100 mm) below the ceiling, 17.6.3.3.1.1 permits the ceiling to be treated as a smooth ceiling.

Where beams project more than four inches (100 mm) below the ceiling, Section 17.6.3.3.1.2 requires the spacing at right angles to the beams to be reduced by one-third of the listed spacing. The selected spacing (S) will be two-thirds of the listed spacing. Where beams are less than 12 inches (300 mm) deep and less than 96 inches (2.44 m) apart from each other on center, Section 17.6.3.3.2 permits detectors to be placed on the bottoms of the beams. **See Figure 3-12.**

Special requirements exist for line-type heat detectors. These detectors can be mounted on sidewalls up to 20 inches (500 mm) below the ceiling, unless otherwise permitted by the manufacturer.

Joisted and beamed ceilings present challenges for detection devices. Detector spacing must be reduced to accommodate for the delay caused by the obstructions.

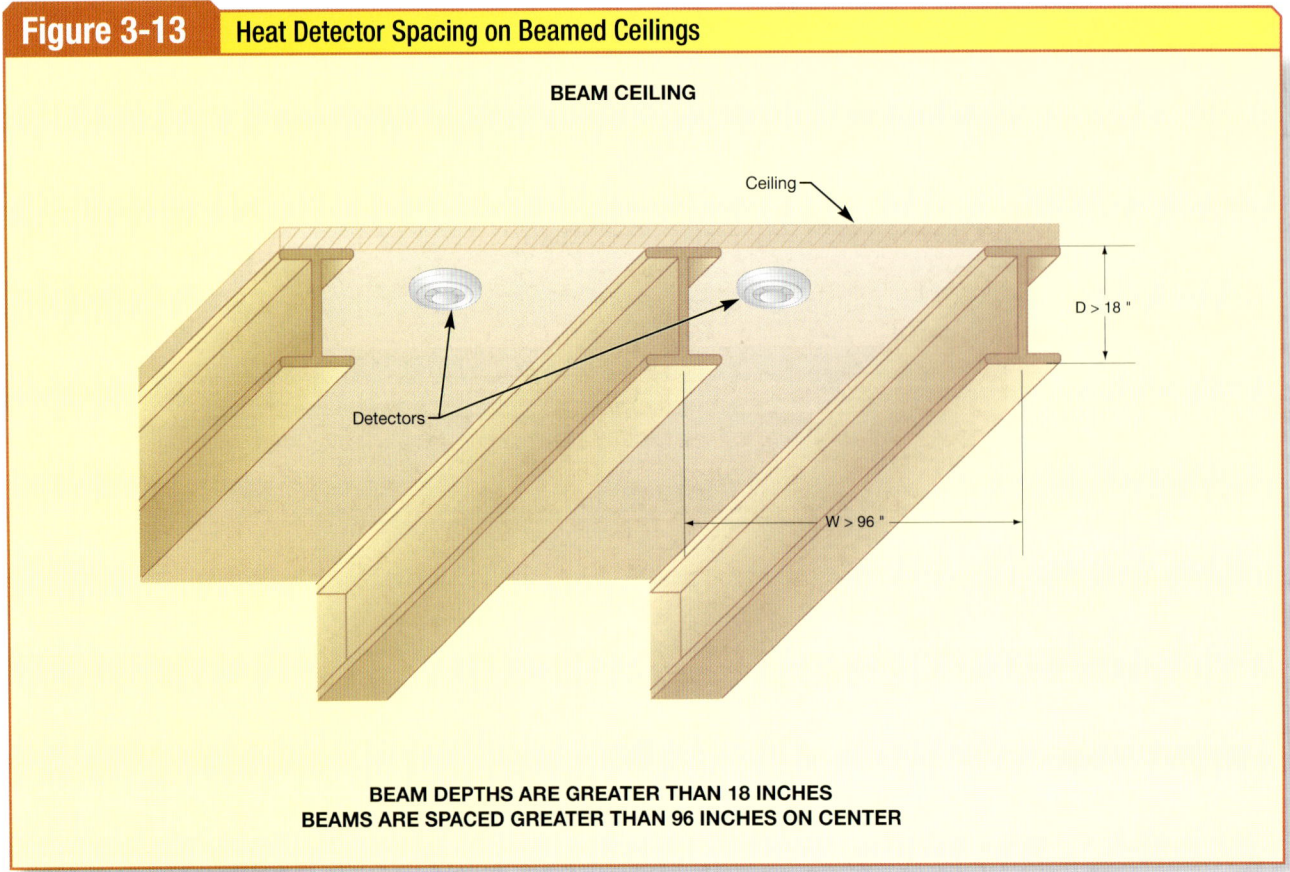

Figure 3-13. When beams are far enough apart, the spaces between them must be treated as separate areas.

Where beams are more than 18 inches (460 mm) deep and more than 96 inches (2.44 m) apart on center, each bay formed by the beams must be treated as a separate area. **See Figure 3-13.**

Section A.17.6.3.3 recommends an evaluation of beam depth, beam spacing, and ceiling height to determine if heat detectors should be placed on the bottoms of the beams or in the bays formed by the beams. If the ratio of the beam depth (D) to ceiling height (H) (D/H) is greater than 0.10, and the ratio of beam spacing (W) to ceiling height (H) (W/H) is greater than 0.40, detectors should be placed in each beam pocket. If either D/H is less than 0.10 or W/H is less than 0.40, heat detectors should be placed on the bottoms of the beams.

High Ceilings. Sometimes, the spacing of detectors is reduced to account for ambient conditions, such as joisted or beamed ceilings. Beams and joists may slow the spread of smoke and heat, thereby resulting in slower response times. Ceilings over 10 feet (3 m) in height may also cause a delayed response of the detection devices because the heat must travel farther to reach a device. Additionally, the fire plume cools as it expands and rises. These delays result in detection at a much later stage in the event, which means the fire will be larger upon detection.

A reduction in spacing will effectively increase the likelihood of early detection by placing detectors closer to potential fires. Therefore, heat detector spacing reductions are also required for high ceilings. When the spacing is reduced, the new spacing becomes the selected spacing (S). The selected spacing (S) will always be less than the listed spacing of the heat detector for non-smooth or high ceilings.

It should be noted that the selected spacing for heat detectors on beamed or joisted ceilings over 10 feet (3 m) is determined by applying both the height reduction and the non-smooth ceiling reduction.

Section 17.6.3.5.1 and Table 17.6.3.5.1 of *NFPA 72* require the application of a correction factor which must be applied to the listed spacing of heat detectors to account for the additional time required for heat from a fire to reach a heat detector on a high ceiling. When the correction factor is multiplied by the listed spacing, the new spacing becomes the selected spacing (S). All points on the ceiling must be within 0.7 times the selected spacing (0.7 S) in order to satisfy the *Code*. This is called the circle of protection.

Table 17.6.3.5.1 of *NFPA 72* contains the correction factors that must be applied to ceiling heights between 10 feet (3 m) and 30 feet (10 m). Correction factors for ceilings over 30 feet (10 m) cannot be extrapolated from this table since there is no test data to substantiate corrections for ceilings over 30 feet (10 m). See Figure 3-14.

Sloping and Peaked Ceilings.
Peaked and shed ceilings present unique spacing requirements. Section 17.6.3.4.2.1 requires a row of heat detectors within 36 inches (910 mm) of the peak of the ceiling. If additional rows of detectors are required, they must be spaced based upon the horizontal projection (plan view) of the protected space.

Where the slope is less than 30 degrees, the height at the peak is used to correct for height (if necessary). For ceiling slopes of more than 30 degrees, the average height is used to determine the correction for height. Refer to Figure A.17.6.3.4(a) in *NFPA 72* for guidance on peaked ceilings.

Section 17.6.3.5.2 does not require the minimum spacing of heat detectors to be less than 0.4 times the height of the ceiling. This spacing applies to all smooth, non-smooth, and high ceilings. This minimum spacing is based on fire plume dynamics.

Smoke Detectors
Smoke detectors are now permitted to be located on ceilings or on walls. The requirements for the location of smoke detectors have changed in the 2010 edition of *NFPA 72*. Smoke detectors are now permitted in the dead air space at the ceiling-wall interface. This change was based upon research conducted over the past several years.

Permitted Ambient Conditions.
Smoke detectors react to particles produced by combustion. However, smoke detectors respond to other stimuli such as moisture, dust, paint spray, lint, engine exhaust, and many other particles commonly found in the environment. Smoke detectors are, therefore, not suitable for use in environments where these particles are normally present. Section 17.7.1.7 requires the selection and placement of smoke detectors to take into account both the performance characteristics and the environment where the detectors are to be placed. Failure to properly select and install detectors may result in nuisance alarms.

Section 17.7.1.8 requires smoke detectors to be installed in areas that do not exceed the following limitations:
- Temperature: 32°F (0°C) to 100°F (38°C)
- Relative humidity: 93%
- Air velocity: 300 feet/minute (1.5 m/s)

Figure 3-14 | Heat Detector Reduction for Height

Table 17.6.3.5.1 Heat Detector Spacing Reduction Based on Ceiling Height

Ceiling Height Above		Up to and Including		Multiply Listed Spacing by
ft	m	ft	m	
0	0.0	10	3.0	1.00
10	3.0	12	3.7	0.91
12	3.7	14	4.3	0.84
14	4.3	16	4.9	0.77
16	4.9	18	5.5	0.71
18	5.5	20	6.1	0.64
20	6.1	22	6.7	0.58
22	6.7	24	7.3	0.52
24	7.3	26	7.9	0.46
26	7.9	28	8.5	0.40
28	8.5	30	9.1	0.34

Figure 3-14. Correction factors must be applied to ceiling heights between 10 and 30 feet high.

Reprinted with permission from NFPA 72®-2013, *National Fire Alarm and Signaling Code*, Copyright © 2012, National Fire Protection Association, Quincy, MA. The information in this table is intended to be used in conjunction with the requirements of this code. It is not the complete and official position of the NFPA on the referenced subject, which is represented only by the standard in its entirety.

Further, Section 17.7.1.9 requires an evaluation of the location of smoke detectors to determine if potential sources of smoke, moisture, fumes, or other sources of nuisance alarms are present. Where these stimuli are present, smoke detectors are probably not the most appropriate choice. Other detectors, such as heat detectors, may be more appropriate for reduction of nuisance alarms.

During Construction. Protection of a building during construction is often required by authorities having jurisdiction. This sometimes requires the use of smoke detectors. Buildup of contaminants will cause increased sensitivity in smoke detectors. Unfortunately, gypsum dust, paint fumes, dirt, and paint overspray can deposit inside smoke detectors and increase detector sensitivity. This almost always results in an increase in nuisance alarms. Unless cleaned, smoke detectors installed during construction will have much shorter lives and will be a source of irritation to the building owner. Mechanical damage is also likely to occur as other trades continue to work in the building.

Section 17.7.1.11 provides requirements for protection of smoke detectors during construction. Section 17.7.1.11 does not prohibit the installation of smoke detectors until the final cleanup of all trades is complete. In some cases, authorities having jurisdiction sometimes want coverage during the construction project. Section 17.7.1.11.1 requires that any detectors installed before final cleanup of all trades must be cleaned and recalibrated or replaced prior to final commissioning of the system.

Detectors installed during construction that are non-operational must be protected from construction debris, dust, dirt and damage. Section 17.7.1.11.2 requires these detectors to be tested to verify proper sensitivity or they must be replaced prior to final commissioning of the system.

Where smoke detectors are not required during construction, Section 17.7.1.11.3 prohibits their installation until the final cleanup of all trades is complete.

High Ceilings. Section 17.7 contains no requirements for the reduction in spacing of smoke detectors for high ceilings. However, Section 17.7.3.1.2(2) requires the designer to evaluate the protected space to determine if stratification will be a factor in the design. Section 17.7.3.1.2 also requires the location and spacing of smoke detectors to be based upon anticipated smoke movement patterns. This section further requires the spacing and location to account for ambient conditions, such as ceiling height, room geometry, ventilation, ceiling shape and surface, and fuel loads. Generally, high air movement areas and ceilings over 15 to 20 feet (5 m to 6.1 m) in height are good candidates for other types of detection besides spot-type smoke detection. Projected beam detection is often a good choice for applications involving high ceilings or large areas.

Smooth Ceilings. A smooth ceiling is defined differently for smoke detectors. Smooth ceilings are those that have obstructions less than 10% of the ceiling height. Section 17.7.3.2.3 provides requirements for the spacing of smoke detectors on smooth ceilings. Unlike heat detectors, there is no listed spacing for smoke detectors. For this reason, Section 17.7.3.2.3.1(1) permits a spacing of 30 feet (9.1 m) to be used as a guide for the spacing of smoke detectors. Section 17.7.3.2.3.1(2) requires a detector to be located within 0.7 times the selected spacing of all points of the ceiling. This section also permits the use of other spacings, based upon ambient conditions, or to protect a specific item, such as a computer mainframe.

Non-smooth Ceilings. Non-smooth (beamed and joisted) ceilings also require special treatment for the spacing of smoke detectors. Section 17.7.3.2.4 contains requirements for the spacing of smoke detectors on non-smooth ceilings. For the purposes of discussion, Section 17.7.3.2.4.1 considers beams and joists to be equivalent for the spacing of smoke detectors. This is unlike the requirements for heat detectors, where beams and joists are treated differently. The spacing requirements for smoke detectors on non-smooth

Fact

Most of the requirements for the spacing of smoke detectors on non-smooth ceilings are based upon studies conducted at the National Institute of Standards and Technology (NIST). NIST is part of the US Department of Commerce, and the Building Fire Research Laboratory conducts many experiments on fire behavior.

ceilings are quite different from those for heat detectors.

For level non-smooth ceilings, Section 17.7.3.2.4.2(1) permits the smooth ceiling spacing of spot-type smoke detectors for beams or joists that are less than 10% of the ceiling height in depth. In this application, smoke detectors can be located on the ceiling or on the bottoms of the beams or joists.

Section 17.7.3.2.4.2(2) provides requirements for beams and joists exceeding 10% of ceiling height (0.1H). For beams and joists exceeding 10% of the ceiling height, and beam spacing equal to or greater than 40% of the ceiling height

17.7.3.2.3.1* In the absence of specific performance-based design criteria, one of the following requirements shall apply:
(1) The distance between smoke detectors shall not exceed a nominal spacing of 30 ft (9.1 m) and there shall be detectors within a distance of one-half the nominal spacing, measured at right angles from all walls or partitions extending upward to within the top 15 percent of the ceiling height.
(2) *All points on the ceiling shall have a detector within a distance equal to or less than 0.7 times the nominal 30 ft (9.1 m) spacing (0.7S).

(Excerpt from NFPA 72.)

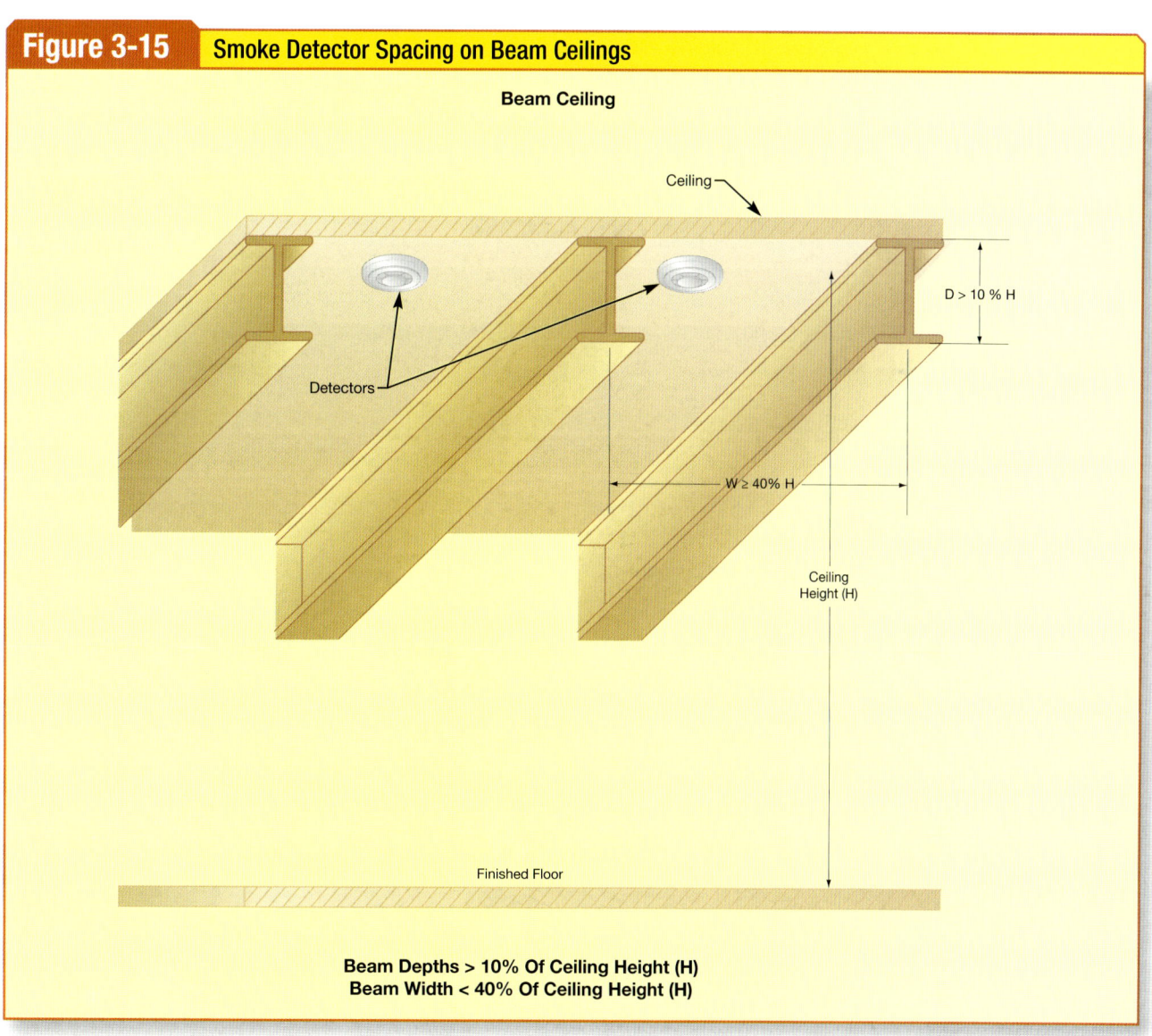

Figure 3-15 Smoke Detector Spacing on Beam Ceilings

Beam Depths > 10% Of Ceiling Height (H)
Beam Width < 40% Of Ceiling Height (H)

Figure 3-15. Non-smooth ceilings require special spacing for smoke detectors.

(0.4H), spot-type detectors must be located on the ceiling in each beam pocket. **See Figure 3-15.**

For joist or beam depths exceeding 10% of ceiling height and for beam spacing that is less than 40% of ceiling height (0.4H), smoke detector spacing must be reduced by 50% (one-half) in the direction perpendicular to the direction of the beams. Reduction of spacing parallel to the beams is not necessary, and the detectors can be installed on the bottom of the beams or on the ceiling. **See Figure 3-16.**

Section 17.7.3.2.4.2(3) covers applications with waffle ceilings or ceilings with intersecting beams. Waffle ceilings with beams or solid joist depths less than 10% of the ceiling height (0.1H) are permitted to use smooth ceiling spacing, and spot-type smoke detectors may be located on the ceiling or on the bottom of the beams.

For waffle ceilings with beams exceeding 10% of the ceiling height (0.1H), the rules are the same as for level, beamed ceiling spacing in Section 17.7.3.2.4.2(2), as described for beamed ceilings. In some cases, smoke detector spacing may be reduced by 50% in both directions, or smoke detectors may be required in each beam pocket. Detectors can be installed on the bottom of the beams or on the ceiling. **See Figure 3-17.**

Section 17.7.3.2.4.2(4) provides requirements for corridors with non-smooth ceilings. For corridors 15 feet (4.6 m) in width or less with ceiling beams or solid joists perpendicular to the corridor length, the smooth ceiling spacing for smoke detectors is permitted, and smoke detectors may be installed on ceilings, sidewalls, or the bottom of beams or solid joists.

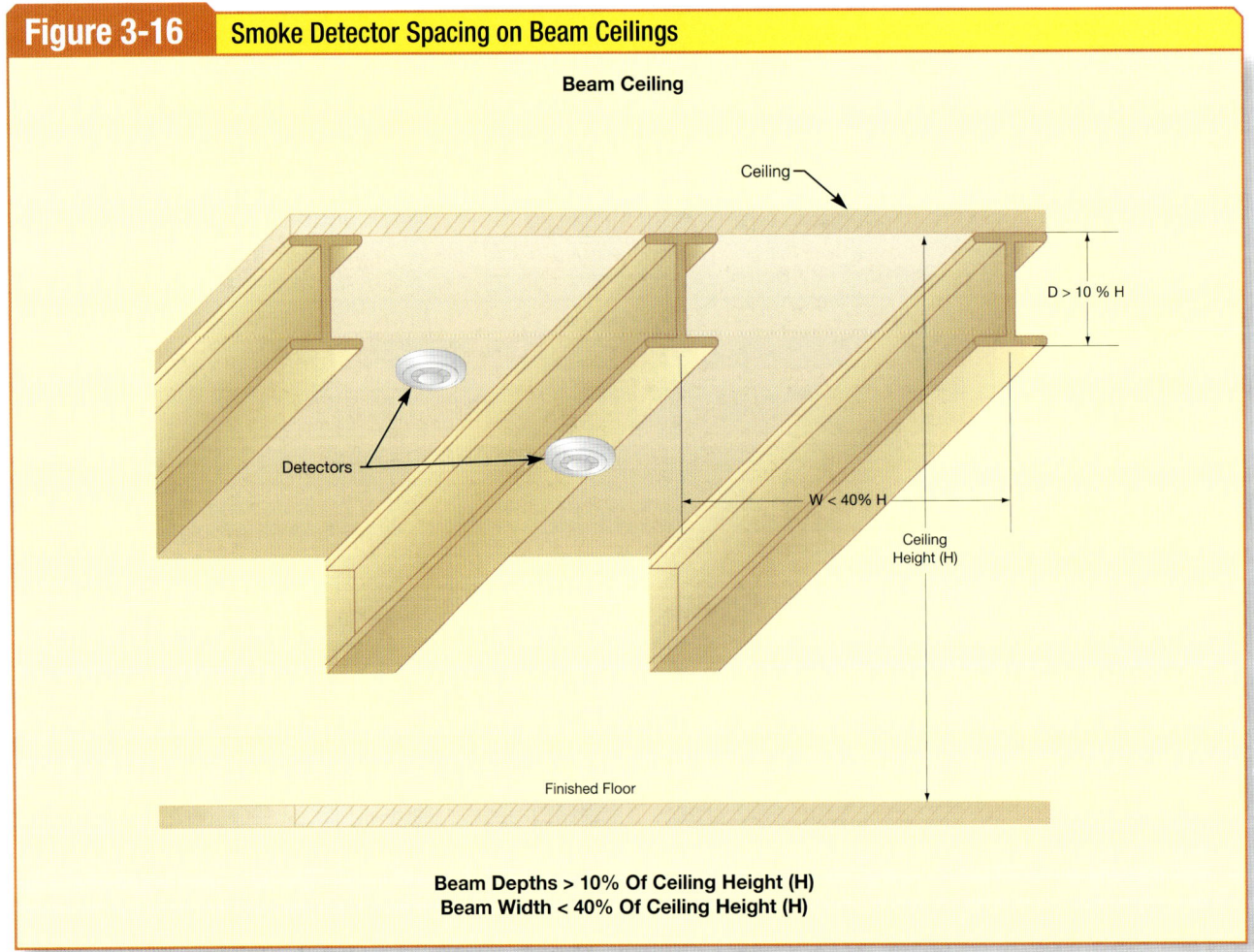

Figure 3-16. Smoke Detector Spacing on Beam Ceilings

Beam Depths > 10% Of Ceiling Height (H)
Beam Width < 40% Of Ceiling Height (H)

Figure 3-16. Joist or beam depth over 10% of ceiling height and beam spacing less than 40% reduces spacing by one half.

Figure 3-17 Waffle Ceilings

Figure 3-17. Waffle ceilings present special issues for smoke detectors.

Small rooms tend to concentrate the smoke, and ceiling surfaces become less significant in the detection process. Section 17.7.3.2.4.2(5) addresses the spacing of smoke detectors in small rooms. This section requires only one smoke detector in rooms of 900 square feet (84 m²) area or less. A smooth ceiling spacing is permitted in these small areas.

Sloping Ceilings. Requirements for sloped ceilings will vary depending on which direction the beams run.

Beams Running Parallel Up the Slope. Section 17.7.3.2.4.3 addresses sloping ceilings with beams or joists running parallel to (up) the slope. For sloped ceilings with beams running parallel to (up) the slope, spot-type smoke detectors must be located in the beam pocket(s). The ceiling height is the average height taken over the slope. Detector spacing is on a horizontal (plan view) projection. The smooth ceiling spacing is used in the beam pockets where more than one detector is required. If the beam depth is less than or equal to 10% of ceiling height (0.1H), spot-type smoke detectors may be spaced at smooth ceiling spacing perpendicular to the beams.

For beam depths more than 10% of ceiling height and beam spacing greater than or equal to 40% of ceiling height (0.4H), smoke detectors may be placed in each beam pocket. For beam depths less than 40% of ceiling height (0.4H), smoke detectors are not required in each beam pocket, but detector spacing must be reduced by 50% in the direction perpendicular to the direction of the beams.

Beams Running Perpendicular Across Slope. For sloping ceilings, with the beams perpendicular to (across) the slope, Section 17.7.3.2.4.4 applies. Spot-type smoke detectors must be located on the bottoms of the beams. The ceiling height is the average height taken over the slope. Spacing is on a horizontal (plan view) projection. The smooth ceiling spacing is used in the beam pockets where more than one row of detectors is required.

If the beam depth is less than or equal to 10% of ceiling height (0.1H), spot-type smoke detectors may be spaced with not more than three beams between detectors and not more than smooth ceiling spacing. For beam depths greater than 10% of ceiling height, detectors are not required to be closer than 40% of ceiling height (0.4H) and cannot exceed 50% of the smooth ceiling spacing.

Section 17.7.3.2.4.6 requires smoke detectors to be mounted on the bottom of joists or beams on sloping ceilings.

Sloping Ceilings with Intersecting Beams. This case is covered by Section 17.7.3.2.4.5. The ceiling height is taken as the average height over the slope, and detectors are placed on the bottoms of the beams. Where the beam depth is less than 10% of the ceiling height, detectors are spaced with not more than three beams between detectors, and not more than the smooth ceiling spacing. If beam depths exceed 10% of ceiling height, detectors are spaced with not more than two beams between detectors, and not more than 50% of the smooth ceiling spacing.

Peaked and Shed Ceilings. For peaked ceilings, Sections 17.7.3.3 and 17.7.3.4 require the first row of detectors within 36 inches (910 mm) of the peak, measured horizontally. Additional rows of detectors must be spaced on a horizontal projection of the ceiling (plan view). For shed-type ceilings, the first row of smoke detectors must be placed within 36 inches (910 mm) of the peak.

Sprinkler Waterflow Alarm-Initiating Devices

Waterflow initiating devices are used wherever automatic sprinkler systems provide coverage in a protected area. Generally, each sprinkler zone has a minimum of one waterflow initiating device. Generally, each floor of a building has waterflow switches installed at each floor test station to detect the flow of water on the zone. However, sprinkler zones are limited to 52,000 square feet in area by *NFPA 13, Standard for the Installation of Sprinkler Systems.* Therefore, large footprint buildings may have more than one sprinkler zone and waterflow switch per floor. The sprinkler contractor is generally responsible for the installation of waterflow initiating devices.

The most common type of waterflow switch is the vane- (paddle-) type switch. This device uses a plastic paddle inserted into the pipe to detect the flow of water. The paddle is connected to an electromechanical sensing device. Waterflow switches are usually conventional devices requiring a signaling line circuit interface to make them addressable. To prevent the device from responding to surges, a delay, called a retard, is used. Most vane-type waterflow devices have preset delays of 30, 60, and 90 seconds, while some have variable settings. Section 17.12.2 requires a maximum retard of 90 seconds, but every effort should be made to use the shortest delay possible without causing nuisance alarms. In some cases, short retards will allow the switch to respond to weekly fire pump tests. The fire pump must be operated to ensure that retards are properly set and spurious alarms do not occur. **See Figure 3-18.**

Pressure switches are also used on larger sprinkler systems, especially where the riser is sufficiently large and vane-type switches are not practical. In this arrangement, the sprinkler system has a clapper (check) valve on the main riser near the point of entry to the building. The clapper valve lifts when water flows into the system because of actuated sprinklers. As the clapper valve lifts, it allows water to flow through a small sensing line into a small tank called a retard chamber. The retard chamber is designed to drain more slowly than it is filled, but it will fill in 90 seconds or less. Once the retard chamber fills, water pushes against the pressure switch and an alarm is generated. The retard is achieved through the retard chamber volume, rather than a time delay built into the pressure switch. The delay is achieved as the retard chamber fills up with water from the clapper valve.

Dry-pipe systems are filled with air instead of water, which could freeze. The air displaces water in the piping but allows oxidation of any components in the pipe,

Figure 3-18 Vane-Type (Paddle) Waterflow Switch

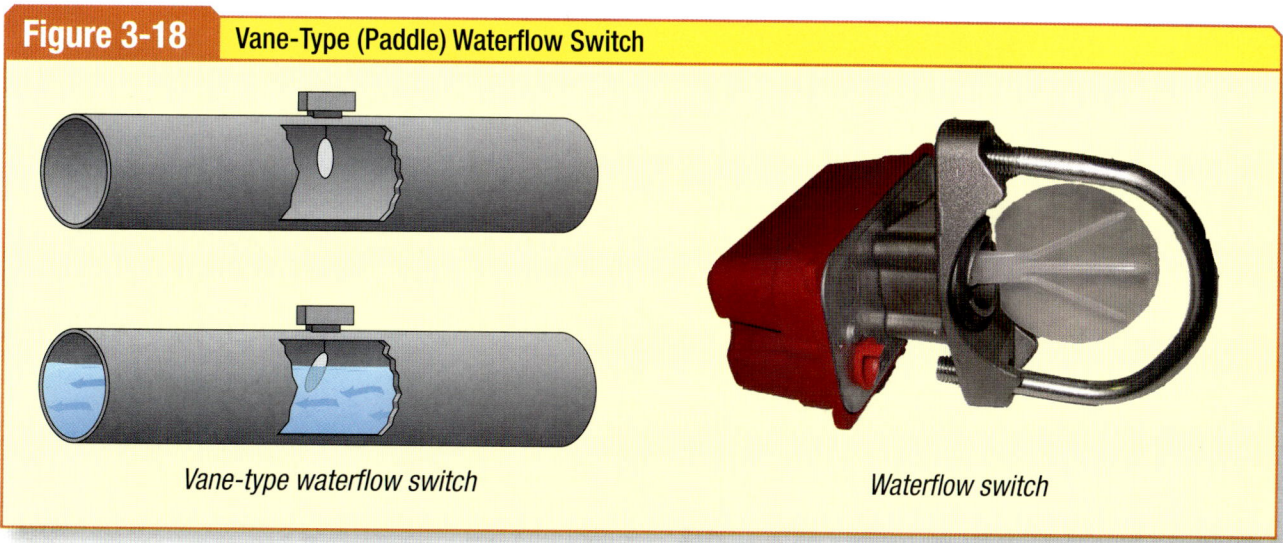

Figure 3-18. *A vane-type waterflow switch detects the flow of water in a pipe. The waterflow switch is a conventional device, usually on a signaling line circuit.*

Retard chambers are installed on sprinkler risers.

such as a waterflow switch. For this reason, dry-pipe sprinkler systems are not permitted to use vane-type waterflow switches because of corrosion issues. Additionally, the sudden inrush of water would rip the paddle from the switch and cause a blockage downstream. Instead, dry pipe sprinkler systems often use pressure switches to detect a sudden increase in water pressure on the dry side of the piping, resulting from an actuated sprinkler. Additionally, pressure switches are used to supervise the air pressure on the system. The air pressure is used to hold the dry valve closed, which prevents the water from entering the dry portion of the system.

Manually Actuated Alarm-Initiating Devices

Manual fire alarm boxes are usually simple switches with convenient operating handles to allow occupants the ability to report fires or other emergencies. Addressable manual fire alarm boxes usually contain a factory-installed signaling line circuit interface to provide an address. Conventional fire alarm boxes can be made compatible with addressable circuits by adding a signaling line circuit interface. This arrangement allows re-use of conventional devices in a retrofit application.

Manual fire alarm boxes may be of the single-action type, requiring only one action (such as pulling the handle) to actuate the box. Manual fire alarm boxes are also permitted to be double-action devices requiring two actions, such as push, then pull. Some manual fire alarm boxes are break-glass type, requiring replacement of the glass after each use. Key-operated (locked) manual fire alarm boxes are permitted by building codes in institutional and detention/correctional occupancies, but only where all staff have keys and are trained in their use. This prevents nuisance alarms caused by

For additional information, visit qr.njatcdb.org Item #1008

Fact

Fire pumps will sometimes actuate vane- (paddle-) type waterflow switches with short retard times. It is necessary to operate a fire pump during testing to ensure this does not happen.

inmates. Key-operated manual fire alarm boxes are not permitted in any other occupancies.

Manual fire alarm boxes may be required for some occupancy types or use groups. Section 17.14 provides requirements for manually operated initiating devices. Manual fire alarm boxes must be spaced at each exit on every floor of a building, with maximum travel distances not to exceed 200 feet (61 m), measured horizontally. The device must be installed within 60 inches (1.52 m) of the door and must not be obstructed by room furnishings or contents. Door openings over 40 feet (12.2 m) require manual fire alarm boxes within 60 inches (1.52 m) of each side of the opening. Good installation practices would dictate that the device be located on the latch side of the door even though there is no code requirement to place it on a specific side.

Manual fire alarm boxes must now be mounted so that the operable part of the device is between 42 inches (1.07 m) and 48 inches (1.22 m) in height above the finished floor. However, the Americans with Disabilities Act (ADA) requires that manual fire alarm boxes be accessible to persons in wheelchairs. The ADA additionally permits only single-action devices because some disabilities will preclude the use of double-action devices. The ADA requires manual fire alarm boxes to be mounted not more than 54 inches (1.37 m) above the finished floor for side reach and not more than 48 inches (1.22 m) above the finished floor for front reach. Mounting the tops of all manual fire alarm boxes at 48 inches (1.22 m) would meet both NFPA 72 and the ADA requirements. Section 17.14.8.2 requires all manual fire alarm boxes to be unobstructed and accessible. Furnishings, plants, and other obstructions are not permitted to prevent access to manual alarm boxes.

Where manual fire alarm boxes are subject to malicious false alarms, a variety of covers are available to reduce tampering. These covers are usually made from plastic and may have a sounding device that emits a loud signal when lifted to access the manual fire alarm box. These devices must be listed for this purpose and are very effective in reducing malicious false alarms.

Fire alarm covers can be used to reduce malicious false alarms.

Even if manual fire alarm boxes are not required by the building codes, Section 23.8.5.1.2 of NFPA 72 requires at least one manual fire alarm box for fire alarm systems. In the case of systems that are connected to a supervising station or that use waterflow detection devices or automatic detection, at least one manual fire alarm box must be provided to initiate a fire alarm signal. This manual fire alarm box must be located where required by the authority having jurisdiction. The exception to this requirement applies where a fire alarm system is installed to provide dedicated elevator recall control and supervisory service.

Supervisory Initiating Devices

Supervisory initiating devices are installed to help ensure suppression systems function properly. Supervisory initiating devices create signals indicating a failure or impairment of a suppression system. These devices may be used to supervise valves in the open position, water reservoir level, dry-pipe air pressure, and room temperature. These devices are generally located and installed according to the particular standard for the suppression system, such as NFPA 13, *Standard for the Installation of Sprinkler Systems*.

Control Valve Supervisory Signal–Initiating Devices. Section 17.16.1 of NFPA

72 contains the requirements for control valve supervisory signal–initiating devices. These switches are commonly referred to as "valve tamper" switches. Section 17.16.1 requires control valve supervisory devices to indicate two distinct signals, normal and off-normal. The switch must indicate the off-normal position within two revolutions of the hand wheel, or one-fifth of the travel distance of the valve, and must restore only when the valve is fully open. The supervisory device must not interfere with the operation of the valve. Some valves have integral switches and do not require an external switch.

Some supervisory devices are designed for large open screw and yoke (OS&Y) valves. They have a switch arm that sits in a milled groove on the valve stem that operates the switch when the valve is moved from the closed position. Some valves have integral switches and are usually called butterfly valves. Devices using "valve laces" are not permitted because they do not self-restore when the valve is open. Therefore, they are not permitted on fire alarm systems.

Some supervisory devices are designed for large OS&Y valves.

Post indicator valves (PIV) are another type of valve that may require a supervisory device. As the name implies, they have a built-in indicator to show valve position (OPEN or SHUT). PIVs are often used for building water supply control and are typically mounted so responders can see the valve position when arriving at the premises. PIVs may contain tamper switches and are generally locked because they are outside and may be subject to vandalism. **See Figure 3-19.**

Pressure Supervisory Signal–Initiating Devices. Pressure-type supervisory devices are designed to monitor the pressure (usually air pressure) of a suppression system, such as a dry-pipe sprinkler system. Section 17.16.2 requires the signals to be self-restoring. Generally, there are high- and low-pressure indicating switches, which are usually set to actuate at +/- 10 psi from nominal pressure. Additionally, pressure switches are used to start and stop the compressor which maintains the pressure.

Pressure switches may also be used to indicate an alarm on a wet sprinkler system. In this case, the switch must be used with a retard chamber to provide the mechanical delay necessary to prevent surges from causing an alarm.

Water Level Supervisory Signal–Initiating Devices. Some water-based suppression systems use reservoirs because of the poor reliability of water supplies or where there is insufficient water available for the

Figure 3-19. Post Indicator Valve (PIV)

Figure 3-19. PIVs are valves which may require supervisory devices because they are outside and subject to vandalism.

application. Where reservoirs are used for sprinkler systems, the level of the water must be supervised. Section 17.16.3 requires a supervisory signal when the level falls or rises three inches (76 mm) from nominal in pressure tanks and 12 inches (300 mm) from nominal in gravity tanks.

Temperature Supervisory Signal–Initiating Devices. Where wet sprinkler systems are installed in areas that are subject to freezing, Sections 17.16.4 and 17.16.5 require the water level and room temperature to be supervised, respectively. Temperatures below 40°F (4.4°C) must indicate a supervisory (off normal) signal and must self-restore above 40°F (4.4°C). If 40°F is chosen as the supervisory alarm point and the temperature drops below 40°F, the operator has enough time to correct the problems before the temperature reaches the freezing point.

Summary

Initiating devices are intended to provide warning of a fire or off-normal suppression system. They provide input to the fire alarm system, which in turn alerts occupants and responders. NFPA 72, National Fire Alarm and Signaling Code, indicates requirements for proper installation of such devices but does not generally require any initiating devices to be installed, with a few exceptions. In most cases, the requirements for initiating devices are found in other codes, such as the International Building Code; NFPA 5000, Building Construction and Safety Code; or NFPA 101, Life Safety Code. Initiating devices can be automatic or manual, conventional or addressable. In all cases, they must be installed properly in order to operate.

Review Questions

1. Which of the following is an alarm-initiating device?
   a. Room temperature switch
   b. Water flow switch
   c. Water level switch
   d. Water temperature switch

2. Which of the following is a supervisory signal–initiating device?
   a. Radiant energy-sensing detector
   b. Tamper switch
   c. Video flame detection
   d. Water temperature device

3. Rate-of-rise heat detectors operate when the temperature increases __?__ per minute.
   a. 8°F to 10°F
   b. 10°F to 12°F
   c. 10°F to 15°F
   d. 12°F to 15°F

4. Which area is a good location for a heat detector?
   a. Attic
   b. Dusty area
   c. Garage
   d. All of the above

Review Questions

5. Which type of detector uses a small amount of radioactive material?
   a. Ionization-type smoke detector
   b. Light obscuration smoke detector
   c. Photoelectric-type smoke detector
   d. Rate-compensated detector

6. A projected beam smoke detector operates on the __?__ principle.
   a. light obscuration
   b. light-refining
   c. light-reflecting
   d. light-scattering

7. Mechanical guards for initiating devices have a tendency to __?__.
   a. cause nuisance alarms
   b. increase sensitivity of the device
   c. reduce sensitivity of the device
   d. none of the above

8. Heat detectors must be located so they are within __?__ multiplied by the listed spacing of all points on the ceiling.
   a. 0.5
   b. 0.7
   c. 0.9
   d. 1.0

9. Open web bar joists are not a factor in heat detector spacing, unless the top chord is more than __?__ deep.
   a. 2"
   b. 3"
   c. 4"
   d. 5"

10. The listed spacing of smoke detectors is __?__.
    a. 15'
    b. 25'
    c. 30'
    d. none of the above

11. For beams and joists exceeding 10% of the ceiling height and beam spacing is equal to or greater than 40% of the ceiling height (0.4H), spot-type detectors must be located __?__.
    a. in alternating beam pockets
    b. in every beam pocket
    c. on alternating beams
    d. on the bottoms of the beams

12. Valve supervisory (tamper) switches must indicate off-normal within one-fifth of the travel distance of the valve or __?__ revolutions of the hand wheel.
    a. 1
    b. 2
    c. 3
    d. 4

Notification Appliances

Fire alarm and signaling system notification appliances are designed and installed to provide warning of a fire alarm, supervisory signal, trouble signal, or other emergency. Mass notification systems (MNS) and emergency communications systems (ECS) may include signals for other emergencies, such as terrorist attacks or tornado warnings. Since most fire alarm control units have a limited number of notification appliance circuits, designers often use power booster panels to provide additional power for strobes. These panels are distributed throughout the building to minimize the length of conductors and voltage drops between the power supply and notification appliances.

Objectives

- » Explain the general requirements of notification appliances
- » Describe various audible signals
- » Describe the requirements for telephone communications systems
- » Explain how a mass notification system works
- » Describe visible signals and how they work
- » Explain the difference between public and private mode signaling

Chapter 4

Table of Contents

General Requirements.................................86
- Mounting 87
- Listing Requirements...........................87
- Mechanical Protection.........................88
- Physical Construction..........................88

Audible Signaling..88
- Bells..88
- Horns..89
- Chimes 89
- Sirens 89
- Speakers 89
- Firefighters' Telephones (In-Building Two-Way Communications Systems) ...89
- Sound Transmission 90
- Mounting Locations..............................93
- Telephone Appliances 93
- Exit-Marking Audible Appliances 94
- Emergency Communications Systems and Mass Notification Systems 94

Visible Signaling and the Americans with Disabilities Act 94
- The Americans with Disabilities Act 95
- Public Mode Visible Signaling 97

Summary 102

Review Questions......................................102

GENERAL REQUIREMENTS

Notification appliances are output devices whose output may be audible, visible, or tactile (sense of touch).

Audible notification appliances can be single-stroke appliances, such as a bell, chime, or horn. Speakers that reproduce pre-recorded or live voice instructions are also audible notification appliances. A piezoelectric buzzer is often used to alert operators of signals at the control unit. Visible notification appliances for occupant warning are usually high-intensity Xenon strobes used to alert the hearing impaired. Additionally, textual displays are often used to provide visual outputs for system operators and responders. Textual displays include liquid crystal display (LCD) panels, cathode ray tube CRT monitors, and light-emitting diode (LED) matrices. A tactile device is the pillow vibrator used to alert hearing impaired persons in the event of a fire.

Generally, *NFPA 101, Life Safety Code*, or local building codes will provide the requirements for the type of notification appliances used and will provide requirements for where they are required. Where notification appliances are required, Chapter 18 of *NFPA 72* provides requirements for their installation and performance. However, it should be noted that *NFPA 72* does contain a limited number of requirements for areas where notification appliances are required.

Chapter 18 of *NFPA 72* is only concerned with the signal reception and intensity and does not specify the type of appliance or the content of the signals. For example, *NFPA 72* requires in-building emergency voice/alarm systems to have a minimum sound pressure level, but it does not specify the content of the voice message for occupants. In-building emergency voice/alarm communications systems are used in high-rise and health care applications where it is impossible or impractical to evacuate all occupants. Aid is directed to selectively evacuate or relocate occupants through the use of voice announcements over a speaker system. The content of the message is not prescribed by the *Code* but is usually specified by the engineer, owner, and the authority having jurisdiction (AHJ).

Sometimes signals are coded, so that responders can react without alerting occupants. There are two basic types of coded signals: non-voice coded and voice coded. Non-voice coded signals were used extensively on older systems installed prior to 1980. They are now installed in limited numbers, particularly in health care occupancies. Coded signals anticipate response by a trained staff. In this case, chimes are used to produce a series of tones to direct trained staff to the fire area. For example, a series of three chimes followed by a single chime could indicate the third floor, north wing. Chimes are frequently used in places where a loud signal might cause undue stress for occupants, such as in a nursing home. Coded signals can also be distributed using voice instruction. An example of a coded voice signal might be, "Paging Dr. Firestone, Paging Dr. Firestone, third floor, north wing."

All notification appliances have a finite area of coverage. The requirements for visible notification appliance spacing are very prescriptive. However, audible notification appliance spacing requirements are non-existent because the *Code* requires a minimum sound pressure level in the protected premises. This is called a performance-based arrangement.

Since most fire alarm control units have a limited number of notification appliance circuits, designers often use power booster panels to provide additional power for strobes. These panels are distributed throughout the building to minimize the length of conductors and voltage drops between power supply and notification appliances.

> **Fact**
>
> In buildings which use total evacuation, most designers are using horns as the audible appliances. They are easier to install, lighter, and usually less expensive than bells, chimes, or other audible appliances.

Notification appliance circuits are polarity sensitive. Therefore, care must be exercised in wiring the field devices to the circuit. Some manufacturers will mark their devices for alarm conditions, while others may mark for supervisory conditions.

Mounting

Section 18.3.5.1 requires notification appliances to be mounted independently of their conductors. The reason for this requirement is the same as for mounting all equipment independently from its conductors: copper is a soft material and will eventually break if under tension. Section 18.3.5.2 requires manufacturer's instructions (found on product cut sheets) to be followed when mounting notification appliances. This requirement is important because appliances like strobes are usually designed to distribute the light in a specific pattern. Mounting in a fashion other than that recommended by the manufacturer will result in a poor distribution of the signal. Visible appliances listed solely for wall mounting are not permitted to be ceiling mounted because they are designed for a different distribution pattern than ceiling appliances.

It is permitted to connect both audible and visible appliances on the same circuit, provided that the devices are compatible with the circuit (usually voltage and current matched). Horns and strobes are frequently used on the same notification appliance circuit. However, speakers generally require a separate notification appliance circuit because amplifiers must provide the signal.

Previous editions of the *Code* permitted silencing of audible notification appliances while visible appliances remained flashing. In this application, it may have been necessary to provide two separate circuits: one for audible appliances and one for visible appliances. Newer technology allows a single circuit to support this feature. However, separate deactivation of audible and visible notification appliances is no longer permitted because the Americans with Disabilities Act (ADA) requires equal access to the signals for all handicapped persons. Turning off the audible signals while strobes continue to flash would discriminate against blind persons.

Fact

Most fire alarm notification appliances are listed for dry locations. Locations, such as freezers, some basements, areas subjected to washing, and any outdoor location are technically considered wet or damp locations. All appliances used in these or other areas subject to moisture must be listed for such use.

Listing Requirements

Section 10.3.1 requires all fire alarm equipment to be listed for its intended purpose, and Chapter 18 reinforces this. Section 18.3.3.1 requires notification appliances that are to be used in special environments to be designed and listed for such use. For example, appliances installed outdoors (such as on a rooftop) must be listed for wet locations. Because ordinary fire alarm appliances are not weatherproof, atmospheric moisture will cause ground faults on the circuit when these appliances are installed in wet or damp locations. In addition, notification appliances used in a dusty or cold environment or those subject to tampering must also be listed for such use. Of course, any notification appliances installed in hazardous (classified) locations must also be listed for such use in order to prevent explosions.

Section 18.3.2.3 requires manufacturers to provide the rated output of the device on the product nameplate. This information is intended to assist designers and installers in the layout, especially when considering coverage and derating factors. Most audible appliances are listed with a sound pressure rating at a distance of 10 feet (3.1 m). This measurement is made at the testing laboratory in an anechoic chamber. Most visible appliances have a minimum visible output measured in candela (cd) on axis and may specify a polar distribution of the light output.

Mechanical Protection

Mechanical guards are sometimes required to protect fire alarm equipment. There are many types of listed guards available for notification appliances, and there are a few important requirements that must be noted. As with initiating devices, notification appliances installed in abusive environments must be mechanically protected. Section 18.3.4.2 requires mechanical guards and covers to be listed for use with the appliance. Section 18.3.4.3 requires any effect on the appliance by the guard to be applied as a derating factor. An example of a derating factor may be that the spacing of a strobe using a protective cover is reduced from 30 feet (10 m) to 20 feet (6.2 m). The manufacturer's installation sheets for the mechanical guarding device will provide derating information.

Physical Construction

Section 18.3.3 contains the requirements for the physical construction of notification appliances. These requirements exist primarily for the manufacturer. The designer and installer must be aware of these requirements and must use notification appliances that are appropriate for the system and environment. Section 18.3.3.2 contains requirements for notification appliance marking as follows:

AUDIBLE SIGNALING

Audible notification appliances include horns, bells, chimes, buzzers, piezoelectric sounders, sirens, and speakers. Any notification appliance may be used to alert operators and occupants of the need for action, provided that it is permitted by *NFPA 101* or local building codes. Although it is best to use a single type of appliance throughout a facility, there is no *Code* requirement to provide this level of consistency.

Bells

Bells are available in sizes from four inches (100 mm) to 12 inches (300 mm). However, six-inch (150 mm) and 10-inch (250 mm) bells are the most commonly used. Bells may be of the continuous vibrating type or the single stroke type. Most are powered by direct current to be compatible with newer technology. Bells usually produce a very wide output throughout the audible frequency spectrum and are generally perceived to have a louder output than horns of the same sound pressure level output.

Bells produce a wide signal in the human hearing spectrum, which makes them easier to hear.

18.3.3.2* Notification appliances used for signaling other than fire shall not have the word FIRE, or any fire symbol, in any form (i.e., stamped, imprinted, etc.) on the appliance visible to the public. Notification appliances with multiple visible elements shall be permitted to have fire markings only on those visible elements used for fire signaling.

A.18.3.3.2 The intent is to prohibit labeling that could give an incorrect message. Wording such as "Emergency" would be acceptable for labeling because it is generic enough not to cause confusion. Fire alarm systems are often used as emergency notification systems, and therefore attention should be given to this detail. Combination audible and visible units may have several visible appliances, each labeled differently or not labeled at all.

(Excerpt from NFPA 72.)

Horns

Horns generally have a continuous vibrating element, and they produce a much narrower frequency response than bells. Horns generally produce an output which is audible to most occupants, unless they have significant hearing loss in the frequency produced by the horn. Horns draw more power than bells, and care should be exercised to ensure the circuit output capacity is not exceeded.

Mini-horns are often used in residential and light commercial applications. They are perfect for bedrooms and small office areas.

Horns and strobes are used where both audible and visible signaling are desired in a single device. Of course, both horns and strobes are available in separate units.

Chimes

Chimes for fire alarm signaling produce a softer output, which is intended to prevent startling occupants. This is especially important in sensitive areas like an outpatient clinic. Chimes produce a relatively narrow frequency response compared to bells.

Sirens

Sirens are generally used in outdoor or high-noise environments. They are usually motor driven and can be operated on either alternating or direct current. Because they are motor driven, they tend to draw high current compared to other appliances like bells. Since the signal tone sweeps from high to low and back, they provide an excellent frequency output, clearly audible even to occupants with moderate hearing loss. They are not practical for all environments, but they work especially well for industrial applications.

Speakers

Speakers are primarily used on emergency voice alarm communications systems where voice instructions must be distributed throughout the premises. Speakers are powered by amplifiers in the main fire alarm system controls or by amplifiers distributed throughout a building. Speakers can reproduce almost any sound, including human speech, chimes, horns, and sirens. Speaker output can be changed by tapping the appliance at different setting levels. Most speakers can be tapped at $1/8$-, $1/4$-, $1/2$-, 1-, and 2-watt settings. High-output settings often cause distortion, making it difficult to understand the message being broadcast. Many older speakers are powered by 70 volts direct current (VDC), but newer systems use 24 VDC supplies. Speakers are considered audible textual appliances because they provide textual (spoken) information.

Firefighters' Telephones (In-Building Two-Way Communications Systems)

High-rise buildings, large footprint buildings, large assembly occupancies, and health care facilities typically have firefighters' telephones (in-building, two-way communications systems) for communications during emergencies. Use of this equipment is generally reserved for first responders. In some cases, handsets are kept at the main controls and are plugged into jacks located throughout the building during the emergency. Other designs utilize telephone stations throughout the building. In either case, the telephones are used to communicate between the main fire alarm system controls and key locations throughout the building, such as elevator lobbies, exit stairs, and the fire pump room. They have proven to be very useful in large fires where fire department radio traffic is heavy. Telephones are also considered audible textual appliances because they provide textual (spoken) information.

Sound Transmission

All audible fire alarm signal measurements are recorded in decibels, A-weighted (dBA). A decibel is a standard unit of sound pressure measurement. The A-weighted scale is used to account for the way humans perceive sound pressure. The transmission of sound throughout a space is highly dependent upon the room furnishings, wall and floor coverings, and ceiling treatments.

In open spaces, sound pressure levels (SPL) generally drop about 6 dBA each time the distance from the appliance is doubled. For example, an appliance rated 85 dBA at 10 feet (3 m) will deliver an SPL of about 79 dBA at 20 feet (6 m), and 73 dBA at 40 feet (12 m). This is a rule of thumb and generally holds true in large spaces where reflection is not a significant factor.

Sound transmission through walls and doors greatly depends on the thickness of the materials, types of materials, and undercut of any installed doors. Dense materials result in higher attenuation (loss) of signals through the wall or door. Concrete or masonry walls will attenuate significantly more sound pressure than a wall constructed of three-eighths-inch drywall on steel studs. Larger door undercuts will allow high frequencies to pass; however, most human hearing loss is in the high-frequency end of the spectrum. A good fire alarm system design will not attempt to penetrate more than one door or wall with an alarm signal. Fire alarm systems in apartments and inner offices should be carefully designed to ensure good signal dispersion in order to meet *Code* requirements. This usually results in the installation of notification appliances in each apartment or area. Of course, the number of appliances will be determined by analysis of the protected space. Walls, doors, and wall/ceiling treatments will all affect the sound transmission in a space.

The fire alarm system must produce signals that are audible above ambient sound pressure levels. However, Section 18.4.1.1 requires visible notification appliances when ambient sound pressure levels are greater than 105 dBA. This requirement exists because adding a fire alarm signal on top of a 105 dBA ambient sound pressure level would cause pain and/or hearing damage.

Audible signaling requirements are performance based. In other words, there is no specific (prescriptive) spacing of appliances provided by the *Code*. However, the minimum SPL must be achieved throughout the space. Designers must understand that room furnishings, wall and floor coverings, and occupant loads all play a significant part in the spacing of appliances needed to achieve the required SPL. Sometimes, appliances must be installed in each room to meet *Code* specifications.

Most manufacturers develop and publish appliance spacing guides for their products. These guides offer practical information on the quantity, spacing, and location of audible appliances. These guides are usually available free of charge from equipment distributors.

Public Mode Signaling. Section 18.4.2 provides requirements for the standard evacuation signal. This signal is called the Temporal-3 pattern and has been required since July 1996. It is only used where total evacuation is planned, such as in a low-rise building. It would not be used where partial evacuation or relocation is planned, such as in a high-rise building or hospital. This signal pattern

Fact

Doubling the output of a device (by changing taps or using high output appliances) will only result in an additional output of about 3 dB. Using a larger number of lower output appliances generally results in a better distribution of the signal.

was chosen because it can be made with any notification appliance and because it is used worldwide as an evacuation signal.

Public mode signaling is designed to alert occupants of a fire or the need to evacuate. Section 18.4.3.1 of *NFPA 72* requires public mode signals to be 15 dBA above the ambient sound pressure level, or 5 dBA above the maximum sound pressure level lasting 60 seconds or more, whichever is greater. The measurement is taken at approximately five feet (1.5 m) above the floor using a sound pressure level meter.

Temporary sound sources, like construction noises, are ignored when taking this sound pressure level measurement. Intervening doors or other barriers should always be closed when taking measurements to ensure the signals are clearly heard after being attenuated. Where the signal in a space is insufficient because of a closed door, additional appliances will be required.

The average ambient sound pressure level is defined as that being measured over a 24-hour period, or the period when the area is occupied by any person. Measurement of the average ambient SPL is time consuming and requires special averaging sound pressure level meters. Obviously, it is not possible to take measurements of the average sound pressure level when a building exists only on paper. For this reason, the *Code* provides guidance in the expected average ambient SPL. **See Figure 4-1.**

These values are expected and are not guaranteed. Adjustments to speaker tap settings or to the number of appliances may be necessary in order to meet *Code* requirements. However, setting speaker taps too high may cause distortion and will reduce intelligibility of the signals.

In some cases, such as theaters or nightclubs, ambient sound pressure levels may be extraordinarily high during evening performances. However, the business may also be open during quieter periods, such as lunch. Sound pressure levels during these periods may be significantly lower, and it may not be possible to design an audible system to satisfy *Code* requirements during performances without causing pain to occupants during quieter times. For this reason, Section 18.4.3.5 permits the fire alarm system to cause a reduction or removal of the ambient noise. For example, it is permitted to use a relay to de-energize stage power supplies when a fire alarm signal is present. Amplifier (stage) power can be de-energized by a fire alarm system control relay to reduce the ambient SPL. Any relays, circuits, or interfaces used to implement this feature must meet the requirements of Chapters 10, 12, 21, and 23 of *NFPA 72*. The audible fire alarm signal can then be designed for the ambient sound pressure level without the higher level noise.

According to Section 18.4.3.5, the fire alarm signal must meet public mode signaling requirements for the reduced ambient SPL. In other words, the signal must be at least 15 dBA above the reduced ambient SPL, or 5 dBA above the maximum SPL after reduction, whichever is greater. Additionally, visible signals are required in the affected areas, even if the local building code or *NFPA 101* does not require them. It should be noted that the provisions of Section 18.4.3.5 require approval of the AHJ.

Audible textual appliances (speakers) must produce signals that have voice intelligibility in addition to the audibility requirements. This means the signals must be clear and understandable. Low intelligibility results in the listeners not

Figure 4-1	Expected SPL by Occupancy Type
Occupancy	**Expected Average SPL (dBA)**
Business	55
Educational	45
Industrial	80
Institutional	50
Mercantile	40
Mechanical Rooms	85
Places of Assembly	55
Residential	35
Storage	30

Figure 4-1. Note that these are expected values which may be used for planning and design. Actual SPL may vary in the field. Source: NFPA 72, Table A.18.4.3.

Reprinted with permission from NFPA 72®-2013, *National Fire Alarm and Signaling Code*, Copyright © 2012, National Fire Protection Association, Quincy, MA. The information in this table is intended to be used in conjunction with the requirements of this code. It is not the complete and official position of the NFPA on the referenced subject, which is represented only by the standard in its entirety.

being able to understand the message. Many variables affect intelligibility, such as echo, distortion, and reverberation. Large, open spaces like atria, covered malls, transportation centers, and sporting venues require special treatment to avoid low intelligibility. Qualified designers can assist with good designs. Intelligibility is required by Section 18.4.10 of *NFPA 72* and can be measured with a quantitative measurement device. These measurement devices are available from a number of manufacturers. The measurement device kit usually includes a source signal generator and meter.

Intelligibility is only required in acoustically distinguishable spaces (ADS), as determined by the system designer. The ADS will most often include, but not be limited to, common areas such as open offices, corridors, stairwells, lobbies, and the like.

Private Mode Signaling. Sections 18.4.3.3 and 18.4.3.4 permit private mode signaling in elevator cars and restrooms, respectively. These areas are usually confined, and using high-intensity audible devices in them would startle occupants. Most designers will place an audible appliance just outside each restroom door with only visible notification appliances inside the restroom. Only low-wattage or low-output appliances are used in elevator cars.

Private mode signaling is only for responders directly concerned with implementation of aid as a result of the signal. For example, a front desk guard sitting five feet from a fire alarm system control unit is expected to pay attention to the system and would not, therefore, need a loud signal to get his or her attention. For this reason, it is permitted by Section 18.4.4 to use lower intensity audible signals for private mode.

Section 18.4.4.1 requires private mode signals to be 10 dBA above the ambient sound pressure level, or 5 dBA above the maximum sound pressure level of the reduced noise lasting 60 seconds or more, whichever is greater. Section 18.4.4.2 permits further reduction of the private mode signaling in areas such as a critical care area in a hospital, where the ambient SPL is low and staff is well trained to respond to signals. Additionally, Section 18.4.4.3 permits the fire alarm system to cause a reduction or removal of the ambient noise, just as is the case for public mode signals. Where this option is chosen, visible appliances must be installed in the affected areas, as required by Sections 18.5 or 18.6. Any relays, circuits, or interfaces used to affect this feature must meet the requirements of Chapters 10, 12, 21, and 23 of *NFPA 72*. Both of the above provisions require permission of the authority having jurisdiction.

Sleeping Areas. Sleeping areas require special treatment to ensure that occupants are awakened in a timely fashion. Recent studies indicate that it takes a signal of at least 75 dBA to awaken the average sleeping person. Of course, the studies assume drugs or alcohol do not impair the sleeping person. Section 18.4.5.1 requires audible signals in sleeping areas to be 15 dBA above the ambient sound pressure level, 5 dBA above the maximum sound pressure level lasting 60 seconds or more, or 75 dBA, whichever is greater. For example, a sleeping room with an ambient of 50 dBA and a maximum SPL of 55 dBA would require an SPL of at least 75 dBA. However, if the ambient SPL were 65 dBA, with a maximum of 70 dBA, the required signal SPL would be 80 dBA.

Sometimes audible notification appliances are not installed in every sleeping room. Additionally, there may be doors or other intervening barriers between the notification appliance and the pillow. Section 18.4.5.2 requires intervening doors, curtains, moveable partitions, or other barriers to be in place when taking SPL measurements. Where the SPL does not satisfy Section 18.4.5.1, notification appliances will be required in the sleeping room.

Many designers and installers have attempted to place notification appliances in corridors of hotels and apartment buildings, in the hopes of providing the required SPL at the pillow. Most walls and doors will attenuate the signal from appliances by at least 25 dBA. This makes

> **Fact**
> Good designers will develop and maintain a log book of different wall and door types, recording the SPL attenuation of each. This will assist in future designs.

it impossible to deliver 75 dBA at the pillow from an 85 dBA appliance located in the corridor. Good design practices dictate appliances in every sleeping room or in the corridor near the door to the sleeping room.

Mounting Locations

Audible notification appliances are typically wall mounted; however, speakers are frequently mounted on ceilings for architectural reasons. When audible notification appliances are wall mounted, they must be mounted so the tops of the appliances are at least 90 inches (2.3 m) above the finished floor and not less than six inches (150 mm) below the ceiling. This ensures that the appliances are above most room furnishings. **See Figure 4-2.**

Ceiling-mounted appliances are permitted in addition to wall-mounted appliances. Ceiling mounting helps to ensure that furnishings like cubicles and filing cabinets will not block the signal from the appliances. Where combination appliances are used, the mounting requirements for visible appliances are used. Section 18.4.8.3 contains requirements pertaining to combination audible/visible appliances.

Telephone Appliances

Telephone appliances are part of a two-way wired communications system, covered by Chapter 24 of *NFPA 72*. These systems were formerly called "firefighters' telephones," but with the introduction of emergency communications systems into Chapter 24 of the *Code*, their use is no longer limited to firefighters. In fact, it is anticipated that these telephones will be used by police and other security personnel during non-fire emergencies.

The installation requirements for telephone appliances are found in Chapter 18 of *NFPA 72*. Sections 18.8.2 and 24.5.1 require firefighters' telephone jacks or stations to be mounted between 36 inches (914 mm) and 66 inches (1.7 m) above the finished floor in a space that is at least 30 inches (760 mm) wide. If telephone stations are accessible to the public, at least one station per floor

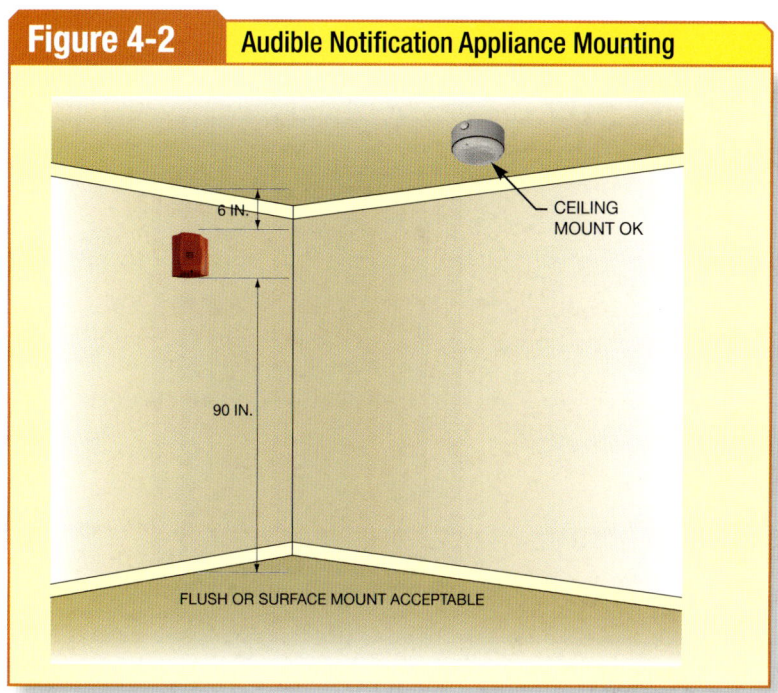

Figure 4-2. The minimum height requirements exist to ensure the signal is not blocked by furnishings located in the room.

must be mounted not exceeding 48 inches (1.2 m) above the finished floor to comply with ADA requirements for wheelchair access.

Telephones or jacks are located in key areas, such as exit stairs, elevator lobbies, and the fire pump room to facilitate communications with the handset at the main system controls.

Exit-Marking Audible Appliances

Exit-marking audible appliances can be used to direct occupants toward an exit, even if they cannot see the exit. Exit-marking appliances use audible signals that operate slightly out of phase, which gives the listener the impression that the signal is providing a direction to follow. This technology assumes the listener can hear equally from both ears. At this time, none of the national building codes requires exit-marking audible notification appliances. Section 18.4.7 of *NFPA 72* provides new requirements for exit-marking audible appliances. The majority of the requirements in Section 18.4.7 defer to manufacturers' recommendations.

Emergency Communications Systems and Mass Notification Systems

Emergency communications systems and mass notification systems are used to alert and inform occupants of a premises, campus, or community. Speakers must be used in order to provide voice instructions. Mass notification systems (MNS) are primarily used by the US Department of Defense (DOD) and other agencies in order to alert occupants of a terrorist attack, weather alerts, or other life-threatening conditions. ECS and MNS requirements are located in Chapter 24 of *NFPA 72*. They cover the following systems:

- One-way emergency communications systems (firefighters' telephones)
- Wide-area (Giant Voice) MNS
- Distributed-recipient MNS

Requirements for in-building fire emergency voice/alarm communications systems are also found in Chapter 24 of *NFPA 72*.

For additional information, visit qr.njatcdb.org Item #1010

VISIBLE SIGNALING AND THE AMERICANS WITH DISABILITIES ACT

The most common visible notification appliances for occupant notification are strobes. Listed system-powered strobes are available in standard ratings of 15, 30, 60, 75, 110, 177, and 185 candela. The higher intensity strobes cover larger areas than lower intensity strobes. These appliances are designed to provide a brilliant flash of white light, much like a camera flash.

Flash rates must be between one hertz and two hertz to prevent seizures in persons with photosensitive epilepsy. The strobe flash is intended to alert persons with hearing impairments; however, they also alert people who do not suffer from hearing loss. Some models have adjustable settings between 15 and 110 cd. A small switch is used to select the output of the device, based upon the size of the area. As a rule, higher intensity strobes will draw more current than low-intensity strobes.

Strobes are mandated by building codes and are located throughout common areas to provide signaling for the hearing impaired. NFPA 72 *provides prescriptive requirements for their spacing and location.*

Liquid crystal displays (LCD) are also categorized as visible notification appliances. These are often found on the operating interface of fire alarm system controls. The LCD provides an alphanumeric display of the system status. This type of display is referred to as a textual visible display because it provides information

Fact

NFPA 72 is now considered as meeting the Americans with Disabilities Act (ADA).

in a text format. LCD displays provide private mode signaling.

Graphic annunciators are frequently used in applications involving large properties because they show the building floor plans and fire alarm zoning. Light-emitting diodes (LEDs) are used to indicate alarm, trouble, and supervisory signals in each zone on the graphic annunciator. LEDs also indicate the type of device actuated, such as waterflow or smoke detector. Graphic annunciators make response faster by providing more information than a visible textual display. **See Figure 4-3.**

Another type of textual visible display used for fire alarm systems is a cathode ray tube (CRT) monitor. Flat panel technology is also used in the same manner as a cathode ray tube (CRT). This type of display is often used in large protected premises systems, high-rise applications, and large proprietary systems, where multiple lines of information or graphics are desired. They are also used in central, proprietary, and remote supervising stations which monitor the protected premises. The CRT display is essentially the same type of appliance as a home computer's CRT, with the exception of the listing for fire alarm use.

Printers are also in the category of visible textual notification appliances. Printers produce a permanent paper record of system signals. Tractor-feed, line printers are usually found in larger systems and supervising stations, where a large volume of signals is expected. However, smaller thermal paper integral printers are often mounted in the control cabinet. Printers are usually connected to the controls through a serial or parallel port, using either an RS-232 or RS-485 connection.

Occasionally, rotating red beacons or red strobes are used to provide occupant notification. However, the *Code* does not specifically cover their use and does not provide spacing rules for their use. These appliances are often used in large industrial applications, where public access is limited and occupants are trained to know what the signal means.

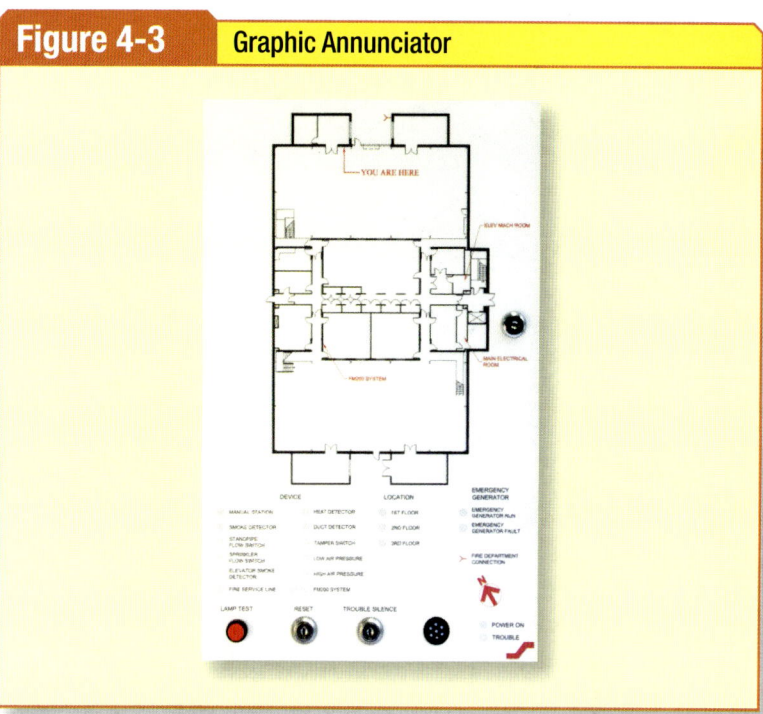

Figure 4-3. Graphic annunciators are required by some building codes to provide a faster response to the emergency. Graphic annunciators usually have indication of alarm, trouble, and supervisory signals by zone or area, and use floor plans for simplicity.

The Americans with Disabilities Act

The U.S. Congress created the Americans with Disabilities Act on July 26, 1990, to provide equal access to facilities and services for all Americans with physical disabilities. The U.S. Architectural and Transportation Barriers Compliance Board (Access Board) was tasked to develop architectural standards for compliance with this new law. These laws are called the Americans with Disabilities Act Accessibility Guidelines (ADAAG), and they provide recommendations for compliance with the ADA. Section 4.28 of the ADA specifically provides guidelines for fire alarm systems.

The ADA cannot be enforced by local jurisdictions as a code since no local government can enforce federal law. However, a few local jurisdictions have adopted local laws with the same language as the ADA, which essentially provides them with the ability to enforce

Fact

Data suggests that visible frequencies of about 6-10 Hz can cause seizures in persons with photosensitive epilepsy. Multiple visible appliances (strobes) in larger areas and corridors must be synchronized to keep the composite flash rate below 2 Hz.

the ADA as a code. Similarly, model code developers have attempted to bring themselves into alignment with the ADAAG by using language similar to the ADAAG in their documents. It must be noted that equivalent facilitation can only be determined by a federal trial after the system is installed.

Most fire alarm designers consider *NFPA 72-2013, National Fire Alarm and Signaling Code,* as equivalent facilitation to comply with the ADA. The requirements found in *NFPA 72* are considered by the Access Board as acceptable in meeting ADAAG requirements.

The ADA, by itself, does not require a fire alarm system. However, where a fire alarm system is installed, it must comply with the ADA. The *International Building Code; NFPA 101, Life Safety Code; NFPA 5000, Building Construction and Safety Code;* or local building codes will mandate where a fire alarm system is required for new construction. However, once a fire alarm system is required, the ADA requires equal access to the system signals by physically challenged persons.

NFPA 72 requires visible notification appliances (strobes) to be listed to nationally recognized product standards. The only nationally recognized product standard used for listing visible notification appliances is UL 1971-4-2002: *UL Standard for Safety Signaling Devices for the Hearing Impaired.* UL 1971 requires a minimum polar distribution of light measured at many points in the protected area to ensure adequate coverage of the signal. The ADA does not specify any product standards or polar distribution but specifies 75 candela strobes spaced so that no point in a space is more than 50 feet from an appliance.

Section 24.4.3.17.3 of *NFPA 72* requires visible notification appliances used for emergency communications systems to be marked with the word "ALERT" on the appliance. If the appliances are used on systems that are exclusively used for fire alarm applications, they must be marked with the word "FIRE" on the appliance. If the existing emergency voice/alarm communications system is to be modified for ECS and fire use, the word "FIRE" cannot be used and must be removed. All appliances must be listed to the requirements of UL 1971.

NFPA 72 assumes a minimum illumination level of 0.0375 lumens per square foot at the floor for visible signaling. The ADA requires a minimum of 0.030 lumens per square foot at the floor for fire alarm signaling. However, recent activities by the Access Board have accepted Chapter 18 of *NFPA 72* as equal to the requirements of the ADAAG.

Reprinted with permission from NFPA 72®-2013, *National Fire Alarm and Signaling Code*, Copyright © 2012, National Fire Protection Association, Quincy, MA. The information in this table is intended to be used in conjunction with the requirements of this code. It is not the complete and official position of the NFPA on the referenced subject, which is represented only by the standard in its entirety.

Figure 4-4 — Wall-Mounted Visible Notification Appliances

Maximum Room Size		Minimum Required Light Output [Effective Intensity (cd)]	
ft	m	One Light per Room	Four Lights per Room (One Light per Wall)
20 x 20	6.10 x 6.10	15	NA
28 x 28	8.53 x 8.53	30	NA
30 x 30	9.14 x 9.14	34	NA
40 x 40	12.2 x 12.2	60	15
45 x 45	13.7 x 13.7	75	19
50 x 50	15.2 x 15.2	94	30
54 x 54	16.5 x 16.5	110	30
55 x 55	16.8 x 16.8	115	30
60 x 60	18.3 x 18.3	135	30
63 x 63	19.2 x 19.2	150	37
68 x 68	20.7 x 20.7	177	43
70 x 70	21.3 x 21.3	184	60
80 x 80	24.4 x 24.4	240	60
90 x 90	27.4 x 27.4	304	95
100 x 100	30.5 x 30.5	375	95
110 x 110	33.5 x 33.5	455	135
120 x 120	36.6 x 36.6	540	135
130 x 130	39.6 x 39.6	635	185

Figure 4-4. This table assumes the appliance is centered on one wall. Source: NFPA 72, Table 18.5.5.4.1(a).

Figure 4-5 Ceiling-Mounted Visible Notification Appliances

Maximum Room Size		Maximum Lens Height		Minimum Required Light Output Effective Intensity): One Light (cd)
ft	m	ft	m	
20 x 20	6.1 x 6.1	10	3.0	15
30 x 30	9.1 x 9.1	10	3.0	30
40 x 40	12.2 x 12.2	10	3.0	60
44 x 44	13.4 x 13.4	10	3.0	75
20 x 20	6.1 x 6.1	20	6.1	30
30 x 30	9.1 x 9.1	20	6.1	45
44 x 44	13.4 x 13.4	20	6.1	75
46 x 46	14.0 x 14.0	20	6.1	80
20 x 20	6.1 x 6.1	30	9.1	55
30 x 30	9.1 x 9.1	30	9.1	75
50 x 50	15.2 x 15.2	30	9.1	95
53 x 53	16.2 x 16.2	30	9.1	110
55 x 55	16.8 x 16.8	30	9.1	115
59 x 59	18.0 x 18.0	30	9.1	135
63 x 63	19.2 x 19.2	30	9.1	150
68 x 68	20.7 x 20.7	30	9.1	177
70 x 70	21.3 x 21.3	30	9.1	185

Reprinted with permission from NFPA 72®-2013, *National Fire Alarm and Signaling Code*, Copyright © 2012, National Fire Protection Association, Quincy, MA. The information in this table is intended to be used in conjunction with the requirements of this code. It is not the complete and official position of the NFPA on the referenced subject, which is represented only by the standard in its entirety.

Figure 4-5. This table assumes the appliance is centered in the space. This table is only valid up to 30 feet. For ceilings over 30 feet, appliances may be suspended or a performance-based design may be used under engineering supervision. Source: NFPA 72, Table 18.5.5.4.1(b).

Public Mode Visible Signaling

Unlike audible notification appliances, the spacing of visible notification appliances is very prescriptive. *NFPA 72* contains tables providing the maximum spacing of visible appliances. Table 18.5.5.4.1(a) contains the room spacing requirements for wall-mounted visible appliances. **See Figure 4-4.** Table 18.5.5.4.1(b) contains the room spacing requirements for ceiling-mounted visible appliances. **See Figure 4-5.**

Table 18.5.5.4.1(a) of *NFPA 72* assumes the visible notification appliance is centered on a wall. Where the appliance is not centered or the room is non-square, there will be other spacing requirements.

Table 18.5.5.4.1(b) of *NFPA 72* assumes a maximum ceiling height of 30 feet. For ceilings over 30 feet high, the appliances must be suspended to a height of 30 feet or lower. It may also be possible to design a performance-based system, using engineering supervision, as an alternative to these prescriptive requirements.

Section 18.5.5 of *NFPA 72, National Fire Alarm and Signaling Code,* provides mounting requirements for visible notification appliances. Section 18.5.5.1 requires all wall-mounted visible appliances to be mounted so that the entire lens of the appliance is between 80 inches (2.03 m) and 96 inches (2.44 m) above the finished floor. This mounting height will help ensure that the

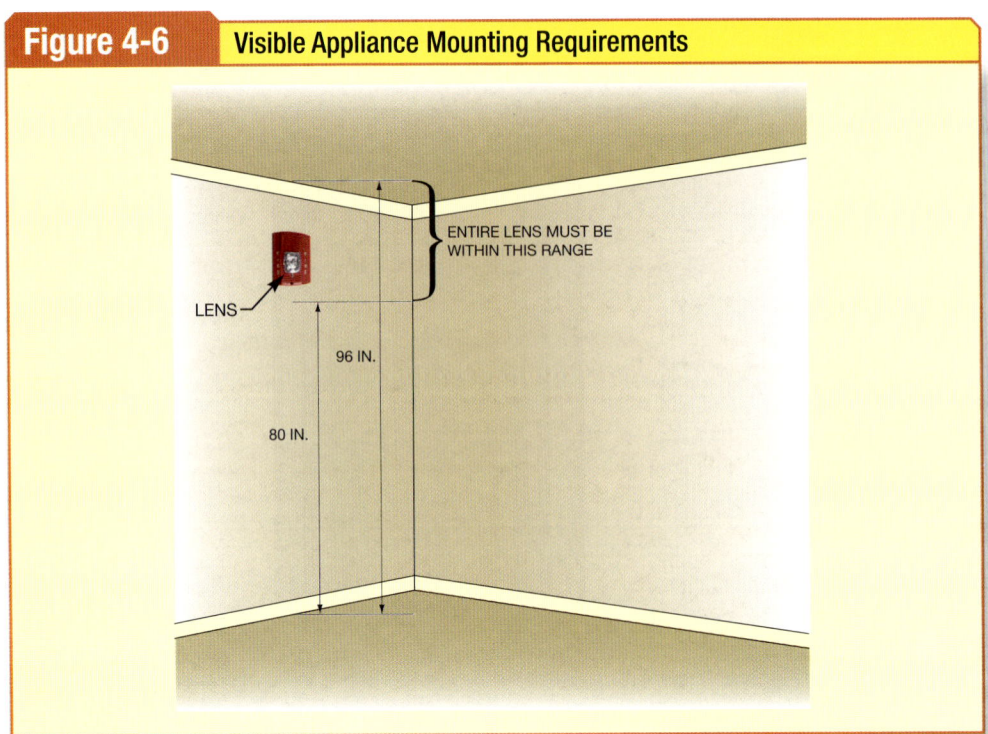

Figure 4-6. The entire lens must be within prescribed limits set by NFPA 72. The strict limits for strobe height relate to the way light intensity attenuates with distance.

appliances are above any furnishings. **See Figure 4-6.** Sometimes, it is not desirable to wall mount visible notification appliances because of architectural concerns, or it is not possible because of the lack of walls. Examples of these cases include ornate lobbies or convention centers with large open areas. In these cases, ceiling-mounted appliances can be used.

The ADAAG requires visible notification appliances to be mounted at least 80 inches above the finished floor, or six inches below the ceiling, whichever is lower. Combination audible/visible notification appliances must also meet these requirements. There is usually some overlap of the mounting height ranges, and mounting appliances in this range can satisfy both ADAAG and *NFPA 72*. It should be noted that *NFPA 72* requires visible notification appliances to be mounted in the same mounting heights used in the testing laboratory. Mounting them higher will attenuate the signal, which reduces the effectiveness of the appliance.

The NFPA spacing tables assume the appliance is centered on one wall of the room, or centered on the ceiling. **See Figure 4-7.**

Where the strobe cannot be centered or the room is non-square, the appliance rating is determined by either using the distance to the farthest wall or doubling the distance to the opposite wall, whichever is

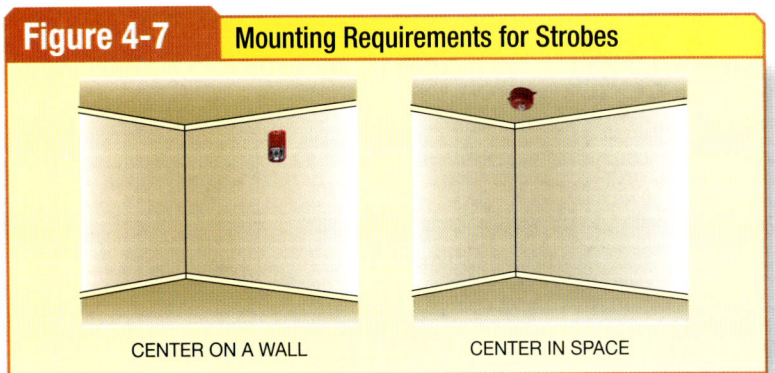

Figure 4-7. In symmetrical areas, the appliance should be centered on either wall or ceiling, as appropriate.

greater. **See Figure 4-8.** The same requirements apply for ceiling-mounted appliances when they are not centered in a room.

It is always best to use the fewest number of visible notification appliances (strobes) in a single area. Multiple strobes flashing out of synchronization in a single viewing area can create a combined strobe rate that can be dangerous to persons with photosensitive epilepsy. It has been anecdotally reported that strobe rates above eight cycles per second (hertz) can cause seizures in persons afflicted with photosensitive epilepsy. Using fewer appliances or synchronizing multiple strobes in a single viewing area will keep composite strobe rates low enough to prevent seizures.

In order to prevent seizures, Section 18.5.5.4.2 requires the use of one of the following:
- A single appliance
- Two groups of appliances in the same room or adjacent space within the same field of view that are synchronized
- More than two appliances or more than two groups of appliances in the same room or adjacent space within the same field of view that are synchronized

Synchronization of appliances is achieved by the use of synchronizing modules or cards supplied by the manufacturer of the controls. Some manufacturers produce equipment or software that synchronizes all appliances on a notification appliance circuit. Other manufacturers make equipment modules for field mounting which require an additional conductor between all synchronized appliances.

Corridors. Section 18.5.5.5 provides requirements for spacing of visible notification appliances in corridors up to 20 feet (6.1 m) in width. Corridors wider than 20 feet (6.1 m) are treated as rooms for the purpose of appliance spacing. Notification appliances in corridors must have a minimum rating of 15 candela; however, higher intensity appliances may be used since they exceed the minimum rating required by the *Code*.

Section 18.5.5.5 requires visible notification appliance spacing in corridors to be such that there are appliances within 15 feet (4.57 m) of each end of the corridor, and not more than 100 feet (30.4 m) between any two appliances. Notification appliances in

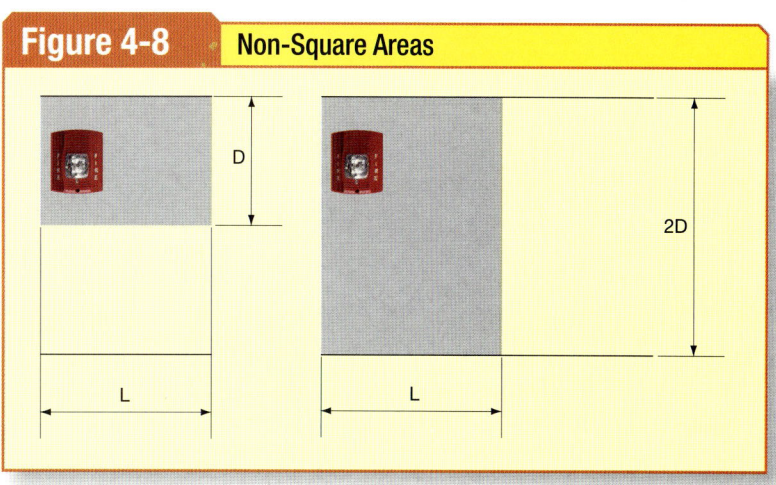

Figure 4-8. Off-center or non-square rooms require special spacing considerations.

 **Fact**

Many manufacturers now produce notification appliances and equipment which can be synchronized by circuit, zone, or system. Be sure to read the manufacturer's instructions before installing equipment.

For additional information, visit qr.njatcdb.org
Item #1009

Synchronization modules are sometimes used to ensure all strobes on a single notification appliance circuit flash in unison.

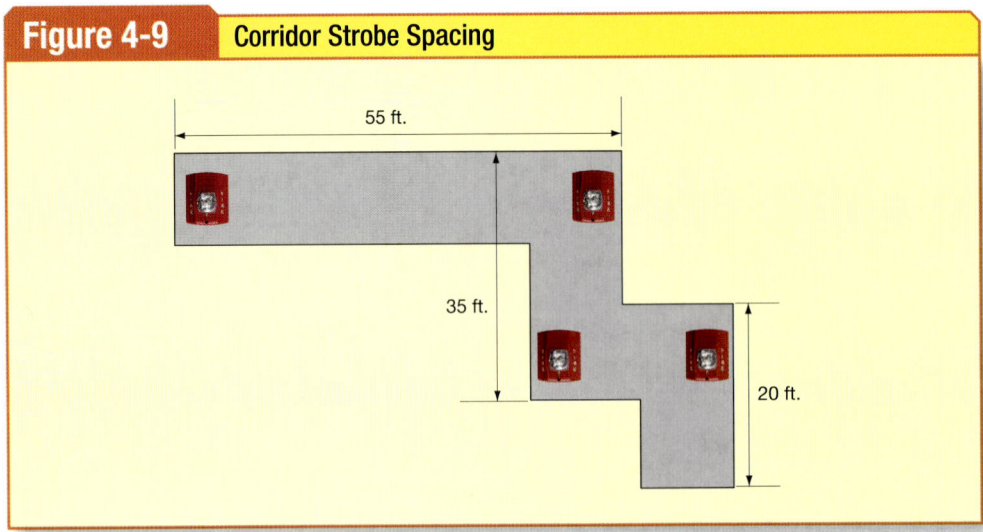

Figure 4-9. Changes in direction, elevation, or viewing path create separate corridors. Note that appliances may be shared between corridors.

corridors may be mounted on the end walls, provided they are not mounted behind obstructions such as an open door. Any time there is a change in direction, a change in elevation, or a door, each corridor section must be treated separately. **See Figure 4-9.** Additionally, when there are more than two corridor notification appliances in a field of view, Section 18.5.5.5.7 requires that they be synchronized. Finally, Section 18.5.5.1 permits room spacing for corridor strobes, as permitted by Section 18.5.4.

Sleeping Areas. Sleeping areas subject to use by persons with hearing impairments must have visible notification appliances and meet special requirements for their mounting. The intense light from visible notification appliances in sleeping areas is actually used to awaken a sleeping person. Most building codes have special provisions for the number of hotel/motel guest sleeping rooms which must be outfitted with visiblrooms that must be outfitted with visible notification appliances. Generally, about 5% to 7% of all guest rooms must meet accessibility requirements.

Section 18.5.5.7 requires visible notification appliances to have a minimum rating of 177 candela where mounted less than 24 inches (610 mm) from a ceiling. Visible appliances mounted more than 24 inches (610 mm) from the ceiling are permitted to have a rating of 110

Strobes must be installed within 15 feet of each end of the corridor and not more than 100 feet between appliances.

Fact

All signaling requirements contained in Chapter 18 of *NFPA 72* assume the occupants are normal, healthy adults. Chapter 18 also assumes the occupants are not impaired by drugs or alcohol.

candela. The higher rating of appliances within 24 inches (610 mm) of the ceiling is to penetrate a smoke layer that might be present during a fire. **See Figure 4-10.**

Additionally, all visible appliances must be mounted within 16 feet (4.87 m) of the pillow. These requirements apply whether the appliance is a stand-alone appliance or integral to a smoke alarm or smoke detector. Section 18.18.5.5.7.1 requires combination smoke alarm/strobe devices to be mounted according to the requirements for smoke alarms, as required by Chapters 17, 18, and 29. However, smoke detectors must be mounted so that the top of the detector is within 12 inches (300 mm) of the ceiling.

Visible notification appliances in sleeping rooms must be located within 16 feet of the pillow.

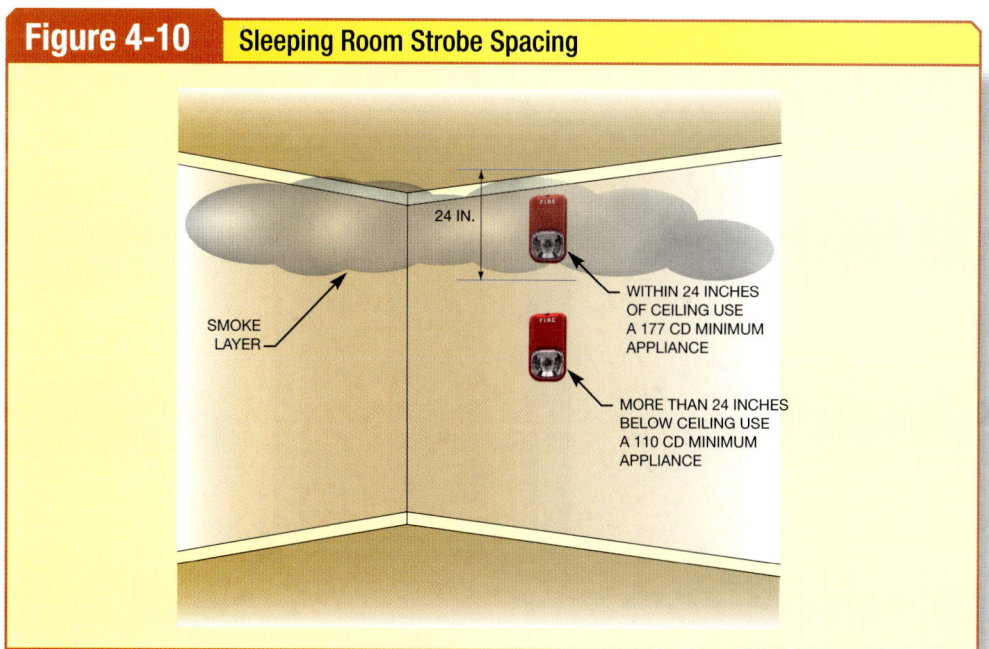

Figure 4-10. A 177 cd appliance must be used where the appliance is mounted within 24 inches of the ceiling in order to penetrate a smoke layer. A 110 cd appliance may be used when mounted at least 24 inches below the ceiling. Any appliance used for sleeping areas must be mounted within 16 feet of the pillow.

Summary

Notification appliances are essential to provide warning to occupants and operators during an emergency. Proper design and installation of notification appliances will ensure that they operate as intended. Appliances may rely on audible or visible signals. The ADA does not require fire alarm systems, but it has special requirements of its own that apply when fire alarm systems are installed.

Review Questions

1. Which of the following occupancy types is a good location for a coded signal?
   a. Convenience store
   b. Hospital
   c. Office building
   d. Warehouse

2. Audible fire alarm signaling is __?__.
   a. determined by AHJ
   b. determined by owner
   c. performance based
   d. prescriptive

3. The average ambient sound pressure level (SPL) is measured over a __?__ period.
   a. 10-minute
   b. 1-hour
   c. 12-hour
   d. 24-hour

4. SPL measurements are always measured using the __?__ -weighted scale.
   a. A
   b. B
   c. C
   d. D

5. Protective covers for notification appliances may require __?__.
   a. derating
   b. more appliances
   c. shorter spacing
   d. all of the above

Review Questions

6. Signaling for occupants of a building is called __?__ signaling.
   a. private mode
   b. public mode
   c. reduced mode
   d. user mode

7. The minimum SPL in sleeping areas must be at least __?__ at the pillow.
   a. 55 dBA
   b. 65 dBA
   c. 75 dBA
   d. 85 dBA

8. Visible notification appliances must be mounted at between __?__ above the finished floor.
   a. 75" and 90"
   b. 80" and 90"
   c. 80" and 96"
   d. 86" and 96"

9. Which of the following is **NOT** a standard rating of a visible notification appliance?
   a. 15 cd
   b. 25 cd
   c. 30 cd
   d. 75 cd

10. A 50-foot-by-50-foot room with a 20-foot high ceiling requires a visible notification appliance rating of __?__.
    a. 75 cd
    b. 95 cd
    c. 110 cd
    d. 185 cd

11. What is the minimum distance that visible notification appliances must be located from the end of a corridor?
    a. 10'
    b. 12'
    c. 15'
    d. 16'

12. The Americans with Disabilities Act Accessibilities Guidelines (ADAAG) require __?__ lumens per square foot as the minimum illumination at the floor.
    a. 0.030 lm/ft^2
    b. 0.033 lm/ft^2
    c. 0.035 lm/ft^2
    d. 0.037 lm/ft^2

Wiring and Wiring Methods

Wiring is one of the most important elements of a fire alarm system. A good wiring installation will usually result in a system that operates reliably for years. However, poor wiring can, and often does, cripple an otherwise good system. Unlike most other low-voltage systems, fire alarm systems are designed and installed to provide life safety. Failure of other electrical systems, such as a paging system, may only result in an inconvenience. However, failure of fire alarm systems may result in loss of life. For this reason, fire alarm system wiring must be held to the highest standards.

Objectives

» Define key terms
» Explain the general requirements of wiring and wiring methods for:
  › Outside circuits
  › Grounding
  › Plenums
» Describe both power-limited and non–power-limited fire alarm circuits
» Explain circuit integrity cable
» Explain Class A and Class X circuit separation
» Perform loop resistance and voltage drop calculations
» Explain firestopping concepts

Chapter 5

Table of Contents

Wiring Requirements for Fire Alarm Systems 106
    General Requirements 106
    Outside Circuits 107
    Grounding 108
    Plenums and Other Air-Handling Spaces 108
    Workmanship 108

Non–Power-Limited and Power-Limited Fire Alarm Circuits 110
    Non–Power-Limited Fire Alarm Circuits 111
    Power-Limited Fire Alarm Circuits 113

Circuit Integrity Cable 116
Class A and Class X Circuit Separation 117
Calculations 117
    Loop Resistance Calculations 117
    Voltage Drop Calculations 118
Firestopping Concepts 120
Summary 122
Review Questions 123

WIRING REQUIREMENTS FOR FIRE ALARM SYSTEMS

Wiring is a large part of any fire alarm system. Sloppy wiring practices are one of the leading causes of fire alarm system failures. Improperly installed conductors may cause insulation damage, which usually results in ground faults. Conduit that is filled beyond capacity can also result in insulation damage as new conductors are pulled. Ground faults may be prevalent when moisture and insulation damage are present. A good installation will not only save the owner money over the life of the system but will also be more reliable.

NFPA 70, National Electrical Code is the most widely adopted electrical code in the world. It is the cornerstone of the electrical industry. It provides many wiring requirements, but it is not a design manual. The *NEC* is adopted in all 50 states of the U.S., Puerto Rico, and has been officially adopted by several other nations, including Mexico, the Philippines, Panama, Venezuela, Costa Rica, Colombia, Peru, and Ecuador. The *NEC* has been translated into Spanish and Mandarin Chinese. It is unofficially used in many other nations such as Thailand, China, much of Africa, and other parts of Latin America. Some jurisdictions do not adopt the *NEC* but use the *NEC* as the basis for a local code. Additionally, the *NEC* is sometimes amended by local jurisdictions to suit unique local restrictions. It is incumbent upon the user to be familiar with any local codes or amendments to the *NEC* before installing any wiring.

NFPA 70, National Electrical Code, applies to all wiring of fire alarm systems. However, *NFPA 72, National Fire Alarm and Signaling Code,* provides the requirements for device selection, performance, and testing and maintenance. Sections 12.2.4 and 12.2.4.3 of *NFPA 72* specifically reference *NFPA 70* for all wiring. It should be noted that *NFPA 72* also contains some requirements for circuit performance, such as Class A separation and survivability. These are not necessarily wiring requirements, but they may have an impact on the type of wiring methods used. *NFPA 72* also contains requirements for the dedicated (individual) branch circuit used to provide primary power for fire alarm controls.

General Requirements

NEC 90.3, "Code Arrangement," states that Chapters 1 through 4 apply generally and Chapters 5 through 7 apply to special occupancies, special equipment, or special conditions. Chapter 7 contains Article 760, *Fire Alarm Systems*. The requirements in Article 760 can be thought of as exceptions or modifications to Chapters 1 through 4 since they contain special requirements for fire alarm systems.

The *NEC* is not a design manual. In fact, the *NEC* will allow wiring that may not permit the fire alarm system to properly operate. For example, the *NEC* allows the use of power wiring (such as THHN) where the manufacturer requires twisted, shielded pairs. Cable capacitance, attenuation, and crosstalk all contribute to poor signal transmission and may be affected by the choice of wiring methods.

Article 760 of the *NEC* comprises four parts. Part I contains general requirements that apply to all fire alarm circuits, Part II applies to non–power-limited circuits, Part III applies to power-limited wiring, and Part IV contains listing requirements.

The scope is contained in Section 760.1, which states that Article 760 applies only to those circuits controlled and powered by the fire alarm system. Article 760 does not apply to burglary wiring, HVAC controls, elevator controls, or communications unless the fire alarm controls provide power or control to those circuits. Other signaling wiring requirements, such as burglary, paging, and HVAC controls, are covered in Article 725.

Section 760.3(E) references Article 725 and clearly states that building control circuits such as HVAC, elevator control, and burglary are not covered by Article 760. However, any circuit powered by the fire alarm system used to control these systems is covered by Article 760. A relay that controls elevator recall features is covered by Article 760 where the relay is powered and controlled by the fire alarm system.

Section 760.3 references several other articles that specifically apply to fire alarm systems. These references are written to bring in other requirements for fire alarm

Fact

The *National Electrical Code (NEC)* was first issued in 1897 to address poor electrical installation practices. The National Fire Protection Association (NFPA) has been the steward of the *NEC* since 1911.

wiring. For example, these include a reference to 300.22, which requires the use of low-smoke producing cables and firestopping. Section 760.3(H) also requires raceways that are exposed to different temperatures comply with 300.7(A). This will require sealing the raceway to prevent condensation from forming in the raceways. 760.3(I) requires fire-rated cables (such as circuit integrity cable) and conductors to be properly supported when run vertically. A new section, 760.24(B), was added in the 2014 edition of *NFPA 70* to provide additional requirements for fastening circuit integrity (CI) cables. This section requires fastening intervals of 18 inches (450 mm) where installed below seven feet (2.1 m) from the finished floor and 24 inches (610 mm) elsewhere.

Section 760.25 requires the accessible portions of abandoned cables to be removed. This requirement is intended to prevent the buildup of combustible cables in a building. Cables are often constructed with flammable insulating materials that, when burned, will produce toxic smoke. Abandoned cables are permitted to remain only where tagged for future use. The intent of 760.3(A) is to prevent fire alarm wiring from contributing to the spread of fire and smoke.

Section 760.30 requires all fire alarm circuits to be identified in a manner that prevents unintentional signals during testing and servicing. This section does not specify the exact means of identification. However, almost all fire alarm specifications will contain additional requirements. These may include painting all junction boxes red in color or placing red stripes on fire alarm conduit. Many jurisdictions also have differing requirements, so be sure to use the method prescribed by the authority having jurisdiction (AHJ).

Outside Circuits

A typical lightning strike raises the ground plane potential to levels that can cause serious damage to circuits. One way to reduce lightning damage is to use optical fiber cables between buildings. Copper conductors require special protection. In fact, Section 12.2.4.2 of *NFPA 72* requires all non–power-limited and power-limited circuits to be provided with transient protection. This section was added to the 2013 edition of *NFPA 72* because the *NFPA 70* does not provide any requirements for protection of fire alarm circuits.

Section 760.32 requires power-limited circuits extending beyond one building to be protected according to Parts II, III, and IV of Article 800 or Part I of Article 300. Non–power-limited circuits must be installed to the requirements of Part I

Fact

NFPA 72, National Fire Alarm and Signaling Code, requires transient protection of all fire alarm circuits which extend outside of a building.

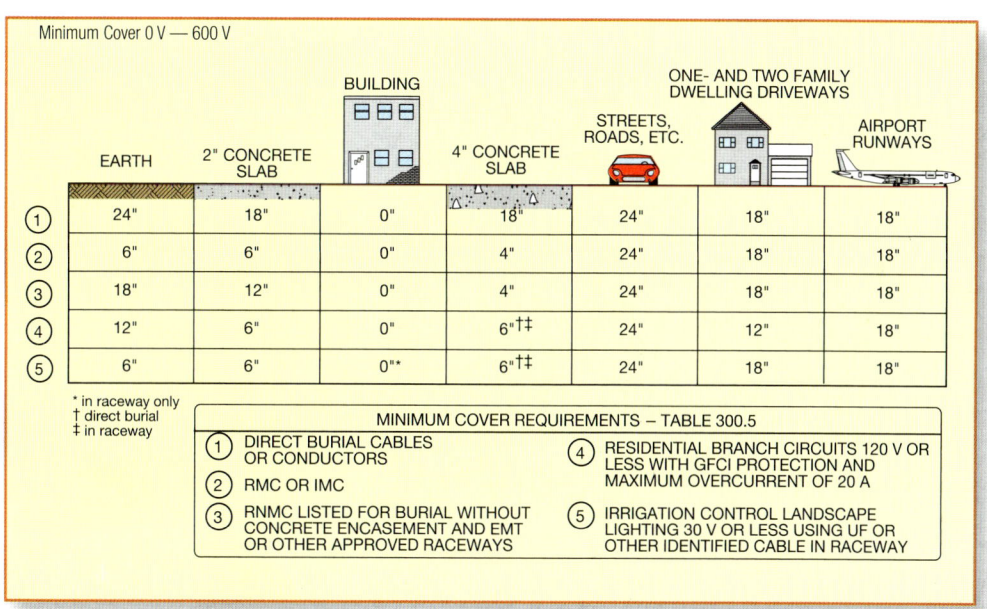

All underground conductors must be appropriately listed for the environment.

of Article 300 and the applicable sections of Part I of Article 225. Parts II, III, and IV of Article 800 apply to above ground (aerial) circuits and require primary (lightning) protection. Primary protection must be installed nearest the point of entry into the building to prevent high voltages from being imposed on internal conductors.

Underground circuits are usually treated as outside branch circuits or feeders. Underground fire alarm conductors must be buried as much as 24 inches (610 mm) below the surface. Table 300.5 provides the minimum burial depths, depending on the type of wiring method used and the type of cover. Other requirements, such as backfill, raceway bushings, and protection from damage can also be found in Section 300.5. Although not specifically required by Articles 225 or 300, primary protection for underground conductors is usually required by many project specifications as a precaution.

Lightning strikes can cause serious damage to low-voltage outdoor circuits, whether run above ground or underground. A typical lightning strike will deliver 10,000 amperes at a million volts or more. The strike can, and often does, raise the ground plane potential to levels that cause serious damage to circuits operating at 24 volts or less.

Grounding

Article 250 provides the requirements for grounding of fire alarm circuits. Generally, the grounding requirements apply to the power supply circuit. The branch-circuit equipment grounding conductor must be grounded at the panelboard from which it is provided power. Most of the field circuits do not have any grounding requirements, but the manufacturer may impose requirements beyond those contained in the NEC. 250.112(I) contains requirements for the grounding of all fire alarm circuits.

Care must be also exercised when using shielded conductors. A shield may act as an antenna, resulting in trouble signals and nuisance alarms if not properly connected. Shields should be continuous over the circuit length, but should only be grounded at the panel or source. Good grounding of the fire alarm control enclosure(s) must be provided, especially when connecting shields to ground.

Plenums and Other Air-Handling Spaces

Section 760.3(B) provides additional requirements for ducts, plenums, and other spaces used for environmental air. This section references Section 300.22, which provides strict requirements for the type of wiring in ducts, plenums, and other spaces used for environmental air. Again, the intent is to limit the type and amount of combustible material in air-handling spaces. In order to accomplish this goal, 300.22 generally requires metal-clad cable or metallic conduit in air-handling spaces. However, these requirements are modified by Parts I and II of Article 760.

Workmanship

Proper workmanship is a large part of a good installation. Sections 760.21 and 760.24 address this issue. Section 760.21 specifically addresses access to electrical equipment behind panels. This section addresses not only access inside equipment enclosures but also addresses spaces above ceiling tiles. Cables and conductors cannot prevent access to the equipment or area. Section 760.24 requires all work to be neat and workmanlike. While the exact meaning of the term "neat and workmanlike" is not defined in the NEC, most installers can identify neat and workmanlike installations. Section 760.24 does require proper support and termination of conductors, except for circuit integrity (CI) cables. New section 760.24(B) was added in the 2014 edition of NFPA 70 to provide additional requirements for fastening CI cables. This section requires fastening intervals of 18 inches (450 mm) where installed below seven feet (2.1 m) from the finished floor and 24 inches (610 mm) elsewhere. **See Figure 5-1.**

The installation must also conform with Section 300.4(D), which contains

---

Fact

NECA 305-2001, Standard for Fire Alarm System Job Practices (ANSI), is a good resource for workmanship methods for fire alarm systems.

Figure 5-1 Workmanship

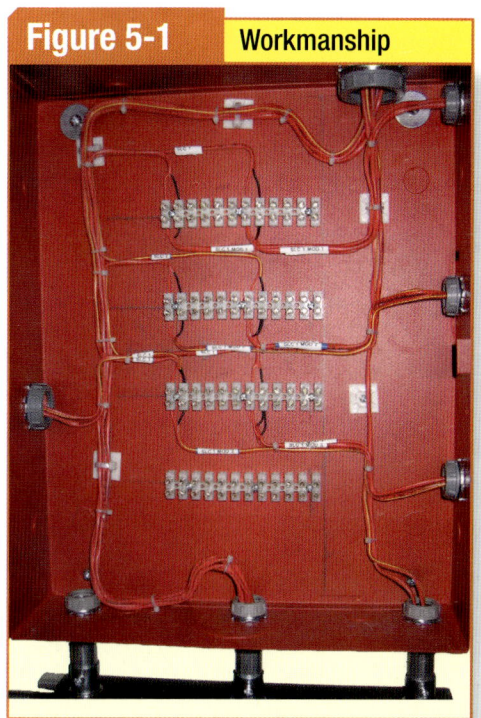

Figure 5-1. Note that all cables are identified and properly secured. This type of quality workmanship will make troubleshooting easier.

requirements for cables in framing members. *NECA 305-2010, Standard for Fire Alarm System Job Practices* (ANSI), may contain additional information regarding workmanship.

Cables. Where cables are used, they must be properly supported. Generally, cables are supported at intervals of 18 inches (450 mm) to 48 inches (1200 mm) and are never permitted to be laid directly on top of ceiling grids or other surfaces. Structural components can be used to support fire alarm cables, provided the cables are not damaged by normal building use. A variety of straps, hangers, staples, and other similar fittings are available and can be used to support cables. Where cables are installed above suspended (drop-in) ceilings, cables must be supported independently of the grid, and support wires independent of the ceiling grid supports are required.

Section 300.4(D) is referenced in Section 760.24, and requires cables and conductors parallel to framing members to be installed at least one and one-fourth inches (32 mm) from the nearest edge of the framing member to prevent puncturing by screws. If this distance is not possible, steel plates measuring one-sixteenth-inch (1.6 mm) thick can be used to protect the conductors. Additionally, intermediate metal conduit, rigid metal conduit, and electrical metallic conduit can be used. It is also permitted to fish cables in a wall that is already constructed.

Even though not specifically referenced by Article 760, Section 300.11 contains important requirements for securing and supporting of cables and raceways. Specifically, Section 300.11(B) does not permit raceways to be used as a means of support for other raceways or cables unless the raceway is identified for such use. Good planning and coordination with other trades is often necessary to meet these requirements.

Terminals. Terminals on fire alarm equipment are usually designed and listed for a certain size and quantity of conductors. Equipment specification sheets provide details for conductor connection to the device or appliance. It is never acceptable to loop wires over a terminal or cut wires from a device having four leads. Looping wires around terminals effectively results in a T-tap. T-Taps result in devices that are not monitored for integrity. It is sometimes necessary to use large conductors in order to reduce voltage drop or loop resistance. However, it is usually not acceptable to connect smaller pigtails to appliances because the terminals do not accept the larger conductor. Shorter runs or smaller loads may be required to reduce the voltage drop rather than forcing larger conductors under a terminal not listed for such use.

Identification. Section 760.30 requires fire alarm circuits to be identified at terminal and junction box locations. This requirement is intended to prevent unintentional tampering or interference with the system. Many AHJ permit junction box covers to be painted red, but some may permit or require other

Figure 5-2. The NEC requires terminal cabinets and junction boxes to be identified, but does not specify a particular method.

methods. **See Figure 5-2.** The NEC does not specify the type of marking, so one must be sure to check the project specifications and contact the AHJ before starting the job.

Hazardous Locations. Any fire alarm wiring in hazardous (classified) locations is covered by Section 760.3(C). Section 760.3(C) references Articles 500 through 516 and Part IV of Article 517 for "Classified Locations." The type of wiring used in these spaces will greatly depend on the class and division of hazardous (classified) location. Wiring in hazardous (classified) areas often requires the use of rigid metal conduit and seals when passing through boundaries. There are explosion-proof fire alarm initiating devices and notification appliances for the instances where detection is needed for hazardous (classified) areas. Generally, the best approach is to install equipment outside the hazardous areas. This will obviate the need for special equipment.

 **Fact**

Manufacturer's literature and device markings will provide an indication the circuit card, device, or equipment is power-limited or non–power-limited. Unless marked as "power-limited," you must assume the circuit is non–power-limited.

Corrosive, Damp, Wet Locations. Corrosive, damp, and wet locations present unique problems for fire alarm systems and are covered by by Section 760.3(D). Section 760.3(D) references Sections 110.11, 300.6, and 310.9 for wiring installed under these conditions. Wiring installed in wet, damp, or corrosive locations must be suitable for the environment. Underground runs between buildings or in concrete slabs that are in contact with the earth are considered to be wet locations. Condensation and seepage will eventually cause ground faults when ordinary conductors (such as THHN) are installed in underground runs, no matter how tightly the conduit is sealed. Failure to remove burrs from conduit cuts may also result in insulation damage and resulting ground faults. Section 300.6 also requires metal raceways to be of materials suitable for the environment in which they are installed.

Some indoor locations, such as indoor pool areas, may be both damp and corrosive. It is possible to install wiring that complies with the NEC in these spaces; however, not all fire alarm equipment may be suited for these environments. For example, smoke detectors are not well suited to pool areas because of the high humidity and corrosive atmosphere present.

Access. Section 760.21 covers requirements for access to electrical equipment. Specifically, fire alarm cables must be installed so their accumulation does not prevent access to equipment. Cables and raceways must be installed so that panels or ceiling tiles can be removed. Access includes the wiring inside conduit and raceways. The internal fill of wiring spaces and conduit are covered by Table 9 of the NEC. Controls must be accessible and must not be installed where obstruction prevents access.

NON–POWER-LIMITED AND POWER-LIMITED FIRE ALARM CIRCUITS

Section 760.41(B) of NFPA 70 requires the primary supply for non–power-limited circuits to be powered by a dedicated (individual) branch circuit. The circuit disconnecting means must be marked in red as "FIRE ALARM CIRCUIT." This mirrors Section 10.6.5.2 of *NFPA 72*. Other requirements in *NFPA 72* include marking the location of the disconnecting means at the fire alarm control unit, limiting access to the disconnecting means, and so forth.

There are two basic types of fire alarm circuits permitted by the *National Electrical Code*: power-limited and non–power-limited. Both types of circuits are still available in the marketplace. Each circuit type has advantages, disadvantages, and special wiring requirements. However, power-limited circuits are quickly becoming the industry standard and are the vast majority of circuits installed today.

Most fire alarm initiating-device circuits, signaling line circuits, and notification appliance circuits manufactured today operate at a nominal 24 volts direct current (VDC). These are most often power-limited circuits. However, installer/maintainers are likely to encounter other circuits known as non–power-limited, which operate at higher voltages and have no limitations on the amount of delivered energy.

Speaker circuits can operate as high as 140 volts and are one of the few non–power-limited fire alarm circuits still manufactured today. Non–power-limited fire alarm circuits are capable of providing voltages up to 600 volts (over 100 volt-amperes) and have sufficient energy to shock and kill or ignite a fire. Non–power-limited circuits are not limited in the energy they deliver, so they must be treated differently from circuits operating at lower energy outputs. They are wired using different methods that are suitable for their higher voltages.

Power-limited circuits generally deliver less than 100 volt-amperes and are inherently limited in the energy they deliver. Tables 12(A) and 12(B) of the *NEC* provide the output limitations of power-limited fire alarm circuits. Table 12(A) provides alternating-current limitations, and Table 12(B) provides direct-current limitations. Power-limited circuits operate at lower voltages and are treated much differently from non–power-limited circuits. Power-limited circuits are very similar to Class 2 circuits in energy output characteristics.

 Fact
Non–power-limited circuits can be wired using *NEC* Chapter 3 wiring methods or by using specialty cables. Specialty cables may require protection and special supporting to protect them from damage.

Article 760 of the *NEC* provides requirements for the installation of both power-limited and non–power-limited circuits. Part II of Article 760 provides requirements for non–power-limited wiring, and Part III provides requirements for power-limited wiring. Other circuits, such as Class 2 and Class 3 (for example, burglary) are covered by Article 725.

The product standards for fire alarm equipment require the manufacturer to provide markings on the product and installation sheets that indicate the energy limitations of the circuit. The laboratory providing the listing for the equipment will determine if the equipment is power limited or non–power limited. This permits the manufacturer to apply markings to the equipment indicating its rating.

Non–Power-Limited Fire Alarm Circuits

Part II of Article 760 applies to non–power-limited fire alarm circuits. Section 760.46 specifically covers wiring methods for non–power-limited fire alarm circuits. Line-side power supply wiring must, of course, conform to Chapters 1 through 4 of the *NEC*. Load-side wiring methods for non–power-limited circuits must also comply with Chapter 3. However, Exception No. 1 to Section 760.46 permits the methods of Sections 760.48 through 760.53 to be used in lieu of Chapter 3 wiring methods. Therefore, non–power-limited circuits may be wired using either Chapter 3 methods or non–power-limited wiring methods. These sections permit special non–power-limited cables to be used, rather than using power wiring methods of Chapter 3.

Non–Power-Limited Cable Types.

Where used, non–power-limited cables may be one of the following types:
- NPLF Non–Power-Limited Fire General Use

- NPLFR Non–Power-Limited Fire Riser Use
- NPLFP Non–Power-Limited Fire Plenum Use

Non–power-limited cable types are designed and tested only for non–power-limited applications. Use for other purposes is severely limited, even though the insulation is rated for 600 volts. Most non–power-limited fire alarm cables are red in color in order to better identify them as fire alarm cables. The listing of non–power-limited fire alarm cables permits only specific fire alarm uses within a building.

Type NPLF cable can be used within a floor but cannot be used for risers (floor to floor), in other spaces used for environmental air, or in plenums. Non–power-limited fire alarm circuits extending from one floor to another must utilize Type NPLFR cable because it is capable of resisting the spread of fire in the vertical direction. Circuits installed in other spaces for environmental air must be wired with Type NPLFP cable. Non–power-limited fire alarm circuits installed in a plenum must be wired using specific Chapter 3 methods (such as MI, MC, and EMT) because Type NPLFP cable is not approved for exposed use in a plenum. In fact, Type NPLFP is only approved for use in other spaces used for environmental air. See Section 300.22(B) for the permitted wiring methods in plenums.

Section 760.176 provides the listing and marking requirements of non–power-limited fire alarm cables and provides guidance on where they can be used. Section 760.176 permits Type NPLFP to be used as a substitute for Type NPLFR, and Type NPLFR as a substitute for Type NPLF. Experienced installers will recognize that it may be more cost effective to use only Type NPLFP cable.

Fact

Non–power-limited cables are plastic jacketed and can be easily damaged by normal building use. When they pass through a floor, they must be protected up to a height of 7 feet (2.1 m). It is always permissible to sleeve fire alarm cables in a raceway.

Non–Power-Limited Wiring Methods. Sections 760.53(A)(1) through 760.53(A)(3) provide the wiring methods specific to non–power-limited fire alarm multi-conductor cables. Non–power-limited fire alarm specialty cables (Types NPLF, NPLFR, and NPLFP) must be installed so that they will not be damaged by normal building use. Generally, these cables can be installed exposed on a ceiling or wall. Exposed non–power-limited fire alarm cables must be adequately supported and protected from damage. Architectural and structural elements such as trim, baseboards, and door frames can be used to protect NPL fire alarm cables. All NPL fire alarm cables must be protected from damage below seven feet (2.1 m) from the finished floor. Exposed cables must be supported at intervals not more than 18 inches (450 mm). However, NPL fire alarm cables can always be fished inside walls for better protection. Where exposed, staples and other approved fasteners are permitted.

Where NPL fire alarm cables pass through a floor or wall, they must be protected by a metal raceway or rigid nonmetallic raceway to a height of seven feet (2.1 m), unless they can be protected using building elements as described above. NPL fire alarm cables in hoistways must be installed in rigid metallic conduit, rigid nonmetallic conduit, intermediate metallic conduit, liquid-tight flexible metallic tubing, or electrical metallic tubing, unless otherwise required by Article 620 for elevator hoistways.

Because the voltage of non–power-limited fire alarm circuits can be as high as 600 volts, terminations or splices of NPL fire alarm circuits must be made in listed fittings, boxes, enclosures, fire alarm devices, or utilization equipment. Some field devices have covers that enclose the terminals. In this case, a back box may not be required. However, back boxes are required to enclose terminals of field devices that do not have approved covers. Flying splices are never permitted on any fire alarm circuit.

Non–power-limited fire alarm circuit conductors must be separated from conductors of all other circuits, with only two exceptions. Class 1 circuits and non–power-limited fire alarm circuits may be mixed in the same cable, enclosure, or raceway, provided that all conductors are insulated for the maximum voltage of any conductor. Class 1 circuits operate up to 600 volts, which makes them very similar to non–power-limited circuits. Non–power-limited fire alarm circuits are permitted in the same cable, raceway, or enclosure as power supply conductors only where connected to the same equipment. These restrictions primarily exist to protect service personnel from coming into contact with higher than expected voltages.

Where Class 1 and non–power-limited fire alarm conductors are installed in the same cable tray or raceway, Section 760.51(A) requires them to be derated in accordance with Section 310.15(B)(3)(a) where the conductors carry more than 10% of their rated capacity. Section 760.51(B) also references Section 300.17 where non–power-limited conductors are installed in the same raceway with power supply conductors. Section 300.17 requires the number of conductors in raceways to be such that heat can be dissipated and that the conductors can be easily removed. Non–power-limited conductors installed in cable trays must conform to Sections 392.22 and 392.80(A). These sections apply mainly to percentage of fill in cable trays.

Only copper conductors are permitted to be used for non–power-limited fire alarm circuits. Section 760.49(A) of the NEC permits conductor sizes of 16 AWG and 18 AWG, provided that they supply loads not exceeding ampacities found in Table 402.5 for fixture wires. Table 402.5 permits 16 AWG to carry eight amperes and permits 18 AWG to carry six amperes. However, conductors larger than 16 AWG must not carry current in excess of the ampacities found in Section 310.15. A variety of insulation types is permitted by 760.49(B), which requires a minimum rating of 600 volts. Any insulation type not provided in Section 760.49(B) must be specifically listed for non–power-limited fire alarm use. Finally, NPL fire alarm conductors can be either solid or stranded copper.

> **Fact**
>
> Power-limited circuits are limited in output. The circuits are designed so they will not cause a fire or create a shock hazard. Tables 12A and 12B in the *National Electrical Code* provide the maximum voltage, current, and power limitations for power-limited circuits.

Power-Limited Fire Alarm Circuits

Part III of Article 760 covers power-limited fire alarm circuits. Power-limited fire alarm equipment must be listed and marked as such, as required by Section 760.121(A)(3). If the manufacturer's product sheet does not classify the device or equipment as power-limited, then non–power-limited wiring methods must be used in accordance with Part II of Article 760. Additionally, Section 760.124 requires equipment to be plainly marked to identify each power-limited fire alarm circuit. Tables 12(A) and 12(B) of the *NEC* provide limitations for power-limited fire alarm circuits. Generally speaking, power-limited fire alarm circuits operate at less than 100 volt-amperes.

Section 760.121(B) of *NFPA 70* requires the primary supply for power-limited circuits to be powered by a dedicated (individual) branch circuit. The circuit disconnecting means must be marked in red as "FIRE ALARM CIRCUIT." This mirrors Section 10.6.5.2 of *NFPA 72*. Other requirements in *NFPA 72* include marking the location of the disconnecting means at the fire alarm control unit, limiting access to the disconnecting means, and so forth.

Power-Limited Cable Types. Section 760.179 provides listing requirements for the use of cables specifically designed for power-limited fire alarm circuits. Where used, special power-limited cables may be one of the following types:
- FPL power-limited fire general use
- FPLR power-limited fire riser use
- FPLP power-limited fire plenum use

Power-limited fire alarm cable types are designed and tested only for power-limited applications. Use of power-limited fire alarm cables for other purposes is severely limited, even though the insulation is rated for 300 volts. Most power-limited fire alarm cables are red in color in order to better identify them as fire alarm cables. However, there are power-limited cables available with different colored stripes, which makes installation and maintenance easier. The listing of power-limited fire alarm cables permits only specific uses within a building. Power-limited fire alarm cables are available in plastic-jacketed or metal-clad versions.

The different types of cables are designed and tested for specific applications. For example, Type FPL cable can be used within a floor, but cannot be used for risers (floor to floor), in other spaces used for environmental air, or in plenums. Power-limited circuits extending from one floor to another must utilize FPLR cable. A riser rating indicates the cable resists the spread of fire in the vertical direction. A new Section 760.135(E)(1) in the 2104 edition of *NFPA 70* now permits FPL, FPLR, and FPLP to be run as risers (floor to floor, when installed in metal raceways. New Section 760.135(E)(2) permits the same cables to be run vertically in plenum communications raceways, riser communications raceways, and general-purpose communications raceways. However, all floor penetrations must be firestopped as required by Section 300.21.

Circuits installed in other spaces used for environmental air or plenums must be wired with FPLP cable. Because FPLP cable is used in plenums, it must be low-flame-producing and low-smoke-producing, and it must produce few toxic gases. Most low-smoke, low-flame cables are made from Teflon, Halar, Kynar, Exar, polyethylene, and polypropylene. Section 760.179 provides the listing and marking requirements of power-limited fire alarm cables. Wise installers will carry Type FPLP for the entire project because it can be used in all applications. **See Figure 5-3.**

Power-Limited Wiring Methods. The line side of power supplies for power-limited circuits must comply with Chapters 1 through 4 for power wiring since they operate at a nominal 110 volts. There are more choices, however, for load-side wiring. Section 760.130 permits either non–power-limited wiring methods or power-limited wiring methods to be used for power-limited circuits. As previously explained, non–power-limited wiring methods can consist of either power wiring methods found in Chapters 1 through 4 or the NPL fire alarm cabling methods permitted by Part II of Article 760. The permitted wiring methods of power-limited fire alarm systems include:
- Chapter 3 methods
- Non–power-limited wiring methods
- Power-limited wiring methods

Where power-limited fire alarm circuit methods are used, the requirements of Sections 760.130(B)(1), 760.130(B)(2), and 760.130(B)(3) must be followed. These sections require the same type of mechanical protection as for non–power-limited fire alarm cabling. Specifically, exposed cables must be protected from damage by using the building features or by fishing the cable in a wall. Where cables are installed below seven feet (2.1 m) from the finished floor, cables must be supported at intervals not exceeding 18 inches (450 mm).

There are restrictions on mixing power-limited fire alarm circuit conductors and conductors of other circuits. In fact, Section 760.136 requires separation of power-limited fire alarm circuits from other circuits such as electric light, power, Class 1, non–power-limited fire alarm, and medium power network-powered broadband communications conductors. However, there are exceptions for connection of power-limited fire alarm conductors to the same equipment as higher voltage circuits. There are additional exceptions for a separation by a listed barrier or a minimum of one-fourth-inch separation within an enclosure. Again, the purpose of separating circuits is to prevent accidental shock hazards and to prevent faults on higher voltage circuits from disabling the fire alarm system.

Section 760.139 permits power-limited fire alarm circuits to be in the same cable, raceway, or enclosure as other power-limited fire alarm circuit conduc-

Figure 5-3 Power-Limited Wiring Methods

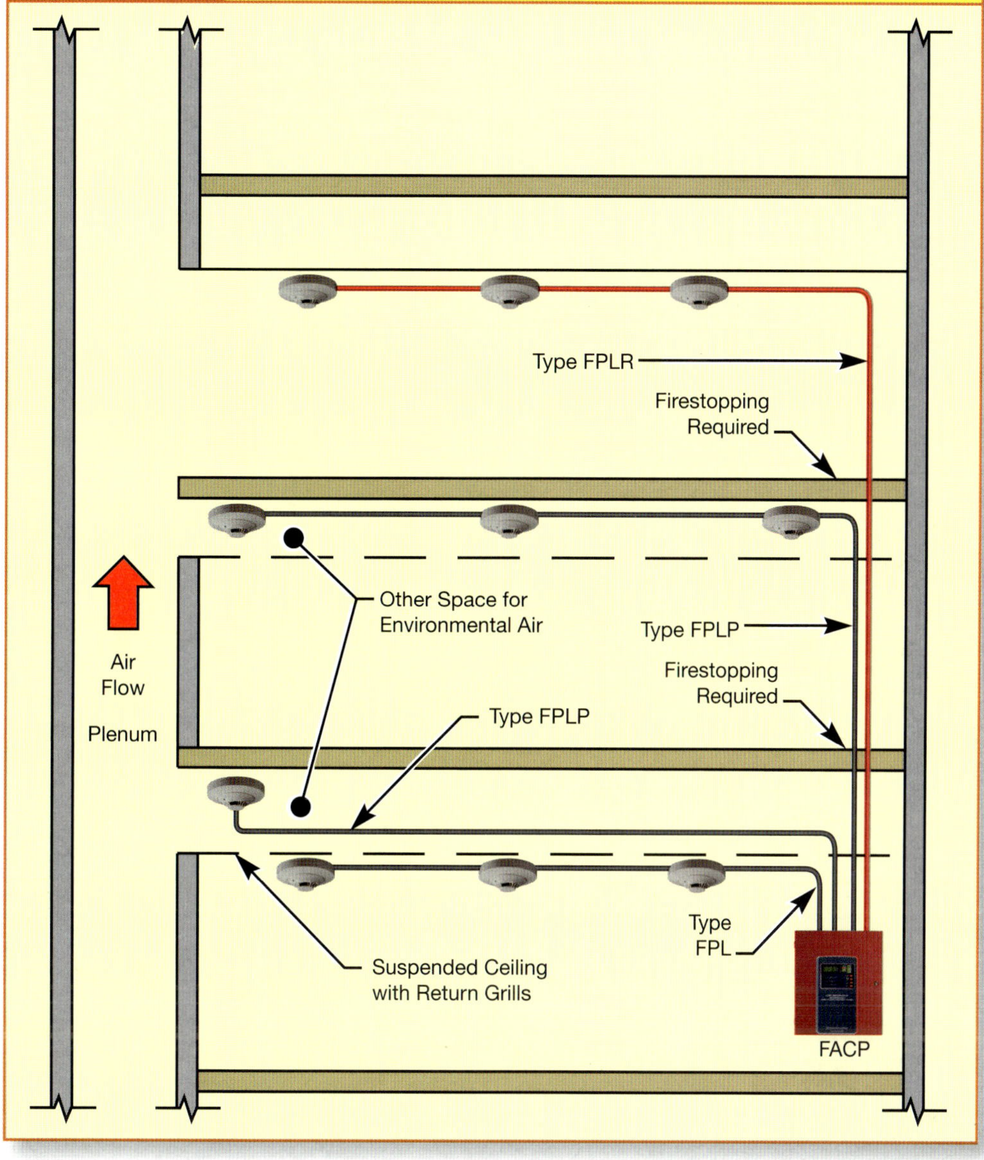

Figure 5-3. Special power-limited fire alarm cables are permitted by the NEC. Cables have specific uses, but substitutes are also permitted.

tors, Class 2 circuit conductors, and low-power network-powered broadband communications conductors.

Power-limited circuit fire alarm cables must be protected in a manner similar to protection for non–power-limited fire alarm circuit cables. Fire alarm cables installed exposed less than seven feet (2.1 m) above the finished floor must be protected from mechanical damage. Building features, such as baseboards, door frames, or other trim, can provide protection. Additionally, power-limited fire alarm cables can be fished in a wall. Raceways are not required when using fire alarm cables but can always be used to provide mechanical protection.

> **Fact**
>
> Table 760.154(D) provides a list of permitted cable substitutions, which include Types CM, CMR, CMP, and CMG. Of course, these cables must also be installed according to their listing.

Where power-limited fire alarm cables pass through a floor or wall, the cables must be protected by a metal raceway or rigid nonmetallic raceway to a height of seven feet (2.1 m), unless they can be protected using building elements as described. Power-limited fire alarm cables in elevator hoistways must be installed in rigid metallic conduit, rigid nonmetallic conduit, intermediate metallic conduit, liquid-tight flexible metallic tubing, or electrical metallic tubing, unless otherwise required by Article 620.

Although stated in Section 300.11(D), Section 760.143 reiterates the fact that power-limited fire alarm circuit conductors cannot be taped, strapped, or attached by any means to any other raceway for support. Sprinkler piping and other piping systems are not permitted as a means of support for circuit conductors.

Power-limited fire alarm circuit conductors tend to be smaller in size because of the low current use of the circuits. Section 760.142 places limits on the minimum conductor size, mainly for pull strength. Multiconductor cables are permitted to have conductors as small as 26 AWG, but single-conductor cables cannot have conductors smaller than 18 AWG.

It should be noted that most power-limited fire alarm cables permitted by Sections 760.154 and 760.179 are not permitted to have insulation ratings marked on the cable, unless required for other uses. This prevents power-limited fire alarm cables from being misused for applications involving a higher voltage. Table 760.154(D) permits substitutions for power-limited fire alarm cables. Generally, communications cables having an equal or higher rating may be substituted for power-limited fire alarm cables.

CIRCUIT INTEGRITY CABLE

In buildings where occupants either relocate or partially evacuate, there exists a need to maintain communications to direct aid or information during a fire. This would not be possible if a fire burned through the circuit conductors, thereby disabling communications to large parts of the premises. Chapter 24 of *NFPA 72, National Fire Alarm and Signaling Code,* requires fire alarm systems that employ partial evacuation or relocation (primarily emergency voice/alarm communications systems) to have survivability. In short, a fire in one evacuation zone cannot cause loss of communications to another evacuation zone.

Section 12.4 of *NFPA 72* defines each level of survivability. Level 0 requires no survivability. Level 1 requires all conductors installed in metal raceways and requires the building to be fully sprinklered, per *NFPA 13, Standard for the Installation of Sprinkler Systems.* Survivability Levels 2 and 3 require not less than two hours of fire resistance. The essential difference between Levels 2 and 3 is that Level 3 requires the building to be fully sprinklered, whereas Level 2 does not.

There are two approved methods of providing this level of survivability using cabling methods: mineral-insulated (MI) cable and circuit integrity (CI) cable. Other methods permitted by *NFPA 72* require a two-hour rated enclosure. MI cable is generally used in fire pump applications and has excellent resistance to fire. However, it requires special fittings and may not meet alarm equipment manufacturers' requirements. CI cable is a special type of

---

12.3.7* Class A and Class X circuits using physical conductors (e.g., metallic, optical fiber) shall be installed such that the outgoing and return conductors, exiting from and returning to the control unit, respectively, are routed separately. The outgoing and return (redundant) circuit conductors shall be permitted in the same cable assembly (i.e., multi-conductor cable), enclosure, or raceway only under the following conditions:

(1) For a distance not to exceed 10 ft (3.0 m) where the outgoing and return conductors enter or exit the initiating device, notification appliance, or control unit enclosures.

(2) For single raceway drops to individual devices or appliances

(3) For single raceway drops to multiple devices or appliances installed within a single room not exceeding 1000 ft² (93 m²) in area

(Excerpt from NFPA 72.)

cable that is used to maintain its insulating characteristics when exposed to fire. CI cable is a plastic-jacketed cable, with a proprietary insulating material that crystallizes when burned. The crystallized material maintains its insulating characteristics, which allows communications (circuit integrity) to continue during a fire

Some CI cables require the use of a metal conduit to provide mechanical protection, while others can be installed without conduit. Almost all CI cable requires a two-hour rated surface on which the cable or conduit can be mounted. The manufacturers' sheets will explain proper installation procedures.

CI cable cannot be spliced between the source (fire alarm controls) and the zone that it serves because there are no terminations approved for two hours of attack by fire. CI cable is available in either non–power-limited or power-limited cable types and is designated by a CI following the cable type (for example, NPLFP CI).

CLASS A AND CLASS X CIRCUIT SEPARATION

Class A and Class X circuits are used where reliability of the fire alarm system is critical. The redundant (return) conductors provide an additional pathway for power and signaling to and from devices and appliances. These circuits must be separated in order to provide this redundancy.

Section 12.3.7 of *NFPA 72* requires all Class A and Class X circuits to be separated.

The annex materials accompanying this section recommend a separation of 12 inches (300 mm) where run vertically and 48 inches (1.22 m) where run horizontally. These recommendations cannot be enforced unless the Code is amended or the project specifications require these values. Some organizations, such as the Federal Aviation Administration (FAA), require minimum separation and enforce A.12.3.7 of *NFPA 72*.

CALCULATIONS

There are two basic electrical calculations required for fire alarm systems: loop resistance and voltage drop. Loop resistance is especially important on the input side, such as initiating device circuits and signaling line circuits. Most manufacturers will specify a maximum loop resistance to ensure the devices will properly transmit signals to the controls.

Voltage drop is important on notification appliance circuits because an appliance operating below its minimum voltage will not provide the required sound pressure level or visible intensity.

Loop Resistance Calculations

Loop resistance is a relatively simple calculation. Table 8 of the *NEC* provides unit resistances of conductors that can be used to determine the overall resistance of a circuit. For example, if a manufacturer specifies a maximum loop resistance of 75 ohms, and 16 AWG solid coated copper wire will be used, the maximum length of the circuit can be calculated.

Table 8 of the *NEC* shows that 16 AWG solid coated copper has a unit resistance of 5.08 ohms per 1,000 feet. The maximum length of the conductors (out and back) will be the combined resistance of the outgoing and return conductors. In this case, the maximum length of the circuit must not exceed 7,382 feet. **See Figure 5-4.**

For additional information, visit qr.njatcdb.org Item #1011

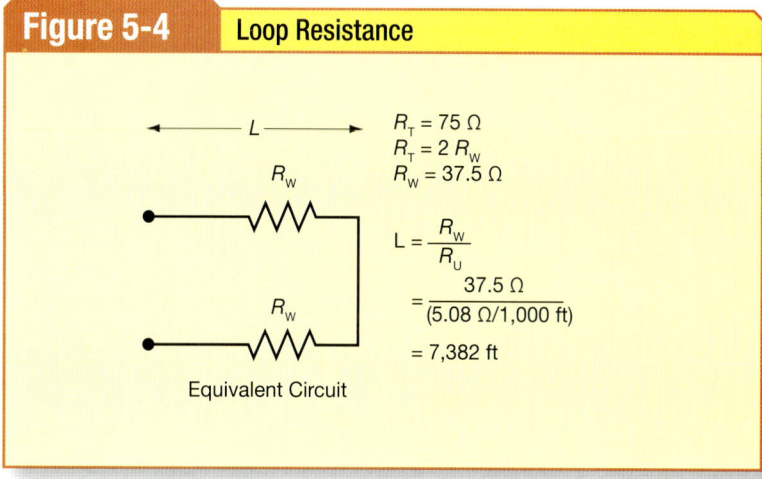

Figure 5-4. Loop resistance calculations are important for signaling line circuits and initiating device circuits. Most manufacturers specify a maximum loop resistance for their products.

Voltage Drop Calculations

Voltage drop calculations are slightly more complex than loop resistance calculations. Generally, voltage at the last appliance on the circuit must not drop more than 15% below the nominal voltage of the system. This value is 20.4 volts for most notification appliances. The resistance of each appliance must be known or calculated. Most manufacturers do not provide any resistance values for appliances. If the resistance is not known, it can be easily calculated using Ohm's Law ($V = IR$), when the voltage and current are known.

For instance, if a notification appliance circuit has solid coated copper 14 AWG conductors and four notification appliances, if the notification appliances each draw 200 milliamperes at nominal voltage, if the manufacturer has not provided any resistance values, and one needs to determine whether a 400-foot long circuit will provide adequate voltage at the last appliance on the circuit, the calculations will be as follows:

First, determine all resistances on the notification appliance circuit, including wire resistances and load resistances. **See Figure 5-5.**

The actual location of appliances on the circuit may not be known. It is conservative to place all loads at the end of the circuit. The resistance of the appliances can be calculated using Ohm's Law. The resistance at nominal voltage can be calculated since the manufacturer provides the load at nominal voltage. The load current at rated voltage is found on the manufacturer's specification sheets for the product being used. In this case, the individual appliance resistance (impedance) is calculated as follows:

$V_T = I_L R_A$, where

R_A is the individual appliance resistance or impedance

V_T is the given voltage of the appliance at load (given as 24 V by the manufacturer)

I_L is the current drawn at 24 V (given as 200 mA by the manufacturer)

Solving for R_A:

$$R_A = \frac{V_T}{I_L}$$

$$= \frac{24}{0.200}$$

$$= 120 \text{ ohms}$$

The wire unit resistance of solid, uncoated 14 AWG is 3.07 ohms per 1,000 feet. In a 400-foot circuit, the resistance of each leg is calculated as follows:

$$R_{WL} = R_U \times L$$

$$= \frac{3.07 \text{ ohms}}{1,000 \text{ ft}} \times 400 \text{ ft}$$

$$= 1.228 \text{ ohms}$$

$$R_W = 2 \times R_{WL}$$

$$= 2 \times 1.228 \text{ ohms}$$

$$= 2.456 \text{ ohms}$$

In order to simplify, all loads are placed at the end of the circuit with all conductor resistance at the front end. This greatly simplifies the calculations. This is a more conservative approach, which will result in safer installations. Combining all load resistances, three resistances remain: the

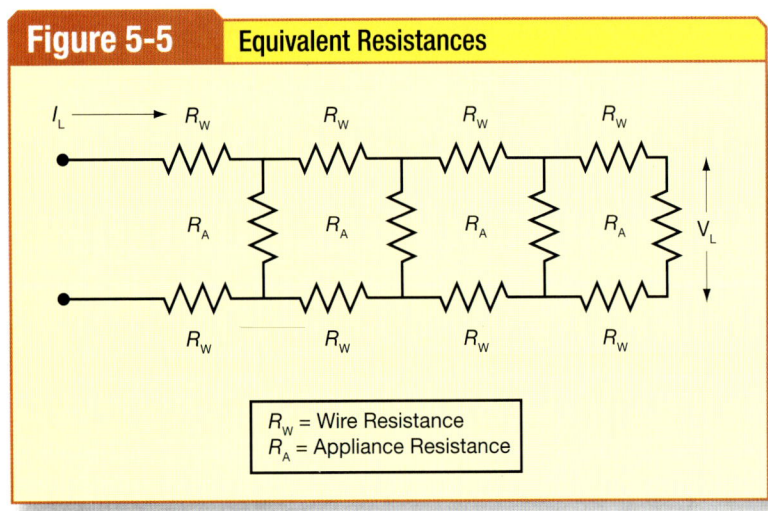

Figure 5-5. Each length of wire and each appliance has a resistance. The manufacturers' product sheets can provide enough information to calculate the resistance of the appliance.

outgoing leg conductor resistance, load resistance, and return leg conductor resistance. **See Figure 5-6.**

The calculated individual appliance load resistance is 120 ohms. This can then be used to further simplify the circuit. In this case, there are four identical notification appliances. The total combined load (appliance) resistance is 30 ohms, and when added to the wire resistance (each leg), the total combined resistance is 32.456 ohms. At 24 volts, the current is calculated using Ohm's Law as follows:

$$I_L = \frac{V_T}{R_T}$$
$$= \frac{24\text{ V}}{32.456}$$
$$= 0.739\text{ A}$$

Using Ohm's Law again, the voltage and resistance of the appliance load is calculated as 22.08 volts. **See Figure 5-7.**

Percentage Voltage at the Load
$$= \frac{V_L}{V_T}$$
$$= \frac{22.18}{24.00}$$
$$= 92.4\%$$

This is within the 15% maximum voltage drop permitted by Sections 10.3.5 and 10.4.3 of *NFPA 72*. If this value exceeded the 15% prescribed by the *Code*, a shorter circuit length, fewer notification appliances, or larger wire would be required. Where larger conductors are necessary, most notification appliances permit up to 12 AWG conductors on the terminals, while some permit only 14 AWG maximum. Therefore, larger conductors may not always be possible.

Manually calculating the voltage drop on notification appliance circuits is time consuming and can be somewhat difficult. Mistakes are easily made. Most

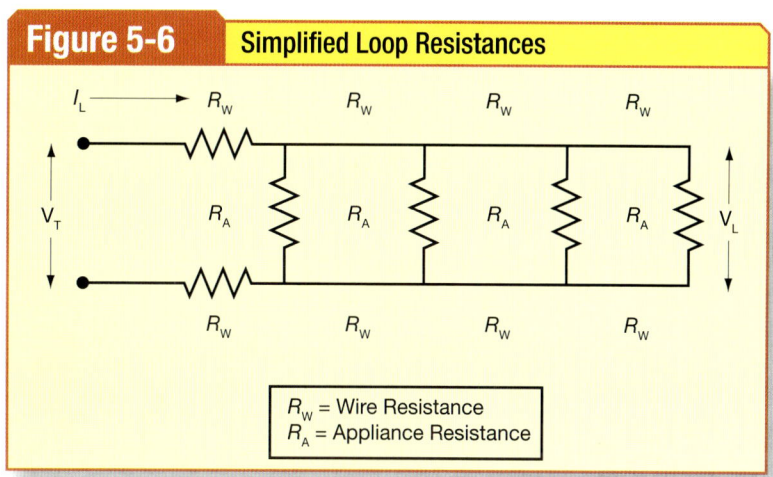

Figure 5-6. All individual wire resistances are grouped at the front end of the circuit, while the appliance resistances are grouped at the load end.

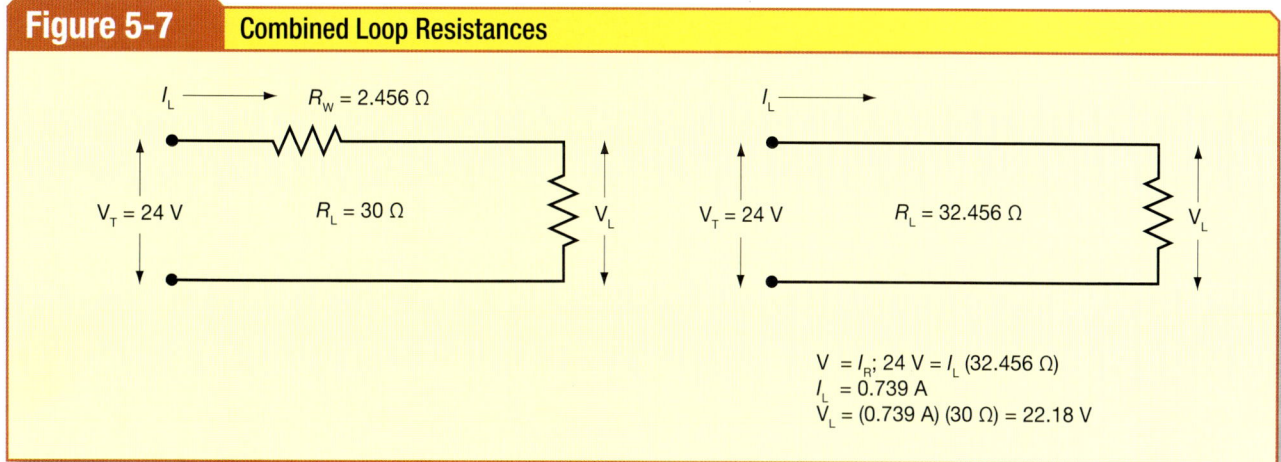

Figure 5-7. The notification appliance resistances and wire resistances are combined into single resistances. All resistances are combined and the voltage at the load can be easily calculated.

manufacturers have already developed spreadsheets to calculate voltage drops, but it is good to understand how these calculations are made.

FIRESTOPPING CONCEPTS

Firestopping is a passive form of fire protection intended to prevent the spread of smoke and fire in premises. Passive fire protection is a simple and effective way of limiting fire exposure. Firestopping, however, is only effective when properly installed and maintained. The *National Electrical Code* requires firestopping for all electrical penetrations in rated walls and floors.

Section 760.3(A) also references Section 300.21, which provides additional requirements for the spread of fire or products of combustion. Specifically, Section 300.21 requires all openings around penetrations in fire-resistant walls, floors, and ceilings (called assemblies) to be sealed using approved firestopping methods and materials (called a system) in order to maintain the fire resistance ratings of the barrier being penetrated. Improper penetrations of cable and conduit can allow heat, smoke, and flames to migrate throughout the building, thereby contributing to the spread of fire and products of combustion. Firestopping is a relatively simple type of passive fire protection and its importance cannot be overemphasized.

Only approved firestopping sealant materials and methods should be used. Firestopping materials must be approved for such use but are never listed. Qualified testing laboratories evaluate and approve firestopping methods and materials from different manufacturers using full-scale fire tests. If the materials pass the prescribed tests, an approval is granted.

Firestopping materials must be applied and used in a prescribed arrangement called a classification. Materials and methods that pass the testing criteria are approved and published in a fire resistance directory. All major manufacturers have field installation manuals based upon UL classifications that can be used to determine what materials and methods are appropriate for a given application. In the rare event that a classification does not exist for a particular application, most manufacturers will assist customers in the development of a new classification.

Firestopping systems are rated in their ability to resist the passage of flame, temperature, and leakage. These ratings are called the F, T, and L ratings, respectively. All ratings are given in minutes, and an F rating of 120 means the assembly resists the passage of flames for two hours. The rating of the penetration must always match the rating of the assembly (barrier). For example, a two-hour rated firewall must have penetrations that are rated for not less than two hours. Penetrations having a rating less than the barrier rating will compromise the rating of the barrier (assembly).

Only approved materials and methods (classifications) can be used to provide proper firestopping. Simply applying a layer of firestopping caulk around a conduit penetration may not be sufficient to prevent the migration of smoke and heat. Most applications for electrical installations are very straightforward when the classifications are followed. Most manufacturers organize their field installation manuals by barrier type, then by penetration type, which makes them easier to find.

Some firestopping products are intumescent, which means they expand when heated by a fire. Intumescent products char and expand to fill gaps around penetrations (between the penetrant and barrier) to prevent the passage of smoke, heat, and flames. Intumescent products include firestopping caulking, pipe wrap, pipe chokes, pillows, paint, and putty. **See Figure 5-8.**

Other types of firestopping materials include ablative or cement-based products. Ablative products usually contain small amounts of moisture and will slowly burn off while providing a cooling effect. An example of these products is firestopping cement which can be poured around floor penetrations.

Fact

Firestopping is required by Section 300.21 of the *National Electrical Code.* A multitude of products are available to comply with the *NEC.*

Figure 5-8   Intumescent Firestopping Products

For additional information, visit qr.njatcdb.org Item #1012

Figure 5-8(a). These products are often used with mineral wool to seal openings around conduit and other static penetrations. Caulking products are not appropriate for dynamic penetrations like wireways.

Figure 5-8(b). These products can be used to seal around conduit and other penetrations in rated walls and floors. Putty pads are used to seal around electrical boxes which penetrate rated walls.

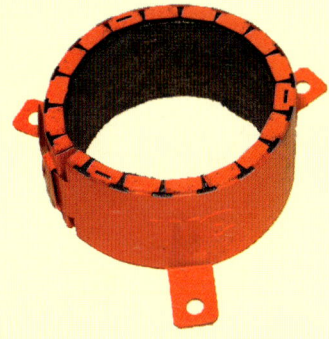
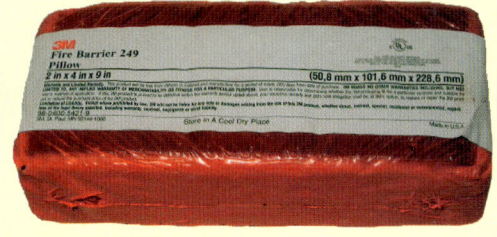

Figure 5-8(c). Pipe chokes are used to seal around conduit and pipes which penetrate rated walls and floors. Pipe chokes typically require no other products to complete the assembly.

Figure 5-8(d). These pillows are commonly used to seal around dynamic penetrations like wireways.

Figure 5-8. Intumescent products can be used to seal around conduit and other penetrations in rated walls and floors.

There are two basic types of firestopping applications: static and dynamic. Static applications work well in penetrations that are not usually moved or expanded, such as a conduit penetration. Static penetrations can be made using a variety of methods, such as firestopping caulk, intumescent pipe wrap, cement-based firestopping materials, and pipe chokes.

Dynamic penetrations, such as a cable tray, permit quick expansion of the penetration but pose problems for more permanent firestopping methods. A firestopping pillow can be used around larger dynamic penetrations. The pillows can be temporarily moved to permit the installation of new conductors. Some pillow firestopping products require a restraining means to ensure that pressures created by the fire do not force the product from its place. Some classifications have maximum size limitations of the firestopping classification.

Holes for penetrations are all too often made much larger than is actually necessary for the penetrant. Before penetrations are sealed using firestopping products, the surrounding area should always be patched using the same materials used in the barrier to minimize the use of firestopping materials. In fact, some UL classifications specify a maximum width of the firestopping product. For example, a penetration that is three feet higher than is necessary for a wireway should not be filled entirely with firestopping pillows. The penetration should be patched using materials to match the barrier, then firestopped using an approved classification.

All firestopping materials must be used exactly as shown in the manufacturer's field guide or UL classifications. For example, firestopping caulk is usually not sufficient as a firestopping assembly by itself. It is almost always used in conjunction with backer rod, mineral wool, or other filler materials. Additionally, intumescent putty pads are generally not acceptable in wireway applications. UL classifications or manufacturers' guides will help determine the proper methods and materials for a particular application.

Additionally, it is never permitted to mix two different manufacturers' products in a single penetration because testing laboratories do not evaluate multiple products in a single classification. However, two different penetrations or systems may use two different manufacturers' products in a rated assembly. In other words, only a single manufacturer's product should be used on one penetration.

Only materials that have been approved as part of a firestopping classification can be used to satisfy *Code* requirements. Fiberglass insulation, high expansion foam, latex caulking, and other materials are not able to withstand the effects of fire and must never be used for firestopping purposes.

Firestopping is not complicated nor is it difficult, but in order for firestopping systems to work, they must be installed strictly according to field guides and UL classifications. Improper use of products or use of non-approved firestopping products will not deliver desired results and will not be *Code* compliant. Finally, most manufacturers are happy to train users on proper techniques and will usually assist the user with unique applications.

Summary

The purpose of the *National Electrical Code* is to prevent fires and electrocutions. However, there are many additional manufacturers' requirements that may restrict wiring methods and materials. The *NEC* also provides related requirements for the reduction of products of combustion. This is accomplished through the use of low-smoke-producing cable insulation and firestopping.

Fire alarm system wiring is a very important part of any fire alarm system. The *National Electrical Code* provides the requirements for the installation of all wiring. Proper installation of the wiring will help ensure that the system operates without faults for many years.

Review Questions

1. Which table in the *NEC* provides power limitations for direct-current power-limited fire alarm circuits?
   a. Table 12(B)
   b. Table 12(A)
   c. Table 11(B)
   d. Table 11(A)

2. Which document contains performance requirements for fire alarm systems?
   a. *NFPA 70*
   b. *NFPA 72*
   c. *NFPA 110*
   d. *NFPA 170*

3. Which article of the *NEC* pertains to the wiring of burglary system controls?
   a. Article 800
   b. Article 760
   c. Article 725
   d. Article 620

4. Section __?__ pertains to wiring in plenums and other air-handling spaces.
   a. 200.11
   b. 200.22
   c. 300.11(D)
   d. 300.22

5. Runs of wiring in underground conduit are considered as __?__ locations.
   a. damp
   b. dry
   c. wet
   d. none of the above

6. Which of the following cable types is appropriate for a non–power-limited circuit?
   a. NPLF
   b. FPL
   c. FNPL
   d. FPLN

7. According to the *NEC*, non–power-limited circuits may operate with voltages as high as __?__.
   a. 70 V
   b. 150 V
   c. 300 V
   d. 600 V

8. Which of the following may be used for power-limited circuit wiring?
   a. Chapter 3 methods
   b. Non–power-limited methods
   c. Power-limited methods
   d. All of the above

Review Questions

9. Power-limited wiring may be combined in the same cable or raceway as wiring of this circuit type.
    a. Class 1
    b. Class 2
    c. Power and lighting
    d. Non–power-limited

10. Where power-limited conductors are installed in the same enclosure as power and lighting conductors, they must have a separation of at least __?__.
    a. ¼"
    b. ½"
    c. ¾"
    d. 1"

11. Where required, survivability means that fire alarm circuits must operate for not less than __?__?
    a. ½ hour
    b. 1 hour
    c. 2 hours
    d. 4 hours

12. A(n) __?__ firestopping material contains small amounts of water that cool the penetration as it is burned.
    a. ablative
    b. classified
    c. intumescent
    d. static

System Interfaces and Safety Control Functions

Fire alarm and signaling systems are designed to detect fires and notify occupants in the event of a fire. In addition, fire alarm systems often interface with a number of other systems and equipment designed to make the premises safer in the event of an emergency. These systems may include suppression systems; fire pumps; elevators; heating, ventilating, and air-conditioning (HVAC) systems; burglary systems; and access control. Fire alarm system interfaces may handle either inputs from suppression system monitoring and supervision systems or outputs, such as HVAC or elevator shutdown.

Objectives

» Explain the connection requirements of combination systems

» Explain how suppression systems are interconnected to work together

» Describe different emergency control functions for such things as:
  › Elevators
  › HVAC systems
  › Door release or door unlocking services
  › Stairway pressurization

Chapter 6

Table of Contents

Combination Systems and Interconnected Fire Alarm Systems 128
    Combination Systems 128
    Interconnected and Networked Systems 130

Connection Methods 131
    Other (Non-Fire Alarm) Interfaced Systems 133

Interconnected Suppression Systems 133
    General Requirements........................ 134
    Sprinkler System Attachments 135
    Fire Pumps 137
    Special Hazards Systems................... 138

Emergency Control Function Interfaces 139
    General Requirements........................ 139
    Elevator Safety Functions 141
    Smoke Control and HVAC Shutdown 151
    Door and Shutter Release 151
    Door Unlocking................................... 152
    Stairway Pressurization 152

Summary 153

Review Questions 153

COMBINATION SYSTEMS AND INTERCONNECTED FIRE ALARM SYSTEMS

Fire alarm systems actually manage other building fire safety features in addition to providing detection and alarm features. Chapters 21 and 23 of *NFPA 72, National Fire Alarm and Signaling Code,* contain a large number of requirements relating to interfaces.

There are many reasons for connecting a fire alarm system to other protective systems in a building. The most obvious one is to detect the actuation of a suppression system used to protect the building and/or its occupants. The requirements for suppression systems depend on building height, area, use group, occupant load, and any special hazards involved. Building designers may also add additional features, based upon the goals of the fire protection system and any special hazards involved. Building codes may also require fire alarm systems to supervise suppression systems where they are installed. Building operators must hear and act upon supervisory and trouble signals from the suppression system in order to maintain them. Alarm signals from suppression systems also indicate a fire condition that requires evacuation. Therefore, interconnection of the two systems is essentially required.

But suppression systems are not the only other systems interfaced with the fire alarm and detection system. Emergency control functions are usually initiated by the fire alarm system and make the building safer during a fire. These emergency functions might include elevator recall, HVAC shutdown, door unlocking, smoke door release service, and smoke control system actuation.

Large properties or additions to buildings sometimes require multiple fire alarm control units because a single control unit may not have the required capacity for the project. Other systems, such as burglary or access control, may also employ separate controls. However, it is not necessary for fire alarms and other interfaced systems to consist of separate stand-alone units. Combination fire alarm and burglary systems are commonplace, especially in smaller commercial and residential applications. Access control is often integral or interfaced with larger fire alarm systems.

Stand-alone main (master) fire alarm control units can be used to provide all fire alarm signaling but are generally not listed for burglary, intrusion detection, or other functions desired by the owner. Use of stand-alone fire alarm systems will result in another stand-alone system where other features are desired. However, *NFPA 72* permits integrated combination systems to be used.

Combination Systems

Combination systems usually consist of a single integrated control unit that performs functions in addition to fire detection and alarm, such as fire alarm and access control. **See Figure 6-1.**

Combination fire alarm systems may include suppression system control, burglary/intrusion, access control, and energy management systems. Section 10.3.1 requires the control unit to be listed for the purpose for which it is used and this requirement dictates a listing for "fire alarm signaling service." Many combination burglary/fire control units are listed only for household use

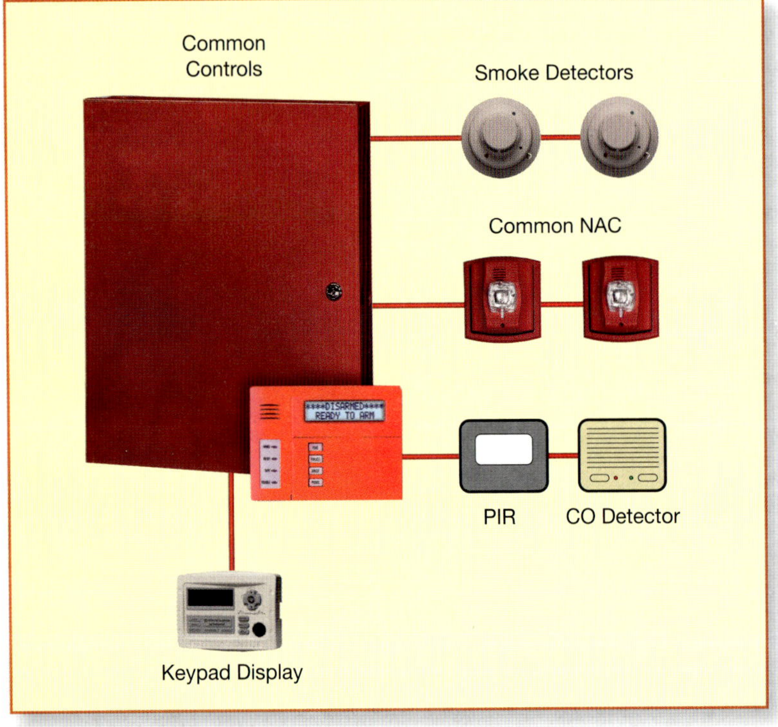

Figure 6-1. Combination systems may include fire alarm, burglary/intrusion, gas monitoring, HVAC, or other systems.

and are not permitted for use in commercial or industrial applications.

One of the most common types of combination systems is an integrated burglary and fire alarm system. The listing of the equipment includes a product evaluation to multiple product standards. NFPA 72 is primarily concerned that the fire alarm equipment is listed for the purpose for which it is used and that the non-fire alarm portion does not interfere with the fire alarm system. Therefore, the equipment must be listed to UL 864, *Standard for Safety Control Units for Fire Protective Signaling Service* when used for commercial use. The *Code* reads as follows:

> **23.8.4 Combination Systems.**
> **23.8.4.1*** Fire alarm systems shall be permitted to share components, equipment, circuitry, and installation wiring with non–fire alarm systems.
> **23.8.4.2** Operation of a non–fire system function(s) originating within a connected non–fire system shall not interfere with the required operation of the fire alarm system, unless otherwise permitted by this Code.
>
> (Excerpt from NFPA 72.)

Combination systems such as burglary and access control are covered by Section 23.8.4 of *NFPA 72*. Section 23.8.4.3 permits non-fire alarm systems to share equipment and wiring with fire alarm systems. The non-fire alarm equipment is permitted to be attached to the fire alarm circuit. However, Section 23.8.4.3.1.1 requires the equipment and pathways to be monitored for integrity; they must be maintained by a single service organization; the installation must meet the requirements of *NFPA 72*; and the equipment must be listed as compatible with the fire alarm equipment or use an interface listed as compatible with the fire alarm equipment. These limitations will help ensure the system operates correctly.

Section 23.8.4.3.2 provides additional requirements for the non-fire alarm portion of the system. Where installed on separate pathways, the non-fire alarm equipment cannot interfere with the fire alarm system. Opens, shorts, and ground faults on the non-fire alarm portion of the system cannot interfere with the monitoring for integrity or control functions of the fire alarm system. Additionally, removal of the non-fire alarm equipment cannot impair the operation of the fire alarm system, as required by 23.8.4.4.3.

Section 23.8.4.6 provides requirements for signal priority on combination systems. On any combination system, fire alarm signals must be clearly recognizable and must be indicated in priority as follows:

1. Life safety signals
2. Property protection signals
3. Trouble signals associated with property protection
4. All other signals

This prevents fire and any other signals from serving the same purpose. Therefore, a burglary/intrusion alarm must sound different from a fire alarm signal. The fire alarm signal must also have priority over all other signals, unless one of those other signals is deemed to represent a more life-threatening situation. An example of this type of situation may be a "duck and cover" bomb threat. **See Figure 6-2.**

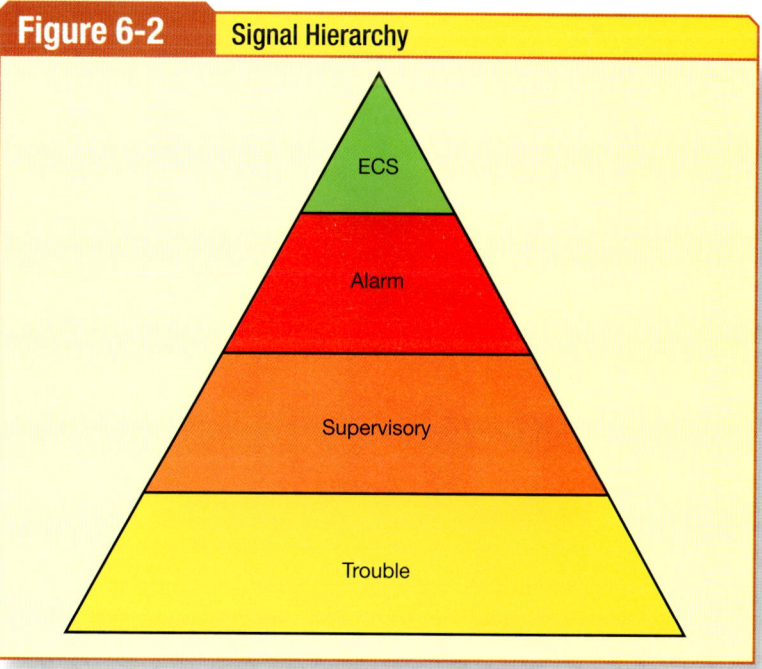

Figure 6-2. *Life-safety signals, such as ECS signals, have the highest priority, while property protection signals, such as supervisory and trouble, have lower priority.*

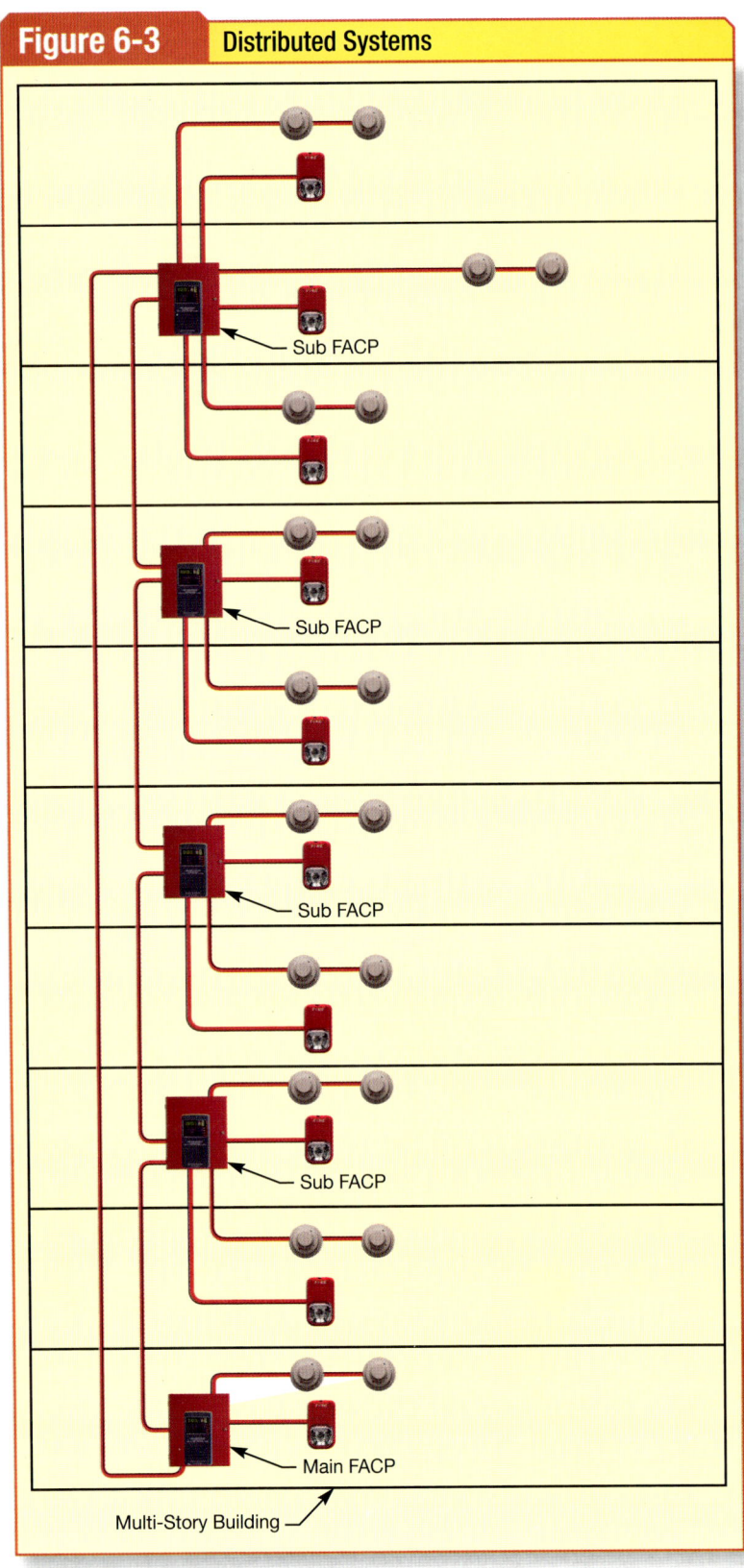

Figure 6-3. Distributed systems are widely used in campus-style and high-rise arrangements because a single control unit is insufficient to cover such a large area.

Keypads or graphical user interfaces on combination systems may be shared or separate and may be capable of displaying both fire alarm and non-fire alarm events. Displays used on combination systems must be capable of quickly displaying fire alarm system information during a fire event. If the authority having jurisdiction determines that the non-fire alarm information is excessive and the fire alarm information is delayed, 23.8.4.7 will require a separate display for the fire alarm system.

Interconnected and Networked Systems

A new Section 23.8.4.8 was added to the 2013 edition of *NFPA 72* to recognize the use of carbon monoxide (CO) detectors. Section 23.8.4.8 requires all CO detector signals to be indicated as a "Carbon Monoxide Alarm." However, the exception to Section 23.8.4.8 permits CO signals to be supervisory, if the building response plan permits it.

Interconnected fire alarm systems may utilize a networked system of multiple fire alarm control units. Large buildings and campus-style arrangements will take advantage of this type of arrangement most often. Systems can be built using fire alarm equipment from one or more manufacturers. Direct or "peer-to-peer" connections are generally possible only when equipment from a single manufacturer's product line is used. However, there are products available from several manufacturers that can integrate very large systems comprising different products. These integration packages use desktop computers and listed proprietary software to allow a single user interface for a multitude of different systems.

It is sometimes necessary to use multiple control units or distributed systems. This is especially true for high-rise, college campus, or multi-building applications. Building additions may also require the use of separate but interconnected controls, especially when no excess capacity is available on the existing system. Cabinet sizes, power supplies, and circuit resistances can also limit the ability of a single control unit to cover large areas. **See Figure 6-3.**

Where multiple fire alarm systems are interconnected in a single protected premises system, Section 23.8.2.1 of *NFPA 72* permits a single system or a combination of subsystems capable of independent operation. Section 23.8.2.2 requires the system to act as a single system without any degradation or loss of system performance under full load.

CONNECTION METHODS

The methods for interconnecting systems permitted by Section 23.8.2.5 include the following:
1. Electrical contacts listed for the connected load
2. Data communications over a signaling line circuit dedicated to the fire alarm or shared with other premises operating systems
3. Other listed methods

Systems from a single manufacturer can often be interconnected through optical fiber cable or copper connections (using a proprietary communications protocol) because they are designed to be compatible with each other.

These interface connections usually involve a common protocol such as RS-232 or RS-485, but communications protocols are usually proprietary. Both RS-232 and RS-485 have length limitations, which usually precludes their use in runs over 20 feet (6 m). It is not normally possible to directly interconnect controls (without using dry contacts) from two or more manufacturers because they "speak different languages." However, there are integration products that can integrate products from several manufacturers. They are used by many designers to integrate multiple systems as a single system.

One method of interfacing controls of different manufacturers is to use dry contacts in the panel being monitored. These contacts are connected to initiating device circuits which are, in turn, connected to the monitoring control equipment. **See Figure 6-4.**

Addressable technology commonly utilizes addressable monitoring modules (on an SLC) located at the controls being monitored. **See Figure 6-5.**

Monitor modules usually have one or two zones connected to the dry contacts in the controls being monitored. The monitor module input circuits are essentially initiating device circuit zones programmed at the main controls to produce alarm, trouble, or supervisory signals. Each zone of a monitor module input is an initiating device circuit (IDC), so

Figure 6-4. Dry contacts are one way of interconnecting fire alarm controls when the controls are incompatible.

For additional information, visit qr.njatcdb.org
Item #1013

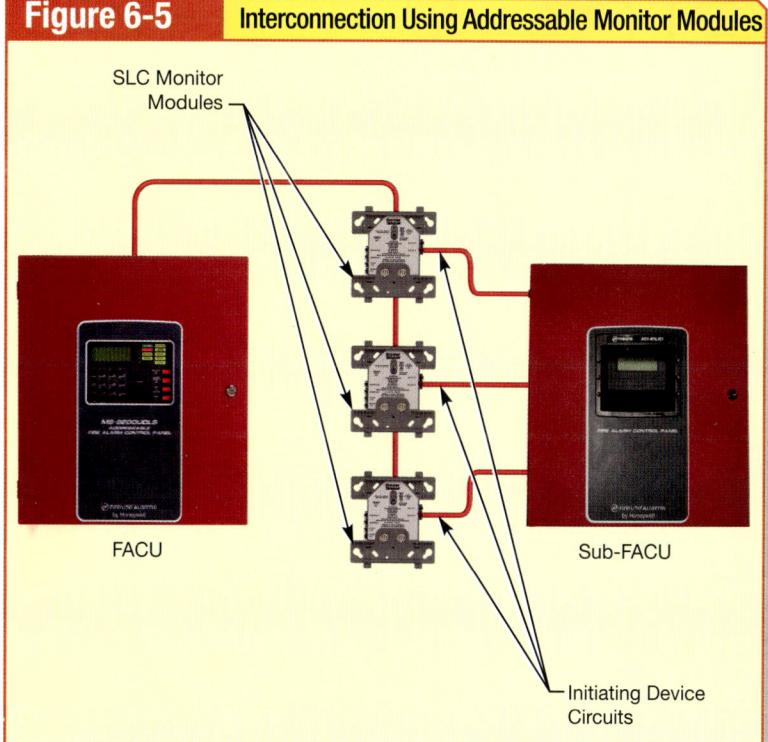

Figure 6-5. Monitor modules are used with addressable technology to monitor and supervise other fire alarm and suppression systems.

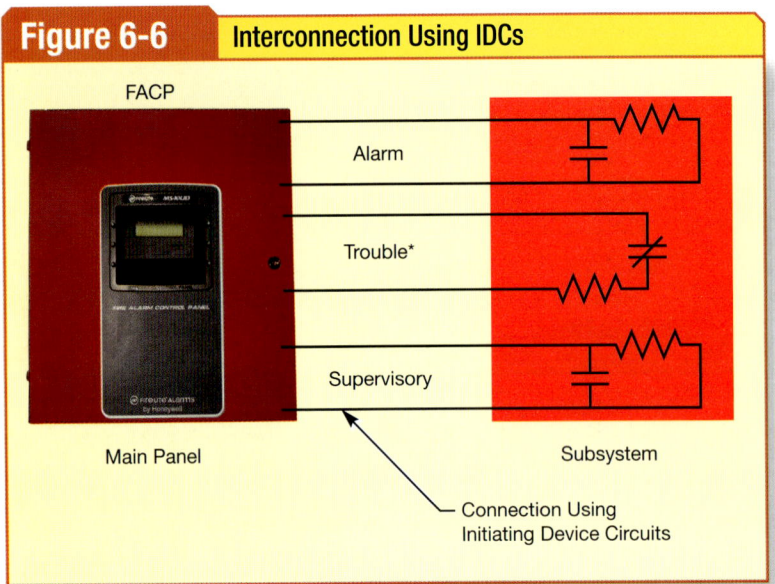

Figure 6-6. The IDCs are used when the main system is conventional, rather than addressable. EOL devices are located at each set of contacts in the monitored controls.

Unless the controls directly communicate using signaling line circuit communications, the signals sent to the main controls from the monitor module cannot provide detail specific to each device or zone on the monitored controls. In other words, they cannot transmit specific information beyond the fact that the panel is sending a signal. It is acceptable to transmit a common signal for the monitored controls. Upon receipt of a signal at the main controls, the operator is required to investigate the signal at the monitored controls in order to determine the nature of the signal. **See Figure 6-6.**

When a single manufacturer's addressable equipment is used, communications between control units usually allow transfer of all data to the main controls. In this case, the status and address of each device can be transmitted to the main controls. **See Figure 6-7.**

Interfaces between system controls and a network are often referred to as gateways. Gateways usually consist of a control unit suitable for the proprietary language of the network, which then is interfaced to subsystem controls. Gateways end-of-line devices are usually required on each input of the module located at the dry contacts in the controls being monitored. This location ensures the wiring between dry contacts and the SLC module are monitored for integrity.

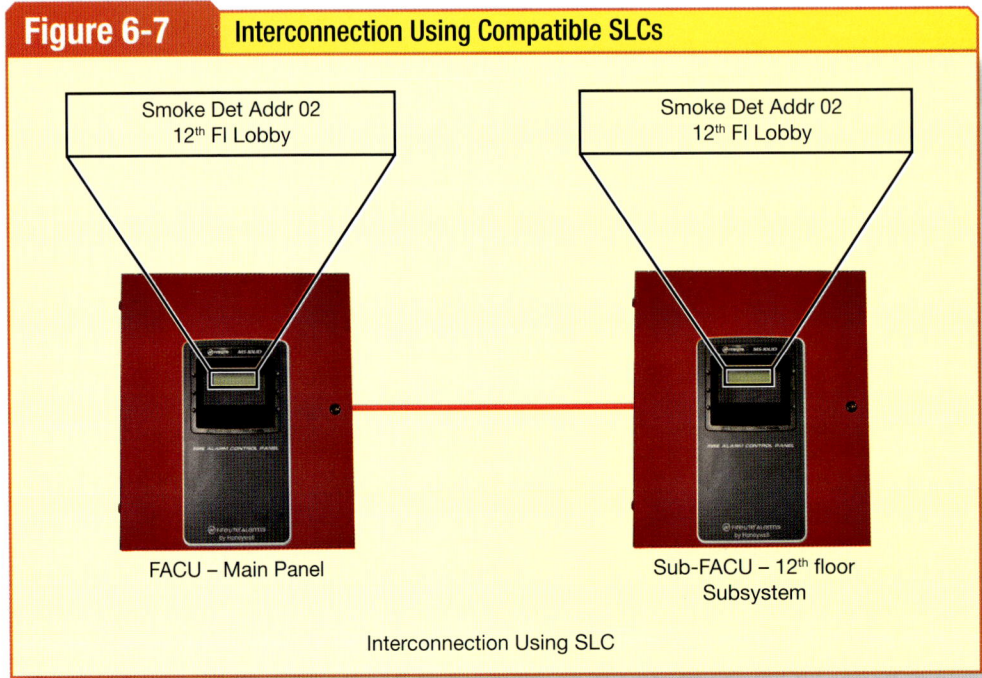

Figure 6-7. Interconnection between two addressable and compatible controls will allow specific device/address information to flow between controls, resulting in a faster response.

may be integral or separate from the system controls, and they may even interface with non-fire alarm systems, such as access control. Section 23.8.2.6.2 requires the gateway to prevent other systems from interfering with the fire alarm system.

A more recent trend in technology is to use building automation control networks (BACNet) and other communications protocols to interconnect the fire alarm system to a network used for other systems, such as HVAC or energy management controls. Listed BACNet gateways are one-way communications portals that permit the transmission of signals over a network but do not permit control of the fire alarm system. Listed BACNet portals may send signals over a local area network or the Internet. Sharing the network reduces the overall building cost but poses significant challenges to the fire alarm system. In all cases, the fire alarm system must have priority over non-fire alarm signals.

All interconnected fire alarm control units must be separately monitored for trouble, supervisory, and alarm conditions, as required by Section 23.8.2.7. However, Section 23.8.2.8 permits signals from interconnected fire alarm control units to be annunciated by zone or by using common signals. Monitoring by common signals simply means that a single signal at the main control unit may represent more than one signal at the interconnected controls. However, all three signals must be annunciated at the main controls.

Silencing or resetting fire alarm signals from outside protected premises without first investigating is a very dangerous practice and is not permitted by the *Code*. For this reason, Sections 23.8.2.9 and 23.8.2.10 require fire alarm signals to be capable of being silenced and reset only at the control unit within the protected premises, unless otherwise approved by the authority having jurisdiction (AHJ).

Other (Non-Fire Alarm) Interfaced Systems

Other systems interfaced with the fire alarm system may include emergency control functions or suppression system interfaces. Fire alarm systems are sometimes integrated with other non-fire alarm systems in order to supervise or control them. The controls for both the fire alarm system and interfaced systems are often separate stand-alone control units. Communications between these systems are usually made through dry contacts or addressable control/monitor modules. Other communications protocols may be available, especially where the same manufacturer produces both the fire alarm and non-fire alarm system controls. However, this arrangement is rarely used in actual practice.

Interconnected non-fire protection systems include:
- Elevators
- Heating, ventilating, and air-conditioning (HVAC) systems
- Access control systems
- Burglary/security systems
- Nurse call systems

Interfaced systems must be designed and installed to ensure the fire alarm system correctly operates. The requirements found in *NFPA 72* are intended to ensure safe and reliable operation of these systems. Chapter 21 provides requirements for emergency control functions and interfaces and will be discussed later in this chapter.

INTERCONNECTED SUPPRESSION SYSTEMS

Interfaced fire suppression systems may include the following:
- Automatic fire sprinkler systems (wet/dry)
- Pre-action suppression systems
- Fire pumps
- Gaseous suppression systems (Halon, FM-200, Intergen, etc.)
- Wet and dry chemical suppression systems

Fact
Most suppression systems have separate standards which provide requirements for their installation. An example is a carbon dioxide (CO_2) suppression system, which is covered by *NFPA 12, Standard on Carbon Dioxide Extinguishing Systems*. These standards contain both installation and testing requirements.

For additional information, visit qr.njatcdb.org Item #1014

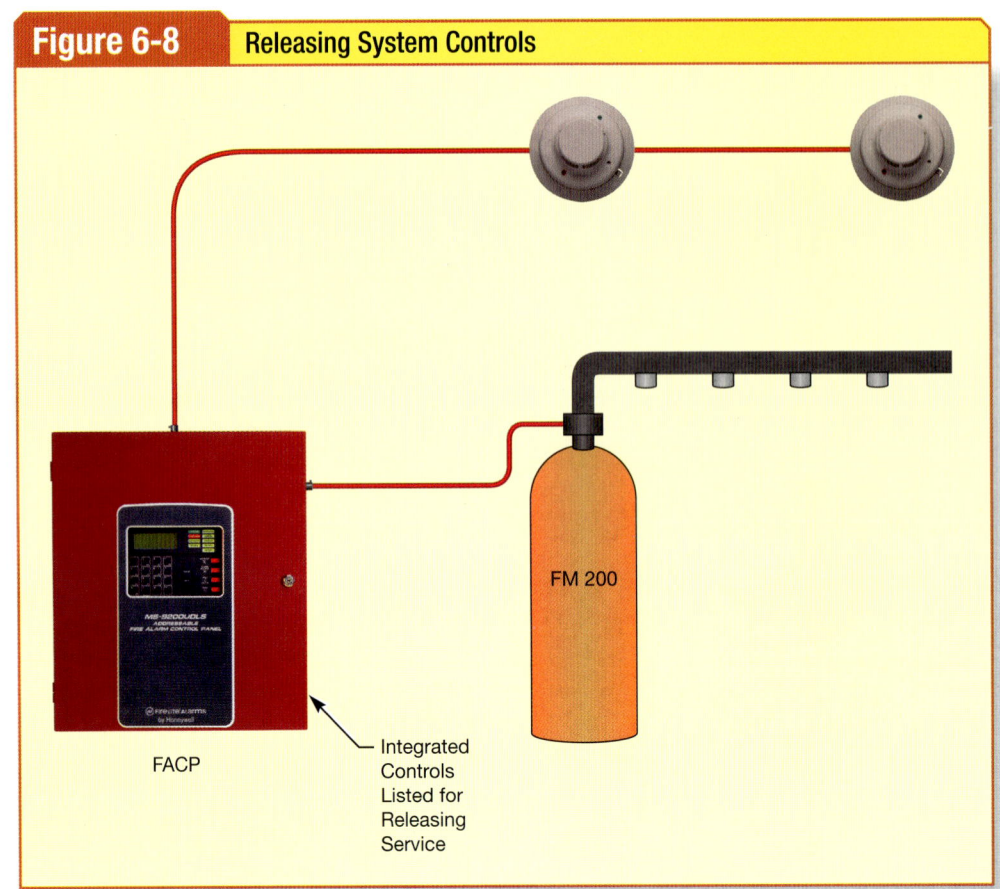

Figure 6-8. Controls used for releasing service must be listed for such use. This arrangement is not often needed because most suppression systems of this nature are pre-engineered.

- Foam suppression systems
- Kitchen hoods

Most of the requirements for interfacing suppression systems are found in Chapter 23 of *NFPA 72*.

General Requirements

Suppression systems may or may not have separate, dedicated control units. Systems with their own detection devices, such as a gaseous suppression system, sometimes utilize separate controls. Most special hazards systems are separate, engineered systems that are covered by other NFPA standards. Chapter 23 of *NFPA 72* contains the requirements for interfacing fire alarm systems with suppression systems. Section 23.11.1 requires control units that serve as suppression system control panels to be listed for releasing service. **See Figure 6-8.**

The main fire alarm system controls may be listed and used for releasing, but separate controls are often used because suppression systems are generally engineered and sold as a separate subsystem. **See Figure 6-9.**

Section 23.11.10 requires separate releasing control units to be monitored by the master fire alarm system for alarm, supervisory, and trouble signals where the premises has a fire alarm system.

Section 23.11.5 requires the installation of a switch to permit testing of the system without actuating the fire suppression system. The switch must be a physical switch, and software is not permitted to cause this disconnect. This requirement exists because of a large number of inadvertent discharges caused by improper lockout during tests.

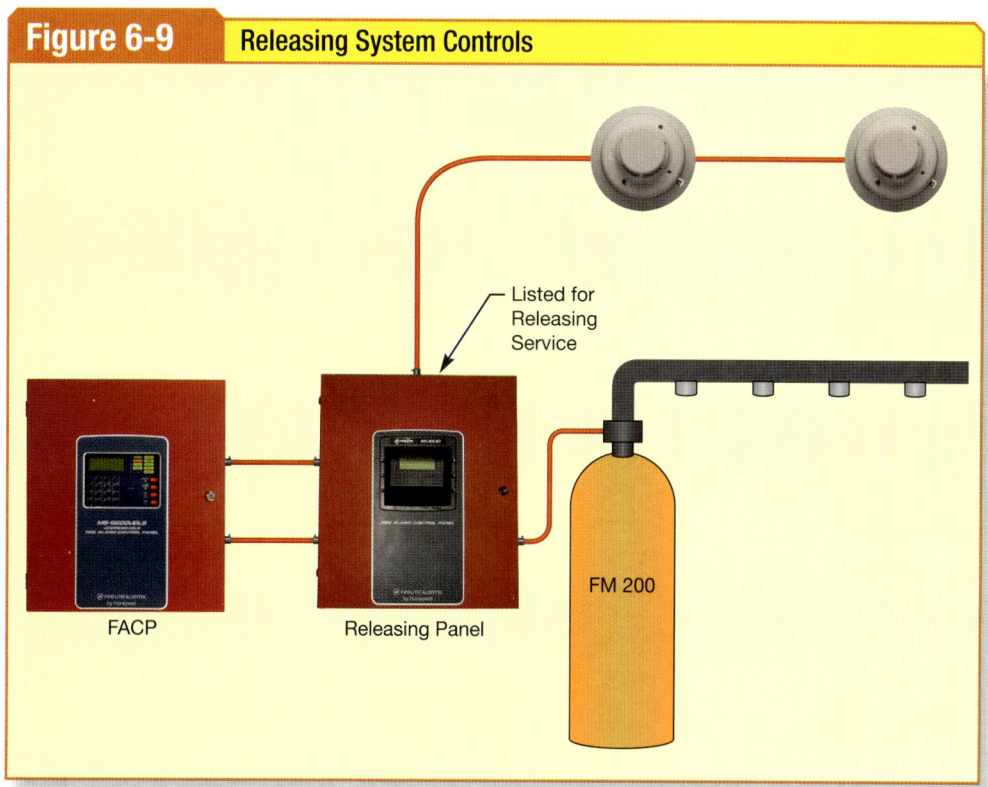

Figure 6-9. Controls used for releasing service must be listed for such use. This arrangement is most often used because most suppression systems of this nature are pre-engineered. Note that the controls are located outside the protected area to prevent damage from a fire or the suppression agent.

Sprinkler System Attachments

Control valve supervisory (tamper) devices are installed to help ensure that suppression systems properly function. Supervisory initiating devices create signals indicating a failure or impairment of a suppression system. These devices may be used to supervise valves in the open position, water reservoir level, dry-pipe air pressure, and room temperature. These devices are generally located and installed according to the particular standard for the suppression system.

Chapter 17 of *NFPA 72* contains the requirements for control valve supervisory devices. It requires control valve supervisory devices to indicate two distinct signals: normal and off-normal. The switch must indicate the off-normal position within two revolutions of the hand wheel or one-fifth of the travel distance of the valve and must restore only when the valve is fully open. The supervisory device must not interfere with the operation of the valve. Some supervisory devices are designed for large open screw and yoke valves. They have a switch arm that sits in a milled groove on the valve stem, which operates the switch when the valve is moved from the closed position. Some valves have integral switches and are usually called butterfly valves.

Waterflow initiating devices are used wherever automatic sprinkler systems provide coverage in a protected area. Generally, each sprinkler zone has a minimum of one waterflow initiating device. Normally, each floor of a building has waterflow switches installed at each floor test station to detect the flow of water on the zone. However, sprinkler zones are limited to 52,000 square feet in area by *NFPA 13, Standard for the Installation of Sprinkler Systems*. Therefore, large footprint buildings may have more than one sprinkler zone and waterflow switch per floor. The sprinkler contractor is generally responsible for the installation of waterflow initiating devices.

The most common type of waterflow switch is the vane- (paddle-) type switch. This device uses a plastic paddle inserted into the pipe to detect the flow of water. The paddle is connected to an electromechanical sensing device. Waterflow switches are usually conventional devices requiring a signaling line circuit interface to make them addressable. To prevent the device from responding to surges, a delay, called a retard, is used. Most vane-type waterflow devices have preset delays of 30, 45, 60, and 90 seconds, while some have variable adjustment. Section 17.12.2 requires a maximum retard of 90 seconds, but every effort should be made to use the shortest delay possible without causing nuisance alarms. If the retard is set too low, weekly tests of the fire pump may cause nuisance alarms. Vane-type switches are never permitted on dry sprinkler systems because the vane would be dislodged by the inrush of water when the system trips. Pressure switches are sometimes used to detect waterflow, especially on dry-pipe sprinkler systems.

Post indicator valves (PIV) are another type of valve supervisory device. As the name implies, they have a built-in indicator to show valve position (OPEN or SHUT). PIVs are often used for building water supply control and are typically mounted so the fire department can see the valve position when arriving at the premises. PIVs may contain a tamper switch and are generally locked because they are outside and may be subject to vandalism. **See Figure 6-10.**

Some buildings have a sprinkler system but no fire alarm system to monitor waterflow. In this case, NFPA 13, *Standard for the Installation of Sprinkler Systems,* requires a mechanical alarm, horn, or siren or a listed electric gong, bell, speaker, horn, or siren. Motor gongs are water operated and require no electric current, but bells or horns may operate on 120 volts alternating current (VAC). NFPA 13 references the *National Electrical Code* for all wiring requirements.

Sprinkler systems do not usually have separate controls. Sprinkler system initiating devices such as waterflow switches are usually connected directly to the fire alarm system as initiating devices. Section 23.8.5.5.2 permits only five waterflow switches to be connected to an initiating device circuit. Additionally, Section 23.8.5.6.2 permits only 20 supervisory switches to be connected to an initiating device circuit. The reason for these requirements is that the system is non-addressable, and large numbers of these devices will delay the response to signals. However, the number of waterflow or valve supervisory switches connected to a signaling line circuit is only limited by the manufacturer's recommendations.

A formerly used and incorrect practice was to install a waterflow and valve supervisory switch on the same initiating device circuit. **See Figure 6-11.**

This practice violates the *Code* because closure of the valve will cause a trouble,

Figure 6-10. *A post indicator valve (PIV) with electrical supervision is often used for water supply control in buildings. Photo courtesy Merton Bunker.*

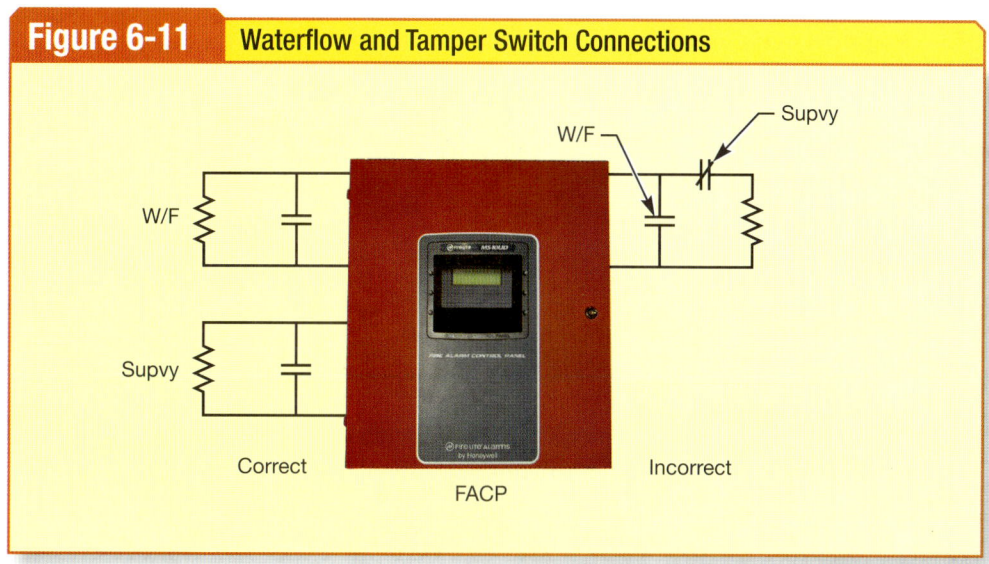

Figure 6-11. The correct method requires one supervisory zone module and one alarm module, unless the manufacturer can provide proper supervision on one module.

rather than a supervisory, signal. When using initiating device circuit technology, two initiating device circuits are required. It should be noted that valve supervisory and waterflow switches are non-addressable and are often connected to addressable systems through the use of dual-input monitor modules, which can annunciate the two signals independently.

Fire Pumps

Fire pumps usually have their own separate controls. Fire pump controls are supervised by the fire alarm system, as required by NFPA 20, *Standard for the Installation of Stationary Fire Pumps*. NFPA 20 requires electric fire pumps to be supervised for the following conditions:
- Pump running
- Loss of phase
- Phase reversal
- Connection to alternate power source (where connection is used)

NFPA 20 requires diesel fire pumps to be supervised for the following conditions:
- Engine running
- Controller main switch turned to the OFF or manual position
- Controller or engine trouble

These signals must be annunciated in a constantly attended location. The fire alarm controls may be used to meet these requirements. **See Figure 6-12.**

Section 23.8.5.9 provides additional requirements for fire pump supervision. It is general practice to transmit a common supervisory signal from the fire pump controller to the fire alarm system controller. A fire pump common supervisory signal is annunciated at the main controls, but operators must immediately

Figure 6-12. Fire pump supervision is required by NFPA 20. The contacts for the required conditions to be monitored are located in the fire pump controller, shown on the right.

Fact

Many jurisdictions have laws or regulations which require persons working on suppression systems to hold a license or certification in that area. Be sure you understand local laws before working on any suppression system interface.

investigate the nature of the signal at the fire pump controller.

Section 23.8.5.9 permits a pump running signal to be either a supervisory or alarm signal. Use of an alarm signal for this purpose is discouraged. If a pump running signal causes an alarm, then precautions must be taken to avoid alarms during the weekly self-test of the fire pump.

Special Hazards Systems

Gaseous suppression, pre-action, and foam systems usually do have controls associated with them but could be controlled by the main fire alarm system instead. Manual kitchen hoods usually do not have system controls but have a simple switch that actuates when the system is manually released.

Most building codes require suppression systems to be connected to the building's fire alarm system where there is a fire alarm system in the building. If no fire alarm control unit exists, connection is required, and the building is sprinklered. Section 23.8.5.5.1 requires the installation of a dedicated function fire alarm control unit designated as:

SPRINKLER WATERFLOW AND SUPERVISORY SYSTEM

Section 23.8.5.7.1 requires the actuation of other suppression systems to cause an alarm or supervisory signal on the fire alarm system, where required by another code or standard. These controls are required to transmit all three signals (trouble, supervisory, and alarm) to the main fire alarm system controls. Section 23.8.5.7.2 requires all suppression system initiating devices to be monitored for integrity. Either the main system panel or the suppression system panel may monitor the field devices for integrity. New Section 23.8.5.7.3 was added to the 2013 edition of *NFPA 72* to require supervision of valves between the suppression system and an initiating device to be supervised for position.

Section 23.8.5.10.1 requires any releasing service fire alarm controls to be connected to the building fire alarm system master fire alarm controls. Section 23.8.5.10.2 requires releasing systems to cause alarm and supervisory signals to be annunciated at the fire alarm control unit.

A closed valve cannot cause loss of a waterflow alarm signal. For this reason, Section 23.8.5.10.4 requires valves that isolate alarm-initiating devices to be supervised by the fire alarm system. This specifically applies to pressure-type alarm-initiating switches installed on sprinkler riser trim kits.

Some supervisory switches are installed outdoors, such as those used on post indicator valves. They are subject to tampering by vandals and arsonists. Section 23.8.5.11.1 requires control units and other alarm and supervisory initiating circuits of suppression systems that are subject to tampering to be designed and installed to create a supervisory signal when opened or removed. Tamper switches on junction boxes are commonly used to create a supervisory signal when the box cover is removed. However, Exception No. 1 to Section 23.8.5.11.1 exempts junction boxes inside buildings and Exception No. 2 permits the use of tamper-resistant screws in lieu of tamper switches.

The means of interconnection between suppression system and fire alarm system controls is generally made through dry contacts or addressable monitor modules. Sprinkler attachments, such as waterflow switches and valve tamper switches, are connected directly to the fire alarm system. Some suppression systems, such as clean agent systems or pre-action systems, have control units that must be interfaced with the fire alarm system. Controls usually have dry contacts suitable for connection to the fire alarm controls through monitor modules or initiating device circuits.

A trouble signal on a suppression subsystem control unit must create a supervisory signal at the main controls because the trouble condition may prevent proper operation of the suppression subsystem. A trouble on the suppression system controls, by definition, is a supervisory signal and must be reported as a supervisory signal rather than a trouble signal. Therefore, it is not permitted to put the suppression system trouble contacts in series with the circuit used for the alarm or supervisory circuit because a trouble on the suppression system controls will cause a trouble signal on the circuit.

Again, it is permitted to annunciate a common supervisory signal at the fire alarm system controls, which means that only one address or circuit is required between the main controls and the suppression system controls.

Pre-action systems are designed to protect areas where wet sprinklers are undesirable. The piping that extends into the protected area has no water until the interlock is actuated. Water flows only after the interlock is actuated and after a sprinkler actuates. Pre-action sprinklers most often use an interlock that requires the actuation of two automatic detectors in order to operate a charging valve. Some systems additionally require loss of air pressure, which is caused by actuation of a sprinkler in the protected area. Once the charging valve is opened, water may flow into the piping. However, water will only flow once a sprinkler is actuated. This arrangement generally uses engineered systems, and the monitored equipment includes detection devices, isolation valves, and releasing control units.

Special hazard suppression systems, such as carbon dioxide (CO_2) systems, Inergen® systems, or FM-200® systems, are sometimes used to protect critical or sensitive equipment. Computer server rooms, uninterruptible power supply (UPS) rooms, switchgear rooms, and other areas with critical or sensitive equipment are good candidates for flooding (gaseous)-type suppression systems. These systems have separate standards and are generally pre-engineered by the manufacturer. Some systems can be actuated manually, while others are automatically actuated.

When discharged, these systems can cause injury or death to occupants. For this reason, the areas protected by these systems must be placarded to warn occupants of these dangers. Manual stations for the actuation of special hazard systems often look like fire alarm manual stations, except they must be marked to indicate they will actuate a system. Occupants and users must pay particular attention to these systems to prevent accidental discharge, which can be fatal or involve extensive clean-up costs.

EMERGENCY CONTROL FUNCTION INTERFACES

Emergency control functions are designed to make the premises safer in the event of a fire. These functions include the following:
- Elevator recall/shutdown
- Smoke control
- Smoke door release
- HVAC shutdown
- Door unlocking
- Stairway pressurization

Chapter 21 of *NFPA 72* provides requirements for emergency control function interfaces.

General Requirements

Fire alarm system interfaces can handle either inputs or outputs. On conventional (non-addressable) systems, inputs are typically provided through initiating device circuits. Outputs are provided through relays powered by the fire alarm system. These output circuits are similar to notification appliance circuits. In some cases, notification appliances have output relay contacts that can be used to control output functions.

Addressable system inputs are usually provided by monitor modules, which essentially contain initiating device circuits as inputs. Addressable system outputs can be accomplished through the use of addressable control modules or relays powered by the fire alarm system. Monitor and control modules are available from most manufacturers in either a single or dual arrangement.

The module address can be programmed to create signals or outputs, based upon the system programming.

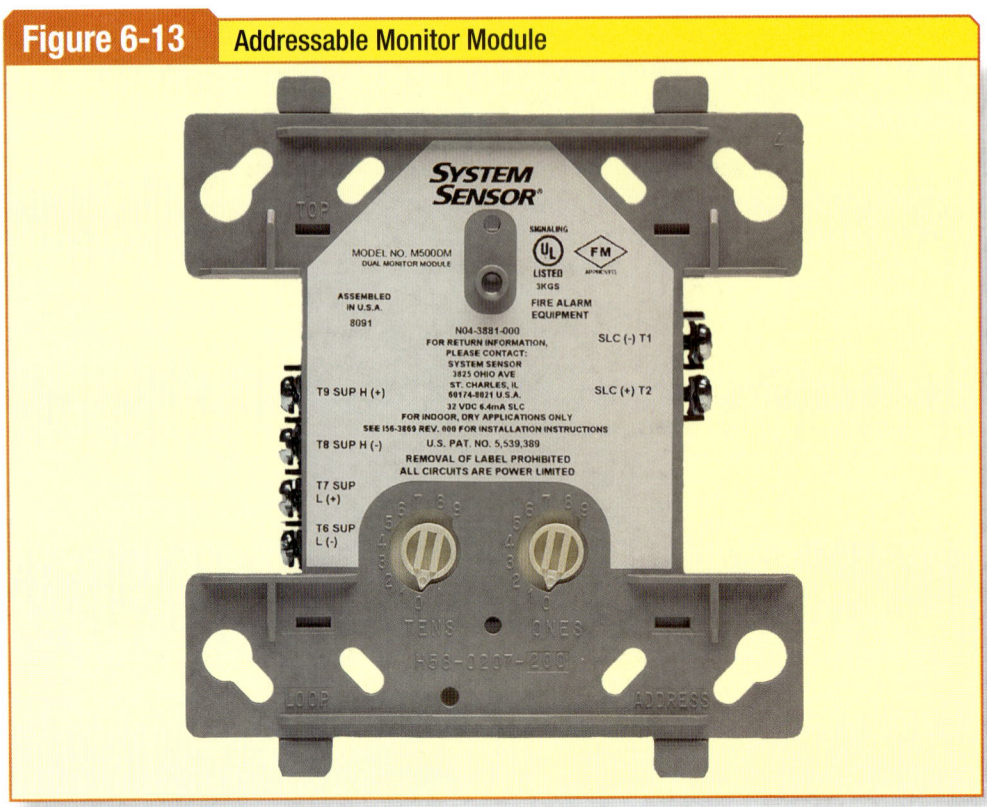

Figure 6-13. Some monitor modules have two inputs (zones), while some have only one. Each input is actually an initiating device circuit.

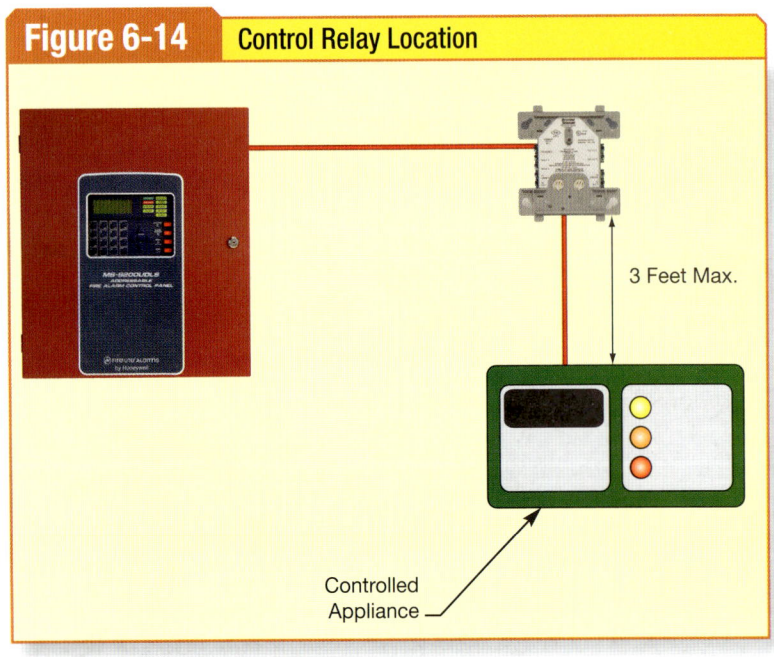

Figure 6-14. The control module or relay must be within three feet (1 m) of the component that controls the emergency function. Ideally, the control element is located within prescribed limits from the component that controls the emergency function to minimize damage to the non-monitored wiring.

Monitor and control modules look very similar. **See Figure 6-13.**

Section 21.2.4 of *NFPA 72* requires all listed control relays or other listed control appliances to be installed within three feet (1 m) of the component controlling the emergency function. This requirement exists to minimize the length of unmonitored wiring between the fire alarm and the interfaced system. **See Figure 6-14.**

The intent of Section 21.2.4 is to locate the control element within three feet (1 m) of the controlled equipment. However, it is not always possible to place the control relay close to the equipment that is controlled. For example, consider an HVAC fan on a rooftop that is controlled by a motor controller on the ground floor. The relay will most likely be located at the motor control center. The intent is to locate the control element within three feet (1 m) of the controlled equipment to minimize mechanical damage to the non-monitored wiring.

Section 21.2.5 requires the control relay to operate within the voltage and current limitations of the fire alarm control unit. Section 21.2.8 also requires the wiring between the fire alarm control unit and the control appliance or relay to be monitored for integrity. Section 21.2.6 requires the wiring between the fire alarm control unit and the control appliance or relay to be a Class A, Class B, Class D, or Class X, per Chapter 12 of *NFPA 72*. Many designers prefer to make all emergency control function circuits Class D. Class D circuits are fail-safe and default to the safe mode of operation.

Addressable devices must be compatible with the proprietary language or protocol used by the panel. Relays or contacts must also be suitably rated for the current and voltage they control, or they are likely to prematurely fail.

NFPA 72 permits several different types of interconnection methods between the fire alarm system and other systems. Section 21.2.10 permits the following:

1. Electrical contacts listed for the connected load
2. Data communications over signaling line circuit dedicated to the fire alarm or shared with other premises operating systems
3. Other listed methods

These methods are the same methods permitted for interconnected fire alarm control units.

Elevator Safety Functions

Elevator safety functions are intended to return elevator cars to a safe level in the event there is smoke present in an elevator lobby. There were many cases where fires impinged on elevator lobby call buttons, shorting them out, and calling elevator cars to the floor that was on fire. Any occupants in the car would generally be trapped without any chance of escape. In fact, two specific incidents in New York during the late 1960s and early 1970s resulted in multiple fatalities. Modern elevators are designed to interface with fire alarms to safely transport occupants to a safe landing in the event of a fire.

They can also disconnect (shunt trip) electrical supplies if sprinklers wet down equipment in the hoistway or elevator machine room.

ASME A17.1-2010/CSA B44-10, Safety Code for Elevators and Escalators, requires Phase I Emergency Recall and shunt trip. *A17.1* is adopted in every U.S. jurisdiction, except Pennsylvania. *ASME A17.1/CSA B44* references *NFPA 70, National Electrical Code,* and *NFPA 72, National Fire Alarm and Signaling Code,* for the electrical and fire alarm installations, respectively.

Firefighters often use Phase II Firefighter's Operation to transport firefighters and equipment, especially in tall buildings. But only the fire service can use Phase II Operation because it requires the use of a special key. Phase I Recall provides the fire service with the added benefit of determining that all elevators are parked in the lobby and that there are no passengers trapped in the hoistway.

Phase I Emergency Recall (Firefighter's Service) Operation. Occupant safety is the most important purpose of Phase I Emergency Recall Operation. This feature causes elevator cars to return to a safe location during a fire so passengers can safely escape. Section 2.27.3.2.3(a) of *ASME A17.1-2010/CSA B44-10, Safety Code for Elevators and Escalators,* requires elevator cars other than those already located at the designated recall level to return nonstop to the designated level (chosen by the authority having jurisdiction) upon actuation of a fire alarm initiating device in an elevator lobby.

Sections 2.27.3.2.3(b) and (c) of *ASME A17.1/CSA B44* require the actuation of a fire alarm initiating device in a hoistway or elevator machine room to cause the car(s) to return nonstop to the designated level. However, actuation of an alarm-initiating device located in the hoistway at or below the lowest level of recall must cause the car(s) to return to the upper level of recall.

Section 2.27.3.2.4(a) of *ASME A17.1/CSA B44* requires that an alarm-initiating device actuation at the designated level

Fact

The designated and alternate recall levels are chosen by the AHJ. The AHJ will usually be the fire marshal or other fire department representative.

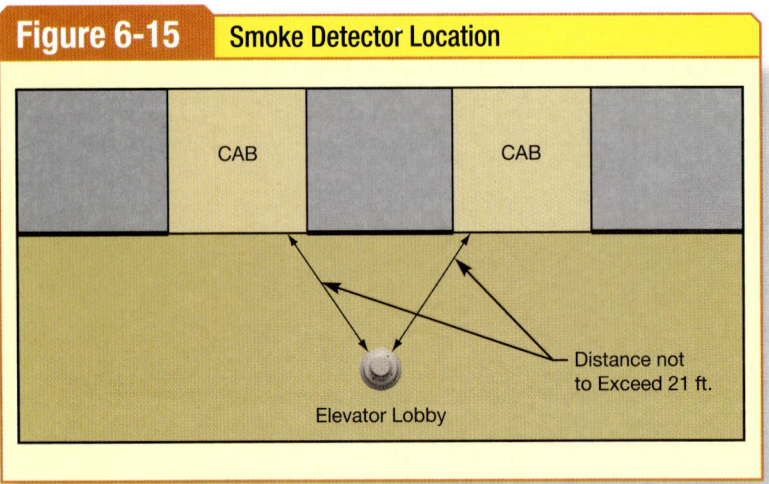

Figure 6-15. Smoke detectors must be located within 21 feet of the centerline of elevator doors. High ceilings may require additional detectors, per chapter 17 of NFPA 72.

must cause the car(s) to return to the alternate level (also chosen by the authority having jurisdiction). It should be noted that unique architecture, such as split-level lobbies or an atrium, might affect the decision in determining the designated level.

Section 21.3 of *NFPA 72*, Elevator Recall for Fire Fighters' Service, provides requirements for fire alarm equipment used for Phase I Emergency Recall Operation. Specifically, Section 21.3.1 requires initiating devices used to initiate fire fighters' service recall to be connected to the building fire alarm system. If there is no building fire alarm system, Section 21.3.2 requires the installation of a fire alarm control unit specifically dedicated for the purposes of elevator control. This panel must be identified as "elevator recall control and supervisory control unit" on system drawings and on the control unit itself. The panel must be located in a lobby or other area where the alarm and trouble signals can be heard.

Section 21.3.3 does not require any other initiating devices beyond those required for elevator control and does not require any notification appliances. *NFPA 101, Life Safety Code* or local building codes will provide the requirements specifying where occupant notification is required. Section 21.3.3 requires only the elevator lobby, machine room, and hoistway smoke detectors to cause Phase I Emergency Recall Operation, unless specifically required by the authority having jurisdiction. The exception permits a waterflow switch to initiate elevator recall upon activation of a sprinkler installed at the bottom of the elevator hoistway (the elevator pit), provided that the waterflow switch and pit sprinkler are installed on a separately valved sprinkler line dedicated solely for protecting the elevator pit and that the waterflow switch is provided without time-delay capability.

In some jurisdictions, actuation of any alarm-initiating device must cause Phase I Emergency Recall Operation. Local codes and amendments can vary from jurisdiction to jurisdiction and must always be consulted.

Smoke detectors must be installed near the elevator bank they serve when used for elevator recall. Section 21.3.5 of *NFPA 72* requires smoke detectors in elevator lobbies to be within 21 feet (6.4 m) of the centerline of each elevator door within the elevator bank under control of the detector. **See Figure 6-15.**

This requirement follows the requirements of Chapter 17 of *NFPA 72* for detector spacing. Non-smooth or high ceilings (over 15 feet or 4.6 m) will require detector spacing to be determined by the requirements of Chapter 17 for those ceiling types. Unenclosed lobbies (such as an atrium) present unique challenges and will require an engineering evaluation to determine spacing. In these cases, consultation with a qualified designer is highly recommended.

Section 21.3.6 does not permit smoke detectors to be installed in non-sprinklered elevator hoistways, unless the detector is used to actuate smoke relief equipment. This requirement exists because hoistways are not always the best environment for smoke detectors. Hoistways are generally

> **21.3.6** Smoke detectors shall not be installed in unsprinklered elevator hoistways unless they are installed to activate the elevator hoistway smoke relief equipment.
>
> *(Excerpt from NFPA 72.)*

ASME A17.1-2010/CSA B44-10

2.27.3.2 Phase I Emergency Recall Operation by Fire Alarm Initiating Devices

2.27.3.2.1 In jurisdictions not enforcing the NBCC, fire alarm initiating devices used to initiate Phase I Emergency Recall Operation shall be installed in conformance with the requirements of NFPA 72, and shall be located

(a) at each floor served by the elevator

(b) in the associated elevator machine room, machinery space containing a motor controller or electric driving machine, control space, or control room

(c) in the elevator hoistway, when sprinklers are located in those hoistways

2.27.3.2.2 In jurisdictions enforcing the NBCC, smoke detectors, or heat detectors in environments not suitable for smoke detectors (fire alarm initiating devices), used to initiate Phase I Emergency Recall Operation, shall be installed in conformance with the requirements of the NBCC, and shall be located

(a) at each floor served by the elevator

(b) in the associated elevator machine room, machinery space containing a motor controller or electric driving machine, control space, or control room

NOTE·(2.27.3.2.2): Smoke and heat detectors (fire alarm initiating devices) are referred to as fire detectors in the NBCC. Pull stations are not deemed to be fire detectors.

2.27.3.2.3 Phase I Emergency Recall Operation to the designated level shall conform to the following:

(a) The activation of a fire alarm initiating device specified in 2.27.3.2.1(a) or 2.27.3.2.2(a) at any floor, other than at the designated level, shall cause all elevators that serve that floor, and any associated elevator of a group automatic operation, to be returned nonstop to the designated level.

(b) The activation of a fire alarm initiating device specified in 2.27.3.2.1(b) or 2.27.3.2.2(b) shall cause all elevators having any equipment located in that machine room, and any associated elevators of a group automatic operation, to be returned nonstop to the designated level. If the machine room is located at the designated level, the elevator(s) shall be returned nonstop to the alternate level.

(c) In jurisdictions not enforcing NBCC, the activation of a fire alarm initiating device specified in 2.27.3.2.1(c) shall cause all elevators having any equipment in that hoistway, and any associated elevators of a group automatic operation, to be returned nonstop to the designated level, except that initiating device(s) installed at or below the lowest landing of recall shall cause the car to be sent to the upper recall level.

(d) In jurisdictions enforcing the NBCC, the initiation of a fire detector in the hoistway shall cause all elevators having any equipment in that hoistway, and any associated elevators of a group automatic operation, to be returned nonstop to the designated level, except that initiating device(s) installed at or below the lowest landing of recall shall cause the car to be sent to the upper recall level.

(e) The Phase I Emergency Recall Operation to the designated level shall conform to 2.27.3.1.6(a) through (n).

2.27.3.2.4 Phase I Emergency Recall Operation to an alternate level (see 1.3) shall conform to the following:

(a) the activation of a fire alarm initiating device specified in 2.27.3.2.1(a) or 2.27.3.2.2(a) that is located at the designated level, shall cause all elevators serving that level to be recalled to an alternate level, unless Phase I Emergency Recall Operation is in effect

(b) the requirements of 2.27.3.1.6(f), (j), (m), and (n)

(c) the requirements of 2.27.3.1.6(a), (b), (c), (d), (e), (g), (h), (i), (k), and (l), except that all references to the "designated level" shall be replaced with "alternate level"

2.27.3.2.5 The recall level shall be determined by the first activated fire alarm initiating device for that group (see 2.27.3.2.1 or 2.27.3.2.2).

If the car(s) is recalled to the designated level by the "FIRE RECALL" switch(es) [see also 2.27.3.1.6(j)], the recall level shall remain the designated level.

2.27.3.2.6 When a fire alarm initiating device in the machine room, control space, control room, or hoistway initiates Phase I Emergency Recall Operation, as required by 2.27.3.2.3 or 2.27.3.2.4, the visual signal [see 2.27.3.1.6(h) and Fig. 2.27.3.1.6(h)] shall illuminate intermittently only in a car(s) with equipment in that machine room, control space, control room, or hoistway.

inaccessible, making maintenance difficult and expensive. Nuisance alarms will occur unless the smoke detectors are properly maintained. However, Section 21.3.8 does not permit smoke detectors in elevator hoistway pits, unless the smoke detector is listed for the environment.

Smoke detectors (or other appropriate automatic fire detectors) are required in sprinklered hoistways because they are necessary to recall cars before sprinkler actuation causes a shunt trip or to provide recall to move cars away from a fire in the hoistway pit if the bottom of the hoistway is sprinklered. Wet elevator brakes and electrical control equipment can cause erratic elevator operation and poses a serious threat to passenger safety. A shunt trip immediately stops the car(s) wherever they are, whether occupants are inside them or not. Phase I Emergency Recall Operation is intended to allow passenger escape before disconnection of the power supply. *NFPA 13, Standard for the Installation of Sprinkler Systems,* provides the requirements for locations of hoistway sprinklers.

When sprinklers are installed in pits, a new Section 21.3.7 was added to the 2013 edition of *NFPA 72* to require automatic detection to initiate elevator recall. Where sprinklers are located above the lowest level of recall, the fire detection devices are required at the top of the hoistway. Where sprinklers are located in the bottom of the hoistway (the pit), fire detection device(s) are required in the pit in accordance with Chapter 17.

Outdoor or non-air-conditioned elevator lobbies, such as in an open parking structure or an unheated elevator lobby, pose significant challenges for fire alarm system designers. These lobbies are damp locations and are unsuitable for smoke detection. *ASME A17.1/CSA B44* requires Phase I Emergency Recall Operation, but smoke detectors are not suited to outdoor environments. Section 21.3.9 permits other types of automatic detection where the environment is not suited to smoke detection. Generally, heat detection is considered an acceptable and affordable alternative to smoke detection in these areas, but any listed automatic

Figure 6-16. *This lamp is located in the elevator cab and illuminates when there is smoke in the elevator machine room or hoistway. It provides a signal to firefighters that the elevators are no longer safe to use.*

detection device (except smoke detection) is permitted under this exception.

Section 21.3.10 requires the actuation of elevator Phase I Emergency Recall Operation alarm-initiating devices to cause an alarm signal on the building fire alarm system. It also requires the indication of the actuated device on any required annunciators. Section 21.3.11 requires the actuation of elevator machine room and hoistway smoke detectors to cause separate and distinct visible annunciation and to alert firefighters that the elevators are no longer safe for them to use. One way this is accomplished is by flashing the fire hat symbol in the car(s) to alert any firefighters that they should exit the car. **See Figure 6-16.**

Section 21.3.12 permits the detectors used to initiate firefighters' Phase I recall to cause a supervisory signal, with permission of the authority having jurisdiction. Where elevator lobby detectors are used for a purpose other than elevator recall, their actuation must initiate an alarm on the system. Most elevator lobby detectors will be used for area detection and must meet this requirement.

Section 21.3.14 requires three separate circuits or control points to be provided between the fire alarm control unit and the elevator controller. These control points or circuits cause Phase I recall for firefighter's service, as required by *ASME A17.1/CSA B44*. In the event of a fire in an elevator lobby, elevator cars must return nonstop to the designated level (determined by the AHJ).

Section 21.3.14.1 requires a fire alarm output to cause elevator recall to the designated level for each of the following conditions:
1. Actuation of initiating device(s) at any elevator lobby except the designated level
2. Actuation of initiating device(s) at the elevator machine room, elevator machinery space, elevator control space, or elevator control room, except where the machine room is located at the designated level
3. Actuation of initiating device(s) in a sprinklered elevator hoistway serving the the elevator under control

Section 21.3.14.2 requires a fire alarm output to cause elevator recall to the alternate level for each of the following conditions:
1. Actuation of initiating device(s) at the designated level elevator lobby
2. Actuation of initiating device(s) in the elevator machine room, elevator machinery space, elevator control space, or elevator control room, where the machine room is located at the designated level
3. Actuation of sprinklered hoistway initiating device(s) at or below the lowest level of recall in the elevator hoistway and the alternate level is above the designated level

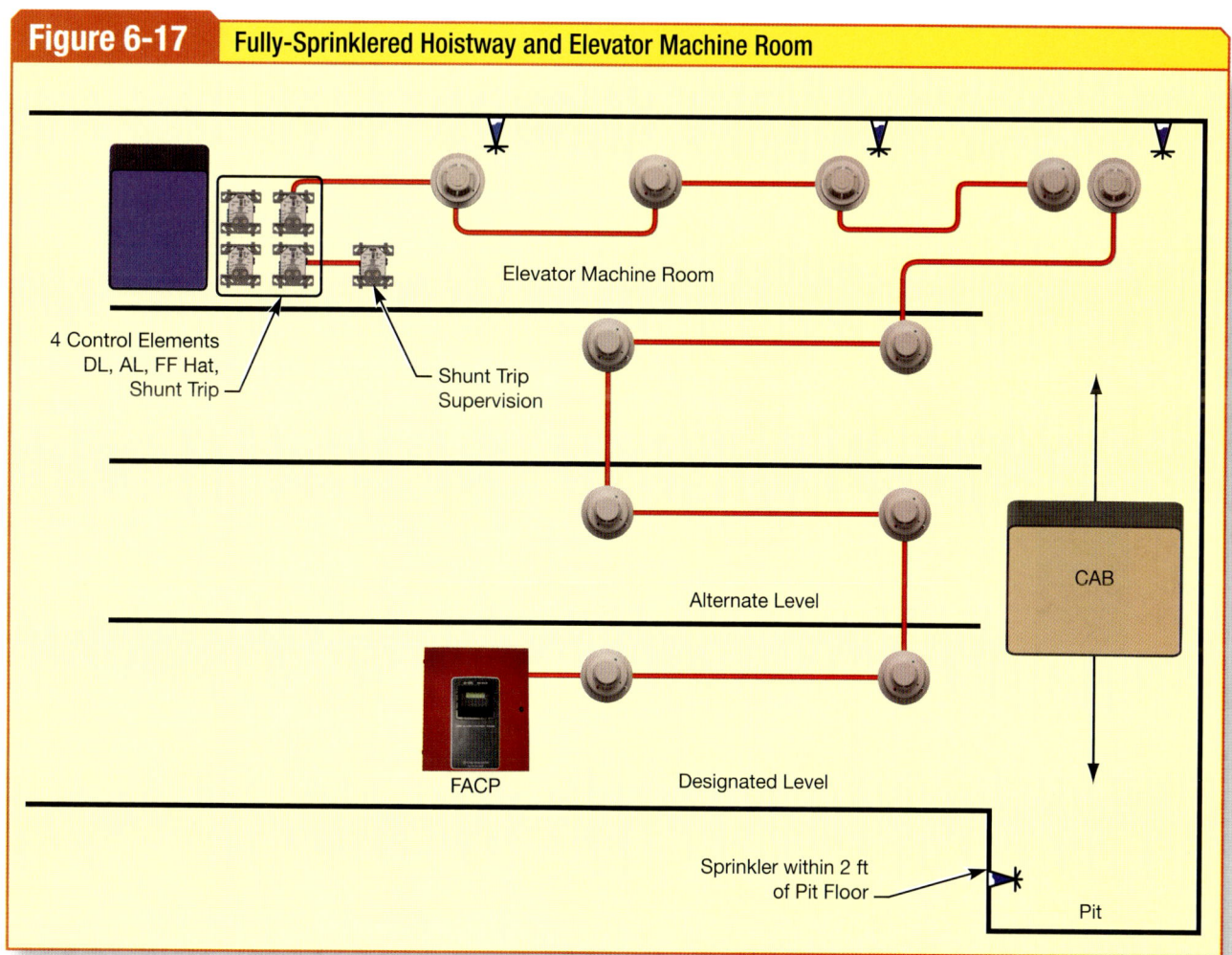

Figure 6-17. *For an elevator with a sprinklered hoistway and machine room, four control circuits are needed, plus one monitor module for shunt trip power supervision.*

Section 21.3.14.3 requires a fire alarm output to cause elevator visual warning (firefighter's hat) for each of the following conditions:
1. Actuation of elevator machine room, elevator machinery space, elevator control space, or elevator control room initiating device(s)
2. Actuation of elevator hoistway initiating device(s)

The third elevator interface circuit is for actuation of the initiating devices in the elevator machine room and hoistway. This circuit is used by the elevator controller to flash the firefighter's hat symbol in the car and is intended to indicate that the elevator(s) are no longer safe to use on Phase II (firefighter's) Operation.

The use of on-cab drivers and controls is becoming more common. This is a fairly recent technology where the driver is on top of the cab and there is no actual machine room. Controls may be located in an elevator lobby, an electrical closet, or a dedicated space near the hoistway. This can complicate the design somewhat because there is no machine room.

Note that addressable systems will use three control points (addressable relays) on a signal line circuit. Although the illustrations show three control circuits, Section 21.4 may require another control circuit for shunt trip.

Only elevator lobby, machine room, and hoistway detectors are required to cause Phase I recall. **See Figure 6-17.**

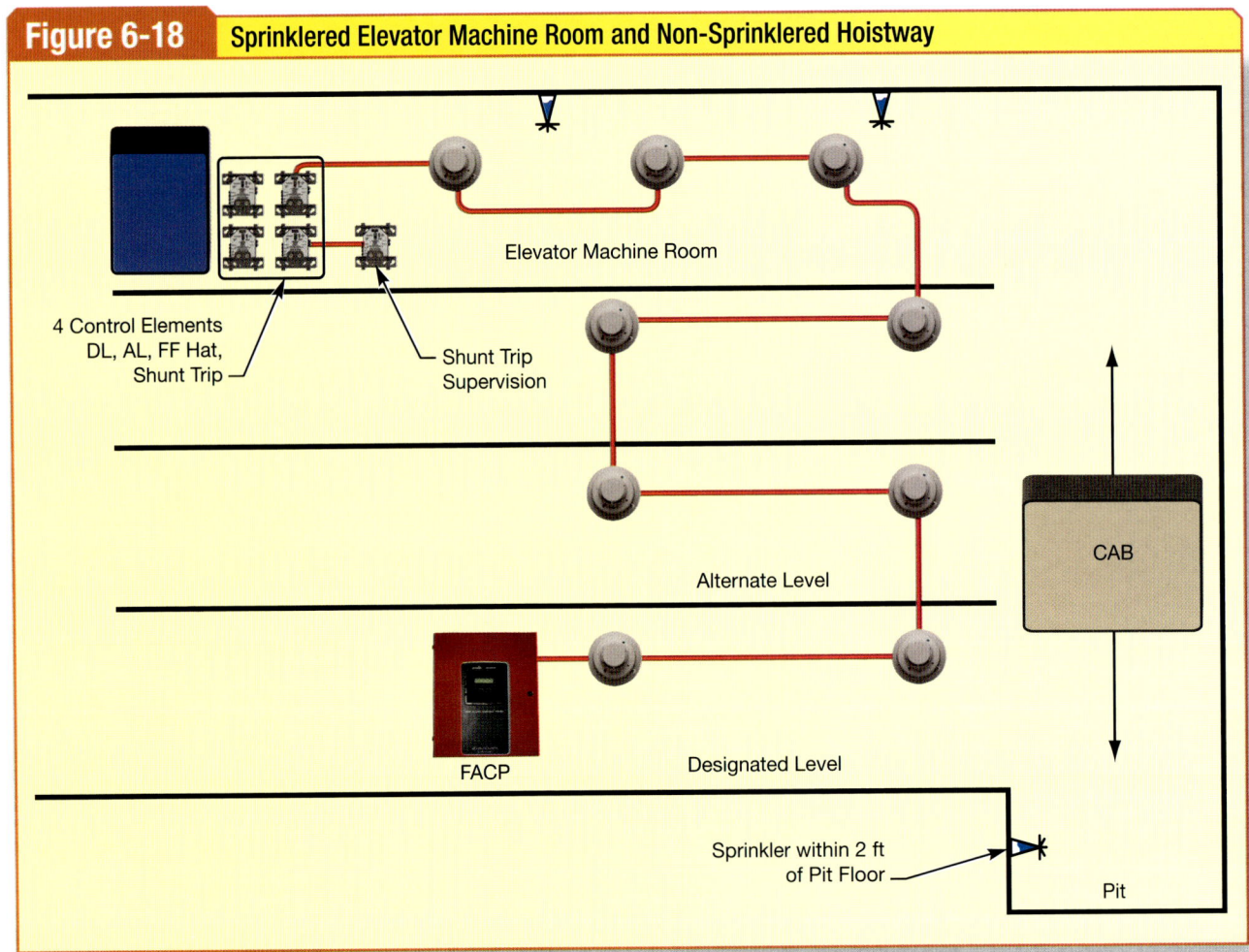

Figure 6-18. For an elevator with a sprinklered hoistway (pit only) and machine room, four control circuits are needed, plus one monitor module for shunt trip power supervision.

Elevator shutdown (shunt trip) is not required for pit sprinklers located less than 24 inches (610 mm) from the pit floor. Sprinklers located at or below 24 inches will not cause braking or electrocution hazards. **See Figure 6-18.**

The requirement to have sprinklers in the hoistway is found in *NFPA 13, Standard for the Installation of Sprinkler Systems*. Section 8.15.5 of *NFPA 13, Standard for the Installation of Sprinkler Systems* provides requirements for elevator sprinklers. Section 8.15.5.6 requires sprinklers at the top of the hoistway only where the hoistway is combustible. Note that this will determine if smoke and heat detectors are required in the hoistway.

Sprinklers are required at the bottom of the pit only where the pit contains combustible fluids. Since most elevators have a hydraulic buffer, pit sprinklers are required. Section 8.15.5.1 requires a sidewall spray sprinkler to be installed so that it is not more than two feet above the floor of the pit. If there are no other sprinklers in the hoistway, other than a pit sprinkler located within two feet of the pit floor, then elevator shutdown will not be required.

Some buildings are not fully sprinklered. In these cases, initiating devices in the hoistway are not permitted by *NFPA 72*, where the sprinkler is located less than 24 inches from the pit floor. **See Figure 6-19.**

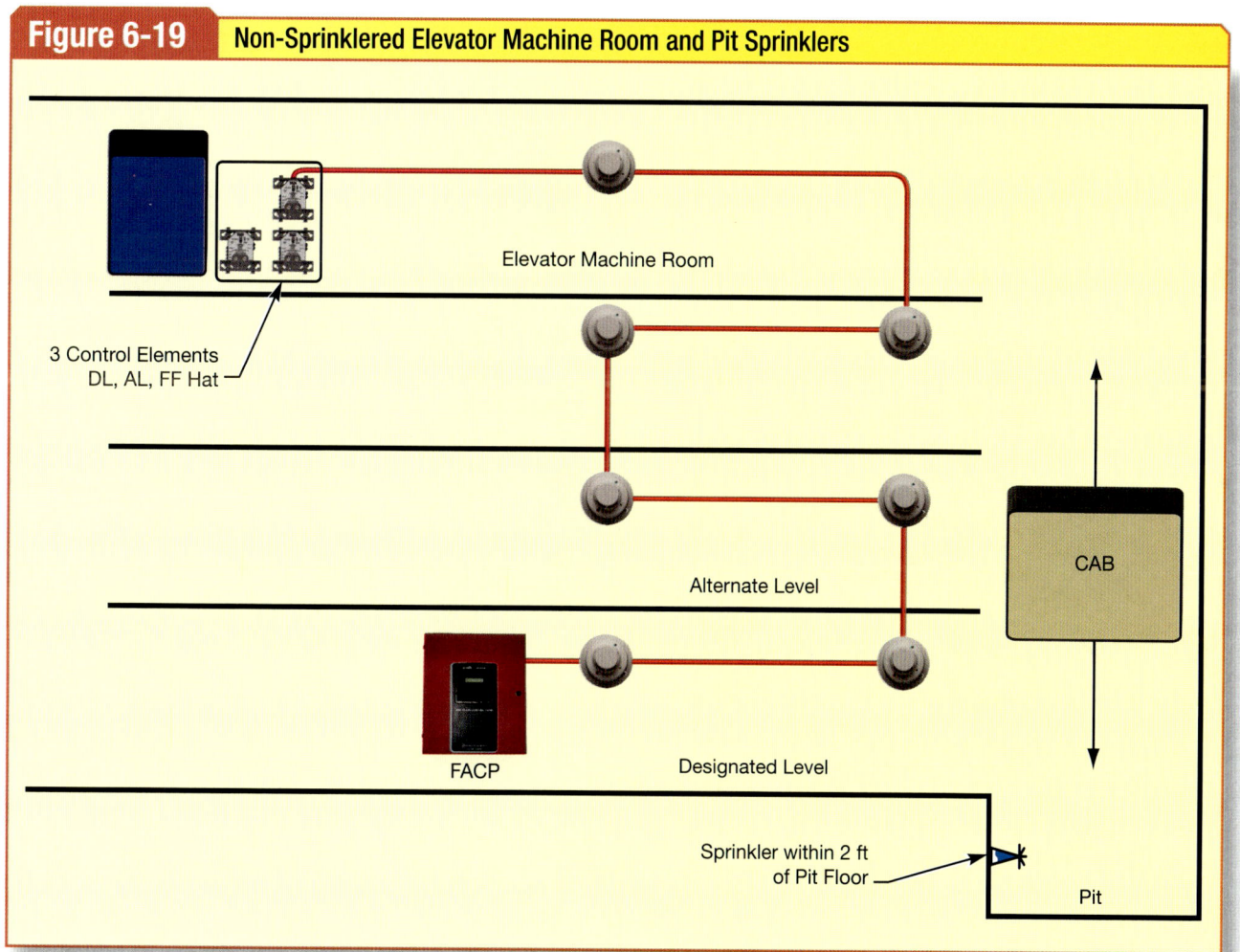

Figure 6-19. *For an elevator with a sprinklered hoistway (pit only), only three control circuits are needed, as there is no shunt trip requirement.*

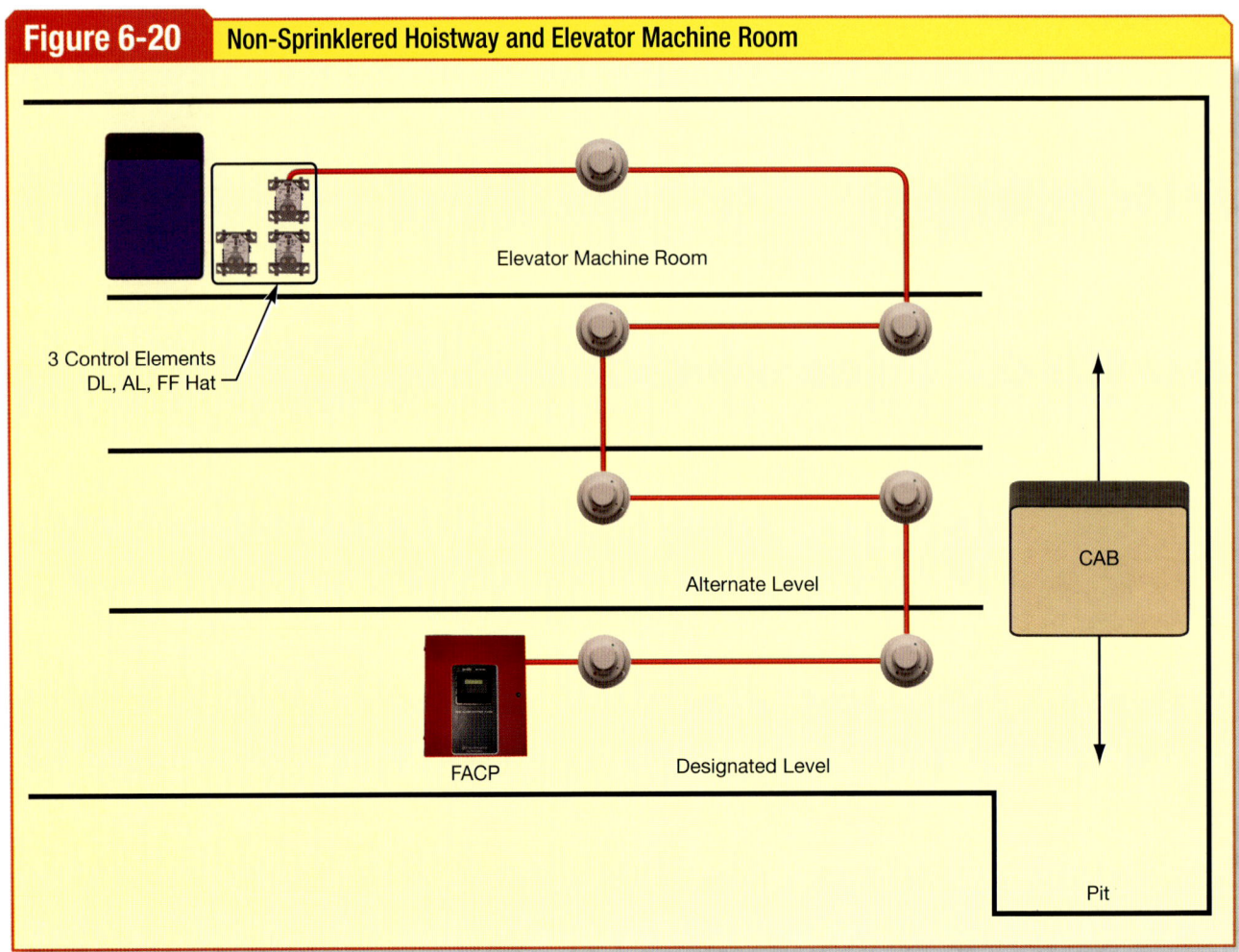

Figure 6-20. For an elevator with a non-sprinklered hoistway and machine room, only three control circuits are needed, as there is no shunt trip requirement.

In other cases, there are no sprinklers in the building. **See Figure 6-20.** Note the absence of a shunt trip relay in the machine room.

The use of initiating device auxiliary contacts to interface with the elevator control circuits is prohibited. This practice violates Section 12.6.1 of the *Code* because the conductors between the detectors and elevator controls are not monitored for integrity. Only listed relays or control modules should be used in lieu of this practice.

Elevator Shutdown (Shunt Trip). Elevator shutdown is frequently referred to as a shunt trip. This feature is intended to immediately stop elevators before water from the elevator machine room or hoistway sprinklers adversely affects brakes and electrical equipment. Wet brakes and/or wet electrical equipment can cause erratic elevator operation or dangerous conditions. Therefore, it is required by 2.8.3.3.2 of *A17.1/CSA B44* to cause disconnection of the main line power supply upon or prior to the application of water from sprinklers located in the machine room or hoistway. Elevator shutdown occurs whether the elevator car is occupied or not and operates regardless of location.

Pit sprinklers located less than two feet (0.65 m) above the pit floor are exempt from causing elevator shutdown because Section 2.8.3.3.4 of *ASME*

A17.1/CSA B44 requires all electrical equipment within four feet (1.3 m) of the pit floor to be NEMA 4 rated, making them resistant to water and the problems associated with wet equipment. Therefore, heat detectors or waterflow switches will typically not be required for elevator shutdown involving pit sprinklers. Some jurisdictions do not enforce the latest edition of *A17.1/CSA B44*, so one must be sure to check the codes before beginning work.

Elevator shutdown can be caused by several different methods. The most common method involves heat detectors located near sprinklers with the understanding that they will operate before the sprinkler fuses (activates). Another method is to use a waterflow switch with no time delay. Systems utilizing heat detectors, controls, and solenoid valves to recall the car(s) can be used to hold elevator shutdown and resulting waterflow until the car stops at a landing. **See Figure 6-21.**

These systems offer greater occupant safety but may cost significantly more than other methods.

Section 2.8.3.3.3 does not permit the use of smoke detectors to accomplish elevator shutdown. Section 21.4.1 of *NFPA 72* requires heat detectors used for elevator shutdown to have a lower temperature rating and a higher sensitivity than the sprinklers in the area. A lower temperature rating will help ensure that the heat detector actuation (and the resulting shunt trip) occurs before water flows. A heat detector must be installed within two feet (610 mm) of every sprinkler, as required by Section 21.4.2. Control modules or relays powered by the fire alarm system can then be used to operate the elevator shutdown (shunt trip) circuit.

Waterflow switches can be used to cause elevator shutdown; however, there must be no time delay between the flow of water and the disconnection of the main line power supply, as the delay will violate *ASME A17.1/CSA B44*. This precludes the use of devices with built-in delay, commonly called retard,

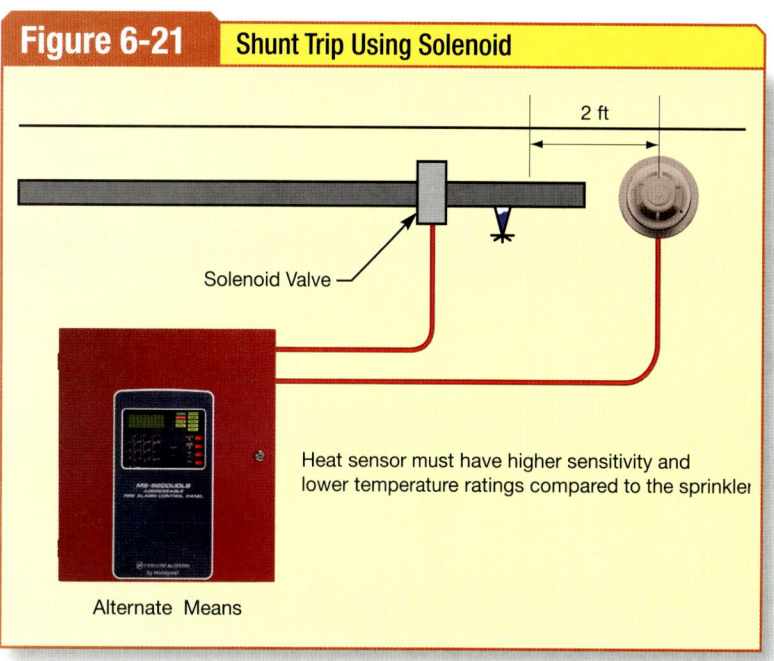

Figure 6-21. *Using a solenoid valve and heat detector is intended for use where the elevator cab would be returned to a landing before shunt trip. This allows occupants to escape, but adds complexity and cost.*

because it can be adjusted to provide a delay in signal transmission to the control unit. There are zero-retard devices available, but they may respond to surges in pressure, causing unwanted shunt trip operation.

It has been the practice of many designers to use heat detector contacts in the shunt trip circuit without connecting them to the fire alarm system. This arrangement avoids the use of a control unit, separate relay, or control module. However, this practice violates 21.4.2 of *NFPA 72* because all fire alarm system wiring must be monitored for integrity. Heat detectors, waterflow switches, or pressure switches must be connected to, and monitored by, the fire alarm system for this reason.

There have been cases where the shunt trip circuit did not have operating voltage, resulting in a failure to shut down the elevator after the flow of water. Both *NFPA 72* and *ASME A17.1/CSA B44* each require control circuits that cause main line power disconnection to be monitored for the

> **ASME A17.1-2010/CSA B44-10**
>
> **2.8.3.3.2** In jurisdictions not enforcing the NBCC, where elevator equipment is located or its enclosure is configured such that application of water from sprinklers could cause unsafe elevator operation, means shall be provided to automatically disconnect the main line power supply to the affected elevator and any other power supplies used to move the elevator upon or prior to the application of water.
>
> *(a)* This means shall be independent of the elevator control and shall not be self-resetting.
>
> *(b)* Heat detectors and sprinkler flow switches used to initiate main line elevator power shutdown shall comply with the requirements of NFPA 72.
>
> *(c)* The activation of sprinklers outside of such locations shall not disconnect the main line elevator power supply. See also 2.27.3.3.6.
>
> **2.8.3.3.3** Smoke detectors shall not be used to activate sprinklers in these spaces or to disconnect the main line power supply.
>
> **2.8.3.3.4** In jurisdictions not enforcing the NBCC, when sprinklers are installed not more than 600 mm (24 in.) above the pit floor, 2.8.3.3.4(a) and (b) apply to elevator electrical equipment and wiring in the hoistway located less than 1 200 mm (48 in.) above the pit floor, except earthquake protective devices conforming to 8.4.10.1.2(d); and on the exterior of the car at the point where the car platform sill and the lowest landing hoistway door sill are in vertical alignment.
>
> *(a)* Elevator electrical equipment shall be weatherproof (Type 4 as specified in NEMA 250).
>
> *(b)* Elevator wiring, except traveling cables, shall be identified for use in wet locations in accordance with the requirements in NFPA 70.

presence of operating voltage. Section 21.4.4 is intended to prevent such an occurrence. **See Figure 6-22.**

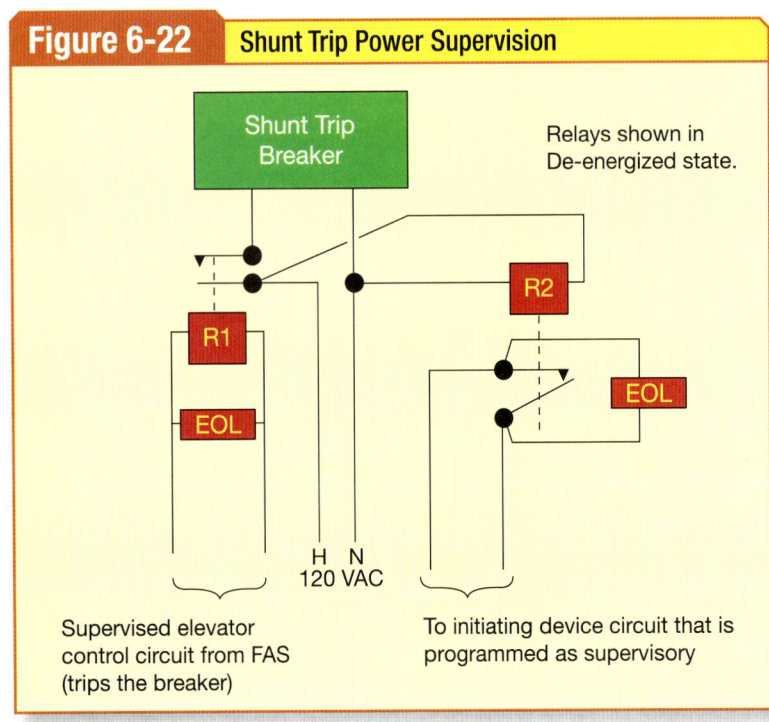

Figure 6-22. Elevator shunt trip power supervision will create a supervisory signal at the controls when shunt trip power is disconnected. Source: NFPA 72, A.21.4.4.

Fire Service Access Elevators. Some building codes now require elevators to be used for first responders during an emergency. Where elevators are specifically designated and marked for use by first responders, Section 21.5.1 requires several conditions to be monitored by the fire alarm system during this use. These conditions include the following:
- Availability of main and emergency power to operate elevators, elevator controllers, and machine room ventilation (if provided)
- Status of elevator cab position, direction of travel, position of landing doors, and whether the cabs are occupied
- Temperature and presence of smoke in associated elevator lobbies and machine rooms

Occupant Evacuation Elevators. NFPA 72 now permits the use of elevators by occupants during a fire emergency. Section 21.6 addresses these requirements. Specifically, these specially designated elevators must be marked for such use and must meet the requirements found in Section 21.5, Fire Service Access Elevators.

Smoke Control and HVAC Shutdown

Smoke control is often provided in assembly occupancies, covered malls, atria, transportation centers, windowless buildings, or other buildings with large occupant loads or large areas. Additionally, special occupancies such as detention, correctional, or health care utilize smoke control because it is impossible or difficult to move occupants. Smoke control systems are designed to provide a clear path of escape for occupants.

Smoke control equipment usually consists of motor-operated dampers and fans controlled by the fire alarm system. Smoke control might be designed to stop or start selected fans to create positive or negative pressures in the building. The pressure differences minimize smoke migration which threatens occupants and causes property damage. One of the most prevalent types of smoke control is stairway pressurization. Fans controlled by the fire alarm system pressurize exit stairways, which minimizes smoke entry into the stairway, thereby allowing occupant egress.

Smoke control is generally initiated by signals from fire alarm system area smoke detectors or duct smoke detectors. The detectors are generally installed throughout the area protected by the system. A signal from one of the smoke detectors typically triggers the fire alarm system programming to send a signal to control modules that actuate smoke control equipment.

During a fire, HVAC fans may spread smoke and toxic gases throughout the building. For this reason, Chapter 6 of *NFPA 90A, Standard for the Installation of Air-Conditioning and Ventilating Systems*, requires smoke detectors to shut down fans of sizes 2,000 cubic feet per minute (CFM) (944 liters per second) or greater. Smoke detection may be required on supply and return sides of the system, depending on size and design. Where fan shutdown is required, Section 21.7 of *NFPA 72* contains requirements for the interface of the two systems.

Section 21.7.2 requires initiating devices connected to the fire alarm system, which causes the operation of fans, dampers, smoke doors, and fire doors, to be monitored for integrity. These devices may or may not be powered by the fire alarm system, such as stand-alone 120 VAC powered devices, but they are required to be monitored for integrity. Section 21.7.4 requires duct smoke detectors when actuated to cause a supervisory signal. But Section 21.7.4.1 allows systems to initiate an alarm signal for a duct smoke detector actuation where the systems are not constantly attended or send signals to a supervising station. A constantly attended location refers to a guard's desk or other location where operational staff are always present. Because not all jurisdictions permit supervisory signals, Section 21.7.4.2 also allows duct smoke detectors to initiate an alarm where required by other codes and standards.

Door and Shutter Release

Door and shutter releasing service is covered by Section 21.8 and is designed to automatically release smoke control doors. The closure of the doors is intended to prevent the flow of smoke from one area of a building to another. Smoke doors are often held open by magnets, which are released when a system smoke detector in the area is actuated. Magnets are safer than wedges because they are automatically released in a fire scenario. Integrated detector/closer units or dedicated stand-alone detectors are sometimes used in place of system-powered area detectors. **See Figure 6-23.**

Where area smoke detectors are used for door releasing service, they must protect both sides of the door being

Figure 6-23. *A smoke detector actuation on either side of the door causes the magnet circuit to be de-energized, thereby releasing the doors.*

> **Fact**
>
> Some circuit breaker shunt trip mechanisms are operated by 24 VDC. If this operating power is supplied by the fire alarm system, then there is no need to supervise the power supply because the fire alarm system already monitors the power supply and provides a battery backup.

controlled. Section 21.8.2 also requires any initiating device used for door release service to be monitored for integrity by the fire alarm system, with the exception of dedicated stand-alone units for door release service. Additionally, Section 21.8.4 does not require a secondary power source for doors that operate (close) on loss of power.

The most common method of door release is accomplished through a set of relay contacts that provide power to magnets holding doors open. Magnet circuits may operate at 24 volts direct current (VDC) or 120 VAC, depending on manufacturer and model. The release occurs when the circuit is opened by the control relay de-energizing the magnet and releasing the door(s). The control relay can be an auxiliary relay on a smoke detector or a dedicated relay. In either event, the contacts must be rated for the applied voltage and current.

Door Unlocking

Door unlocking requirements were added to the *Code* following the MGM Grand Hotel fire in Las Vegas in 1980. In this fire, many occupants died when they were trapped in stairways that filled with smoke (79 people died in this fire). Stairway doors were locked for security purposes, trapping occupants in the exit stairways. The stairways were not pressurized and filled with toxic smoke from the fire on the lower floors.

Section 21.9.1 requires any device or system used to actuate exit door locking or unlocking devices on doors in the direction of egress to be connected to the fire alarm system serving the premises. Section 21.9.2 requires all exit doors in the direction of egress controlled by the control system to unlock in the direction of egress immediately upon receipt of a fire alarm signal. These requirements are intended to prevent occupants from being trapped. Security often trumps fire protection in the post 9-11 world. These provisions only require unlocking in the direction of egress.

Door locking systems typically operate with both primary and secondary power supplies. Loss of primary power must cause all doors to unlock because secondary power may also be lost, which keeps doors secured and which may trap occupants. Therefore, Section 21.9.3 does not permit secondary power to be used to maintain doors in a locked position, unless unlocked within 10 minutes of loss of primary power.

Section 21.9.5 requires all exit door unlocking to occur concurrently or before the actuation of all public mode notification appliances. Finally, Section 21.9.6 requires all doors that are controlled (and unlocked) by the fire alarm system to remain unlocked until the fire alarm system is manually reset.

Stairway Pressurization

Another "lesson learned" from the MGM Grand fire was the need for stairway pressurization. This scheme is designed to pressurize stairways so that smoke cannot enter the enclosure. This allows occupants to escape without becoming overcome by smoke and toxic gasses. Many building codes require stairway pressurization for high-rise buildings. High-rise buildings are generally defined by these codes as having occupiable floors more than 75 feet (23 m) above the lowest level of fire department access. Where required, the general requirements for emergency control function interfaces in Chapter 21 will apply.

Summary

Fire alarm and signaling systems monitor, supervise, and control other systems used to make buildings safer. These systems may include suppression systems and emergency control functions. Emergency control functions and interfaces make the building safer in the event of a fire. They prevent the circulation of smoke throughout a building, prevent occupants from being trapped in elevators, and prevent occupants from being trapped behind locked doors. Many of these functions were developed following tragic fires, such as the MGM Grand fire in 1980. Chapter 21 of *NFPA 72* now contains requirements for all emergency control function interfaces, and Chapter 23 contains many other requirements for interfaces. These requirements are among the most important in the *Code* and must be followed to ensure all system components correctly operate.

Review Questions

1. Which of the following standards applies to commercial fire alarm system controls?
    a. UL 684
    b. UL 862
    c. UL 864
    d. UL 846

2. The AHJ may require __?__ when information displayed on a combination system annunciator is excessive.
    a. a faster system processor
    b. a telephone at the control unit
    c. larger batteries
    d. two separate displays

3. Which of the following protocols is permitted to interconnect HVAC or other systems to a network shared by the fire alarm system?
    a. ABCNet
    b. ACBNet
    c. BACNet
    d. CABNet

Review Questions

4. Fire alarm control units that control a suppression system actuation must __?__.
    a. be listed for releasing service
    b. be located in the protected space
    c. have a supervising station connection
    d. have an emergency generator back-up power supply

5. The maximum delay (retard) on a waterflow switch is __?__ seconds.
    a. 15 seconds
    b. 30 seconds
    c. 60 seconds
    d. 90 seconds

6. Fire pump installations are covered by *NFPA* __?__.
    a. 12
    b. 17
    c. 20
    d. 25

7. How many fire alarm control outputs are required for an elevator with a sprinklered machine room and a non-sprinklered hoistway?
    a. 1
    b. 2
    c. 3
    d. 4

8. A control relay or control module must be located within __?__ of the controlled circuit or device.
    a. 1'
    b. 2'
    c. 3'
    d. 4'

9. The maximum time delay between the flow of water from a machine room or hoistway sprinkler and the elevator shunt trip must not exceed __?__ seconds.
    a. 3 seconds
    b. 2 seconds
    c. 1 second
    d. 0 seconds

10. *NFPA 90A* requires HVAC systems with a capacity of __?__ CFM to have supply-side smoke detection.
    a. 1,000 CFM
    b. 1,500 CFM
    c. 2,000 CFM
    d. 2,500 CFM

Advanced Detection Topics

All fires can be detected because all fires produce changes in the ambient environment. These changes include, but are not limited to, heat, light radiation, smoke particles, carbon monoxide (CO), and carbon dioxide (CO_2). Some plastics may also produce hydrogen chloride (HCl) and other toxic gases. Most of these by-products occur in a measurable quantity when combustible materials are burned. However, the ability to quickly detect the fire will depend on, among other variables, the quantity of the products, placement and quantity of detectors, and the type of detectors used.

Objectives

- Describe basic topics in fire science, including:
  - The fire triangle
  - The chemistry of fire
  - Products of combustion
  - Smoke detectors and standardized tests
- Explain the need for and uses of specialized detection devices, such as:
  - Smoke detectors in high air-movement areas
  - Radiant energy fire detectors
  - Video image smoke detectors
- Explain smoke control applications, such as:
  - Stair pressurization
  - Door releasing services

Chapter 7

Table of Contents

Basic Fire Science 158
    Fire Triangle 158
    Chemistry of Fire 158
    Products of Combustion 158
    Smoke Detectors and Standardized Tests 159

Advanced Detector Applications 163
    Detector Selection 163
    Suitable Ambient Conditions 164
    High Volume Air-Movement Areas 166
    Air-Sampling Smoke Detectors 168
    High Ceilings and Projected Beam Detectors 170
    Radiant Energy Fire Detectors 171
    Video Image Smoke Detection 173

Smoke Control Applications 174
    Heating, Ventilating, and Air-Conditioning Systems 175

Smoke Detectors for Door Release Service 179

Summary 181

Review Questions 181

BASIC FIRE SCIENCE

A basic understanding of fire science is necessary to better design and install fire alarm initiating devices. To better understand the best type of detection for a fast response, it is necessary to understand how fires burn. Fire, or combustion, is a self-sustaining chemical reaction where molecular bonds between chemical compounds (fuels) are broken. Material is thermally decomposed and reduced into other compounds. Initially, the bonds are broken when the materials are heated by a source, such as a match or electrical spark. The amount of energy to break the bonds will vary from one material to another. Wood, for example, requires much more energy to ignite than gasoline because the chemical bonds in wood are much stronger than those of gasoline.

Fire Triangle

In order for combustion to occur, there are three things that must be present: oxygen, fuel, and ignition source. This is known as the "fire triangle," and all three of these elements of the triangle must be present in order for combustion to occur. The word "fuel" refers to the combustible room contents, which include solid or liquid combustibles and flammables. **See Figure 7-1.**

Removal of one side of the triangle will prevent combustion from happening. This principle is the basis for almost all fire prevention programs in use today.

There have been many reported cases where fires have self-extinguished because they lacked proper oxygen to sustain combustion. This phenomenon is the basis for many gaseous suppression systems, such as carbon dioxide suppression systems.

Chemistry of Fire

Oxidation is the type of chemical reaction involved with all fires. Many combustible materials, such as kerosene, gasoline, propane, and methane, have hydrogen-hydrogen bonds as part of their chemical makeup. Hydrogen-hydrogen bonds are weaker than hydrogen-carbon bonds and they are more easily broken by heating. Materials like wood and paper have hydrogen-carbon bonds, which require more energy to break. Hydrogen-carbon bonds are very strong but readily break when sufficiently heated, thereby releasing the contained energy. The hydrogen-oxygen and carbon-oxygen bonds are more stable and are the result of the combustion process. Therefore, combustion tends to create simpler stable compounds like water (H_2O), carbon monoxide (CO), and carbon dioxide (CO_2).

The breaking of chemical bonds releases energy, which appears in the visible spectrum as flames or sparks. It also creates other products of combustion, such as heat. Heat can be radiated in the visible or invisible spectrum. Infrared light is invisible but produces heat. Heat can be transferred by radiation, conduction, or convection.

Products of Combustion

All fires create products of combustion that can be detected. However, not all combustion creates flames. A condition called pyrolysis exists where a material is heated to the point where the material is ejected from the surface as smoke particles, but flames are not present. Pyrolysis often happens just before flaming occurs. An example of this is an overloaded conductor that smokes because of the heat produced by excessive current flow. Flames are not present, but smoke is created.

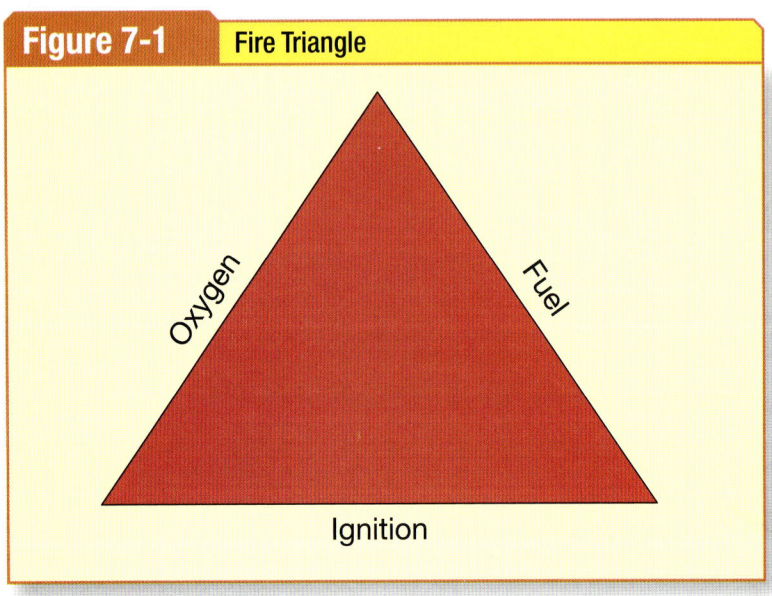

Figure 7-1. Combustion requires all three of these elements in order to occur.

The type and amount of combustion product will be determined by the fuel type and oxygen available. Most fires produce heat, radiation (light), and particles of combustion (smoke). It is these products of combustion that the fire alarm system is designed to detect.

Heat is produced when the materials chemically (exothermically) react to form other chemical compounds. The chemical reaction generates heat as the complex compounds of the fuels are broken down and formed into simpler compounds. Heat generally is transferred to the air in the space being protected, which increases the temperature in the space. This heat release contributes to the consumption of more material, which results in a sustaining chain reaction.

The combustion process also produces radiated energy. Radiated energy from fires can be visible, infrared, or ultraviolet. The wavelengths produced by the fire are dependent on the fuel being burned. Radiant energy is measurable, provided that the detector can see the signature and that it is designed to detect that wavelength.

Fires may exist in the flaming state, but they can also smolder for prolonged periods without flaming. Burning sparks and embers may not actually produce flames, but they emit radiated light energy, which can be detected. The involved fuel(s), oxygen levels, room geometry, fuel configuration, and other variables will determine the type of combustion (fire). Some fires produce significant quantities of heat, with little or no smoke. This may be common in fires involving flammable liquids such as alcohol. However, many fires do produce smoke in a measurable quantity. This is because there may be other fuels burning and producing smoke in the protected space.

Smoke particles may be large or small, depending on the type of fuel and the oxygen content. Visible smoke particles usually range from about 0.01 microns (10^{-8} m or 0.4 millionths of an inch) to about 5,000 microns (5×10^{-3} m or 0.2 inch) in width, depending on the materials and type of combustion. By comparison, dust particles tend to be approximately 1 micron to 1,000 microns in size.

Flaming fires tend to produce smaller smoke particles (about 1.5 microns or smaller) because the energy from the fire creates sufficient energy to break the chemical bonds and produce smaller compounds. However, "cold" fires (smoldering fires) and pyrolysis tend to produce larger particles (about two microns or larger) because less energy is delivered to the smoke particles. Additionally, smoke particles tend to combine as they move farther from a fire. This process is called agglomeration, and it is commonly found in smoke that has been transported away from a fire.

Smoke Detectors and Standardized Tests

Particle size can play an important part in the detection process. Ionization-type smoke detectors are well suited to detecting particles one micron and smaller. This makes them slightly better for detection of flaming fires. Photoelectric-type smoke detectors are slightly better for the detection of larger particles (0.3 micron and larger), which makes them better for detection of smoldering fires, aged smoke, or transported smoke.

Smoke density is measured in units known as percent per foot obscuration (%/foot obscuration). Testing laboratories use a standard testing tool for smoke obscuration when evaluating smoke detectors, and detector sensitivity is measured in the same units.

Some smoke detector test devices use a light source and photocell receiver placed five feet apart. Smoke in the test room or smoke box is measured using the photocell test device, and the level of obscuration is compared to the detector undergoing the test. Most commercial smoke detectors are capable of detecting smoke quantities between 0.5%/foot obscuration and 4%/foot obscuration. Most modern analog addressable detectors can send the actual quantity of smoke being sensed by the detector.

Underwriters Laboratories conducts tests to ensure smoke detectors and smoke alarms meet certain criteria. Two product standards provide requirements for smoke alarms and detectors. UL217 provides requirements for smoke alarms

Fact

Most fires will produce smoke in measurable quantities, which can be detected. However, flammable liquid fires may initially produce more heat than smoke. Detection must be matched to detect the "signature" as quickly as possible, while not producing a nuisance alarm.

Fact

Actual fires usually produce ample quantities of black smoke because fuels found in buildings today contain a large amount of plastic. Product standards only publish test results of the light gray smoke test used to determine the relative sensitivity of a detector.

and UL268 provides requirements for system-powered smoke detectors. UL tests involve the use of a smoke box set up for a laminar flow of smoke from 30 feet/min to 150 feet/min (9.1 m/min to 45.7 m/min). Smoke is produced by a one-eighth-inch diameter cotton wick, which produces a light gray smoke in the chamber. The wick is kept at 10% relative humidity and a temperature of 113°F (45°C).

Each detector subjected to the test must be energized for at least 16 hours prior to tests. No fewer than a dozen detectors from a lot are used during tests of a particular model. In order to become listed, detectors must actuate at not less than 0.5% and before reaching 4.4% obscuration of light gray smoke. For the black smoke test, the wick is soaked in kerosene and detectors must respond between 0.5% and 12.9% obscuration. Detectors provided with sensitivity test features (electrical or mechanical) must respond to a threshold not exceeding 6% obscuration of light gray smoke. Smoke in the test box is measured by a measuring ionization chamber (MIC), which is very sensitive and acts as the benchmark for the test.

UL also uses a full-scale test room for smoke detector tests. This room is 36 feet (11 m) by 22 feet (6.7 m) with a 10-foot- (3 m) high smooth ceiling. The detectors are placed 17.7 feet (5.4 m) from the test fire on a diagonal. Several photocell units and a MIC are used to measure the smoke in the room, and they are located between the detectors under test and along the walls near the detectors under test. **See Figure 7-2.**

During the full-scale fire tests, several different types of fires are lit and the detectors under tests must respond within prescribed time limits in order to be listed. The tests include a shredded newspaper test, a wood fire test (using dried fir strips), a gasoline (30 ml) fire test, and a polystyrene (29 g) test. Responses to each fire scenario are as follows:

Test	Minutes from Ignition
Paper	4
Wood	4
Gasoline	3
Polystyrene	2

Figure 7-2 — UL Smoke Detector Test Setup

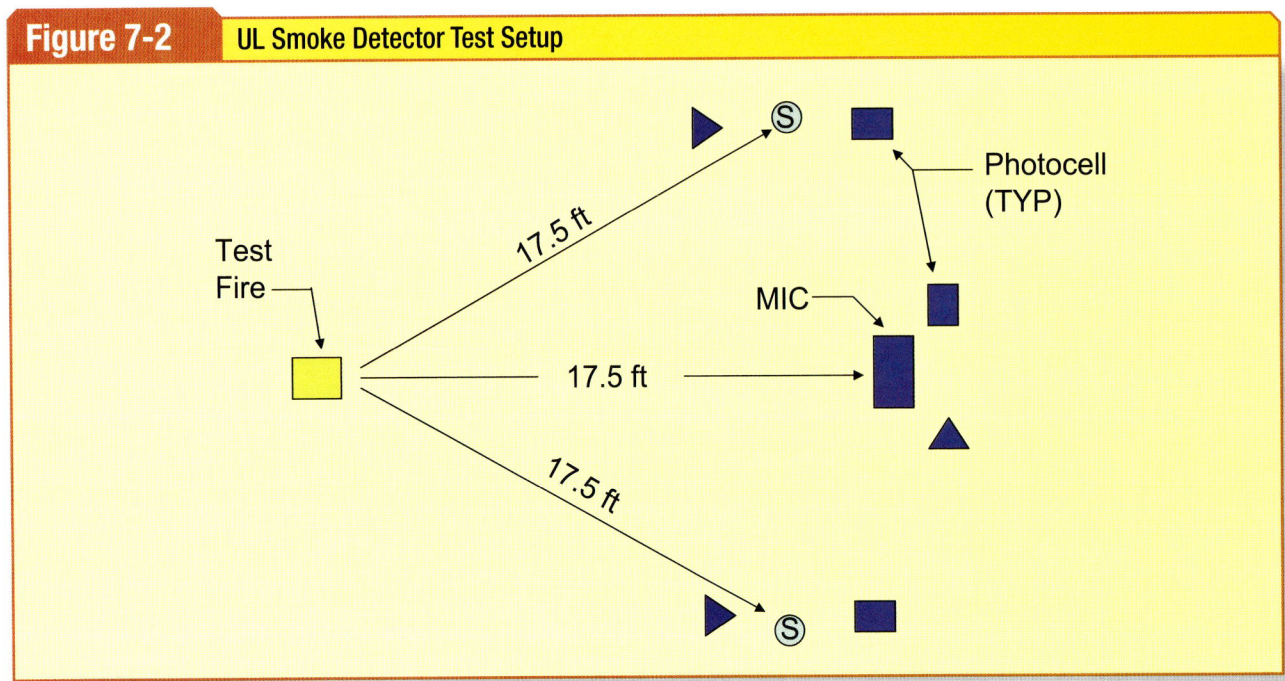

Figure 7-2. Smoke detectors are subjected to a series of tests, but have no listed spacing. NFPA 72 permits 30 feet (9.1 m) as a guide for detector spacing.

The UL test room is also fitted with vents and fans to ensure the smoke is drawn towards the detectors under test.

Other tests conducted by UL include a dynamic load test and a wind stack test. The dynamic load test is a simple test where the test subject is energized for a period of not less than 30 days. Units that fail to operate during this time are not listed. The stack effect test is designed to ensure that pressure from the conduit system does not prevent a response by the detector. A small box is used, and a pressure is applied to simulate a downdraft. The detector must respond within two minutes to a burning cotton wick inside the box.

Underwriters Laboratories also tests heat detectors to determine their listed spacing. A room of slightly larger proportions is used for heat detector testing, but the ceiling height is 15 feet, 9 inches (5.1 m). In the center are four 165°F (74°C) sprinklers, spaced at 10 feet (3 m). The heat detectors under test are spaced diagonally at 10-foot (3 m) intervals. A pan fire is located in the center of the room. **See Figure 7-3.**

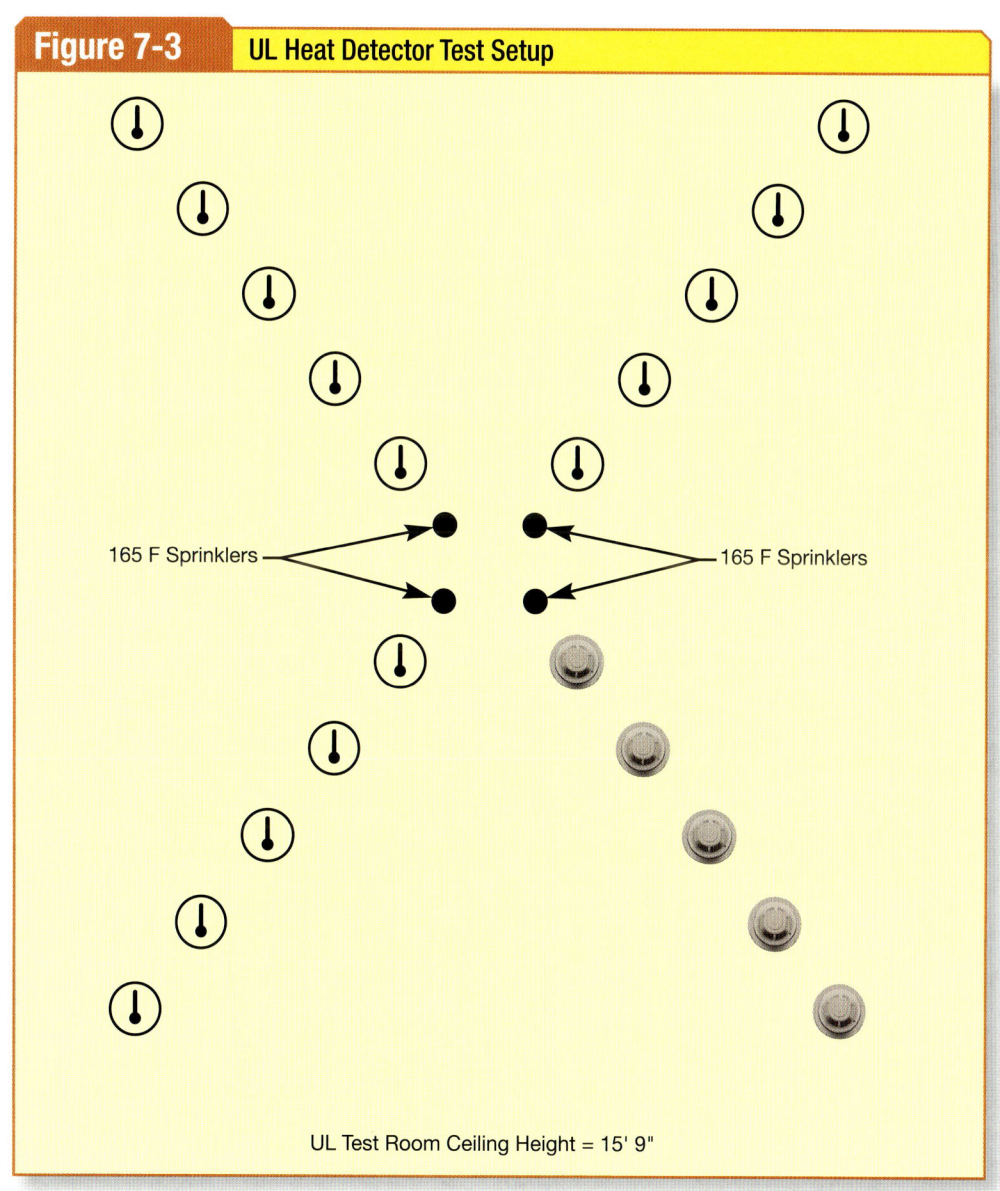

Figure 7-3. *Heat detectors are subjected to a series of tests to determine their listed spacing.*

The pan fire, consisting of alcohol and Heptane, is adjusted to cause a sprinkler actuation within two minutes +/- 10 seconds. The last heat detector to actuate before a sprinkler will determine the listed spacing. Because of room size limitations, UL can test detectors with listed spacings up to 50 feet (16 m). However, Underwriters Laboratories Canada (ULC) can test heat detectors with listed spacings up to 75 feet (24.2 m) because the test room is larger. Equipment listed by ULC is often suitable for use in the United States because ULC is owned by UL and many devices are cross listed for use in the United States.

The relative quantity of smoke detected by smoke detectors is small. A light haze at the ceiling is usually sufficient to actuate a detector. However, the type and color of smoke may play an active part in determining when the detector actuates. Photoelectric-type smoke detectors rely on reflected light to cause an actuation. Larger particles help reflect more light, resulting in a faster actuation of photoelectric-type smoke detectors.

Ionization-type smoke detectors will respond faster to small particles (less than one micron in width) produced by flaming fires. Projected beam-type detectors utilize a light-blocking principle, so darker smoke will result in a faster actuation. However, total blockage of the beam will almost always result in a trouble signal.

Fires in large spaces will produce a plume that rises in a conical shape above the fire. The plume expands as it rises. The smoke and other products of combustion rise because the heat from the fire makes them buoyant compared to the surrounding air. It is usually the plume that transports the products of combustion to the detectors. However, the plume entrains, or mixes with, colder air as it rises. **See Figure 7-4.**

Smoke and products of combustion tend to cool and expand as they move farther from the fire. If the plume temperature cools to match the ambient temperature, the plume will cease to rise. This phenomenon is known as "stratification."

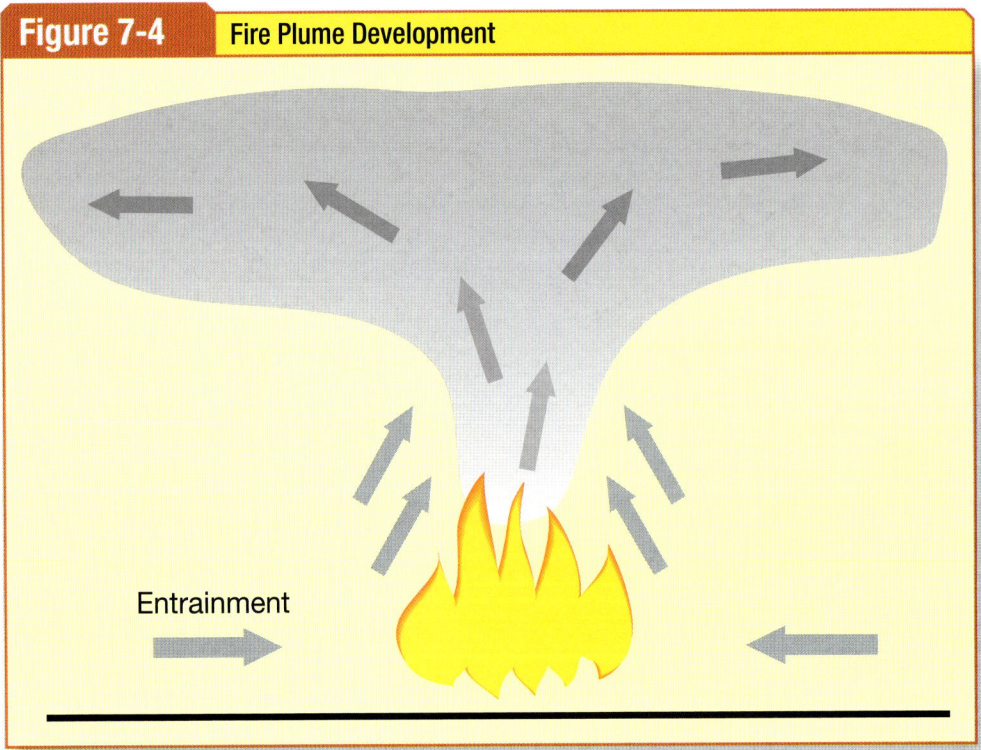

Figure 7-4. As the hot products of combustion rise from the fire, they expand and pull in (entrain) cooler air from the surrounding area.

Stratification is often observed on a cold, calm day, when smoke from a chimney rises somewhat before spreading out or stratifying. **See Figure 7-5.**

Stratification usually occurs in large open spaces, like an atrium. Understanding this concept is important because products of combustion may never reach the ceiling (and hence, the detectors) when stratification occurs. The exact behavior of a fire plume will depend on many variables, such as the size and growth of the fire, room size, ambient temperature, and fuel loads. Stratification is not the only problem with which fire alarm system designers must contend. Building ventilating systems or building construction also contributes to smoke dispersion. For example, special structures such as towers can create a stack effect inside the building that can move smoke far from a fire.

ADVANCED DETECTOR APPLICATIONS

Automatic detectors are currently available to detect smoke, heat, and light radiation. They also have been developed to detect gases like carbon monoxide (CO), carbon dioxide (CO_2), and hydrogen chloride (HCl). However, CO, CO_2, and HCl detectors are not yet widely used for fire detection. Additionally, CO detectors for fire detection should not be confused with CO detectors used to prevent poisoning in dwelling units. In addition to these metrics, some manufacturers are currently producing detectors that sense infrared light. Sensing these different products of combustion helps eliminate nuisance alarms caused by false stimuli.

Detector Selection

The proper selection and installation of detectors will greatly improve the chances of detecting a fire while there is still time to react. An additional concern is the prevention of nuisance alarms. Improper placement of smoke detectors is the greatest cause of nuisance alarms. Frequent nuisance alarms in a building tend to create a "cry wolf" syndrome among occupants. Improper selection or installation of detectors can lead to nuisance alarms or non-operation of the detectors. Occupants must have a high degree of confidence in the system, so proper selection, design, and installation of system detectors is essential.

Time is a critical factor in detection system design. Most flaming fires will double their output approximately every two minutes. The time required for a detector to detect its particular signature will vary greatly depending on fire growth rates, location and spacing of detectors, ventilation, ceiling height, room size, and temperature. A fast response may be desirable for certain applications, like actuation of a fire suppression system. Radiant energy fire detectors, such as flame and spark/ember detectors, have extremely fast actuations (usually measured in milliseconds). Radiant energy detectors are

Figure 7-5. Stratification occurs when the plume reaches the same temperature as the surrounding air. Stratification can prevent or delay detection if it occurs below the detection devices. In this graphic, the plume has stratified three times during the fire growth.

generally used in conjunction with suppression systems for this reason.

Smoke is almost always present in measurable quantities during the development of a fire scenario. Smoke in small quantities is an irritant. In higher concentrations, it can be toxic. Most fire deaths are caused by smoke inhalation rather than heat exposure. For these reasons, detection system design in most applications begins with smoke detectors. Many building codes specify the locations and type of detection required, and some building codes require smoke detection as the primary means of detection.

Suitable Ambient Conditions

However, smoke detectors are not always appropriate for the ambient environment or the type of fire being detected. *NFPA 72* contains warnings about the ambient environment and tries to steer the user from making improper choices. These sections of the *Code* are often ignored, which leads to future problems for the owner and contractor.

Section 17.7.1.7 requires the selection and placement of smoke detectors to take into account the performance characteristics of the detector and the location where it will be installed. The intent of this section is to prevent nuisance alarms caused by application of smoke detectors.

Section 17.7.1.8 of *NFPA 72* does not permit smoke detectors to be installed where any of the following conditions exist:

1. Temperatures below 32°F (0°C)
2. Temperatures above 100°F (38°C)
3. Relative humidity exceeding 93%
4. Air velocity exceeding 300 feet/minute (1.5 m/second)

Smoke detectors used in conditions exceeding these limits will cause nuisance alarms, or they may not actuate in a fire. In areas where these conditions exist, alternate types of detection are more appropriate and must be used. An alternate means of detection in these areas generally includes heat detection.

Areas not suited to smoke detection include outdoor locations, walk-in freezers, kitchens, bars and lounges where smoking is permitted, parking garages, attics, wet basements, rooms containing swimming pools or spas, and high air-movement areas. Either line-type or spot-type heat detection may be used, depending on the application. Line-type heat detectors are often used to protect conveyors, wireways, and cable trays because they protect long rectangular areas.

There are smoke detectors listed for ambient conditions outside the limits of Section 17.7.1.8. These specifically listed smoke detectors are designed for harsh environments and can be used in temperatures up to 120°F (49°C) in higher humidity. By contrast, there are smoke detectors available and listed for cleanroom-type applications. Smoke detectors designed for cleanrooms must only be used where dust and moisture are strictly controlled and should not be used in other applications, such as office areas.

Section 17.7.1.9 requires an evaluation of potential ambient sources of smoke, moisture, dust, fumes, or electrical influence in order to minimize nuisance alarms. For instance, ionization detectors in altitudes greater than 3,000 feet (1,000 m) may result in nuisance alarms because of their technology. Ionization-type smoke detectors contain a radioactive source that ionizes the air in the sensing chamber. There is less oxygen at high altitudes, which means that there are fewer ionized air particles to conduct the sensing chamber current. Therefore, installation of ionization detectors in altitudes greater than 3,000 feet (1,000 m) may result in nuisance alarms. Other factors that may result in nuisance alarms include humidity, diesel exhaust, aerosols, chemical fumes, and dust. Additionally, it is impossible for smoke detectors to distinguish between "friendly" smoke (from cigarettes in a nightclub, for example) and "hostile" smoke (from a real fire). Section A.17.7.1.9 provides a list of potential causes of nuisance alarms for smoke detectors.

Section 17.7.3.1.2 requires the designer to consider and account for factors which will affect the detector response. These factors include, but are not limited to the following:

1. Ceiling shape and surface
2. Ceiling height

Fact

Smoke detectors installed in freezing temperatures often freeze and "go to sleep." Detectors in high heat will often cause nuisance alarms because electronics fail at high temperatures.

3. Configuration of fuels (contents) in the protected area
4. Ventilation
5. Ambient temperature
6. Ambient humidity
7. Atmospheric pressure
8. Altitude

These considerations must not be dismissed without carefully understanding the consequences of placing detectors in questionable conditions. All of these factors will either cause nuisance alarms or will delay the detector response. Neither of these conditions is desirable. The *Code* requires the user to consider factors that result in nuisance alarms and non-operation. Section 17.7.3.1.2 does not definitively require the user to do anything beyond consideration of the conditions. Prudent designers will heed this advice, but foolhardy designers will ignore these warnings. Great care must be exercised when designing systems in buildings with unique characteristics such as those covered by Section 17.7.3.1.2.

When smoke detectors are inappropriate for an application, the *Code* may permit other types of detection (such as heat detection) instead. Heat detection is inherently slower than smoke detection and may not be appropriate for some applications, such as when a smoldering fire is expected. Local building codes may require smoke detection for certain occupancies or applications. These requirements usually do not conflict with *NFPA 72*.

Underfloor spaces that are not used as a plenum or air-handling space are sometimes protected by smoke detectors. Where smoke detectors are installed under raised floors, they must be oriented according to Figure A.17.7.3.2.2 of *NFPA 72*. **See Figure 7-6.**

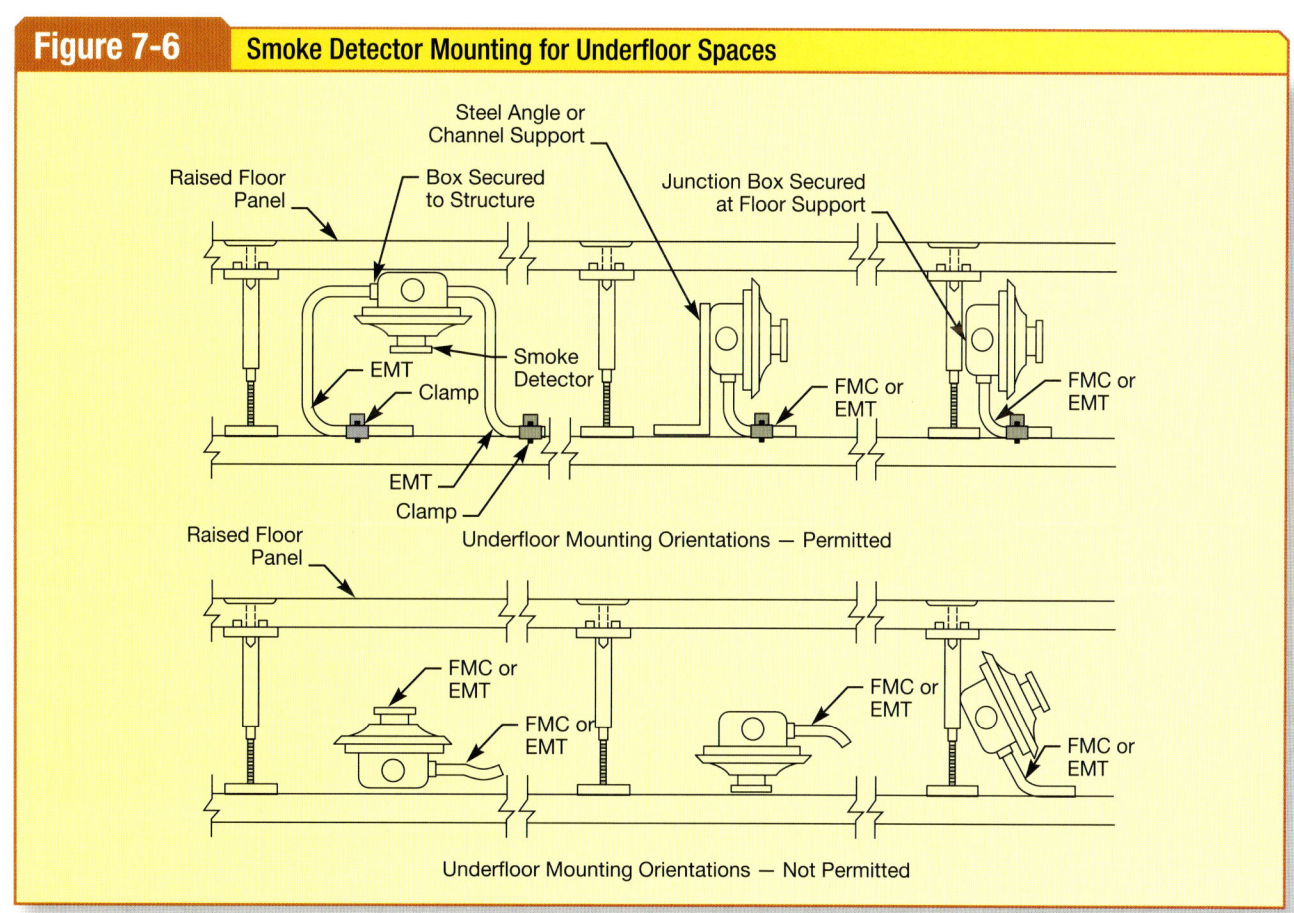

Figure 7-6 Smoke Detector Mounting for Underfloor Spaces

Figure 7-6. *Where smoke detectors are installed under raised floors, they must be oriented according to Figure A.17.7.3.2.2 of NFPA 72.*

Placing detectors with the sensing chamber facing up will result in damage or nuisance alarms caused by dust and dirt from the floor above the detector. Additionally, Section A.17.7.4.1 does not recommend smoke detector placement in direct air streams or within three feet (1 m) of a supply diffuser. Annex materials are not enforceable as *Code*, however.

Section 10.4.4 of *NFPA 72* also requires each fire alarm control unit to be protected by a smoke detector. However, the first exception permits use of a heat detector where conditions are not appropriate for smoke detection. It must be noted that conditions inappropriate for smoke detection are usually not appropriate for control units. A good remedy for this problem is to relocate the controls to a suitable location or provide air conditioning in the space. A second exception was removed from the 2010 edition of the *Code*, and once again use of a smoke or heat detector to protect control units is required even in fully sprinklered buildings.

> **10.4.4*** In areas that are not continuously occupied, automatic smoke detection shall be provided at the location of each fire alarm control unit(s), notification appliance circuit power extenders, and supervising station transmitting equipment to provide notification of fire at that location.
>
> *Exception: Where ambient conditions prohibit installation of automatic smoke detection, automatic heat detection shall be permitted.*
>
> (Excerpt from NFPA 72.)

Control units, by definition, include any system elements that monitor inputs and control outputs. Section 10.4.4 specifically applies to control units, notification appliance circuit power booster supplies, transmitters, and other critical elements. However, monitor modules, control modules, and relays are not typically considered as control units and do not require protection under this section.

A.10.4.4 provides guidance for protection of control equipment in situations where a high ceiling exists. Specifically, A.10.4.4 states: "Where total smoke detection coverage is not provided, the Code intends that only one smoke detector is required at the control unit even when the area of the room would require more than one detector if installed according to the spacing rules in Chapter 17. The intent of selective coverage is to address the specific location of the equipment."

Detectors are frequently used to actuate suppression systems. Suppression system actuation must be fast; otherwise, the fire can grow to a size that cannot be controlled. Again, the first line of defense may be smoke detection, but this may not be enough. Smoke detection may have an excessive time delay or may be inappropriate for the application.

An example of an inappropriate location for a smoke detector is a paint spray booth. In these, a radiant energy-sensing fire detector might be the better choice. Radiant energy-sensing fire detectors may provide a fast response. Unfortunately, they may also be susceptible to sunlight or other sources of radiated energy because they are designed to detect those frequencies.

It is necessary to select the most appropriate type of detection to achieve the fastest response while preventing nuisance alarms. Understanding the type of fire expected and the environment will help the user create a system that meets these goals. A balance of response time, nuisance alarm prevention, and cost is generally achieved in good design. Most suppression systems that use detection are engineered systems covered by other standards and are not part of the fire alarm system. For example, clean agent suppression systems, like Inergen© are covered by *NFPA 2001, Standard on Clean Agent Fire Extinguishing Systems*.

High Volume Air-Movement Areas

High volume air-movement areas, such as computer rooms protected with large volume air-conditioning equipment, require special treatment to ensure smoke detectors cause an alarm. Large air exchange rates can dilute smoke, making it very difficult to detect using spot-type smoke detectors. Areas with more than eight air changes per hour are generally considered high volume air-movement areas.

High volume air-movement areas require an adjustment of detector spacing to ensure a timely response. Table 17.7.6.3.3.2 and Figure 17.7.6.3.3.2 of *NFPA 72* contain the reduced coverage requirements for smoke detectors used in high volume air-movement areas. However, Table 17.7.6.3.3.2 and Figure 17.7.6.3.3.2 are not permitted for above-ceiling or under-floor spaces. The reduced area of coverage is provided in square feet or square meters, which is the only time smoke detector coverage is given in terms of area. The square root of the reduced area of coverage is the horizontal spacing between detectors. **See Figure 7-7.**

In order to properly evaluate the spacing, the number of air changes per hour or minutes per air change must be known. These values can be calculated once the room volume and the volume of air supplied per minute are known. The room volume is calculated by multiplying the length by width by height. The total flow of air into the room must also be determined by adding the flow from each diffuser. The changes per hour for a given room can be calculated by the following formula:

Figure 17.7.6.3.3.2 in *NFPA 72* shows that the required spacing of spot-type detectors for 34.3 air changes per hour

---

$$\text{Air changes per hour} = \frac{\frac{60 \text{ min}}{\text{hr}}}{\frac{\text{Room volume CF}}{\text{Total diffuser flow CFM}}}$$

For example, if a rectangular room with a 10-foot-high ceiling, 50 feet long by 35 feet wide has 20 ceiling diffusers, each with a rated capacity of 500 cubic feet per minute, air changes would be calculated as follows:

Room Volume $= H \times L \times W$

$= 10 \times 50 \times 35$

$= 17{,}500$ cubic feet

Flow of air into the room $=$ (Number of diffusers) $\times$ (flow from each diffuser)

$= 20 \times 500$

$= 10{,}000$ cubic feet/minute

$$\text{Air changes per hour} = \frac{\frac{60 \text{ min}}{\text{hr}}}{\frac{\text{Room volume CF}}{\text{Total diffuser flow CFM}}}$$

$$= \frac{\frac{60 \text{ min}}{\text{hr}}}{\frac{17{,}500 \text{ cubic ft}}{10{,}000 \text{ CFM}}}$$

$= 34.3$ air changes/hr

Figure 7-7 — High Air-Movement Areas

Minutes per Air Change	Air Changes per Hour	Spacing per Detector ft²	m²
1	60	125	12
2	30	250	23
3	20	375	35
4	15	500	46
5	12	625	58
6	10	750	70
7	8.6	875	81
8	7.5	900	84
9	6.7	900	84
10	6	900	84

Figure 7-7. Table 17.7.6.3.3.2 of NFPA 72 contains the reduced coverage requirements for smoke detectors used in high-volume air-movement areas.

Reprinted with permission from NFPA 72®-2013, *National Fire Alarm and Signaling Code*, Copyright © 2012, National Fire Protection Association, Quincy, MA. The information in this table is intended to be used in conjunction with the requirements of this code. It is not the complete and official position of the NFPA on the referenced subject, which is represented only by the standard in its entirety.

is approximately 230 square feet. Since detector spacing is always a linear measurement, the final step of the calculation is to take the square root of 230, which results in a maximum linear spacing of 15.17 feet. **See Figure 7-8.** Note that Figures 7-7 and 7-8 contain the same data.

Air-Sampling Smoke Detectors

Spot-type detection is not appropriate for underfloor or above-ceiling high-volume air-movement areas. These areas are best suited to air-sampling-type smoke detection or projected-beam smoke detectors. Air-sampling smoke detectors generally use a network of pipes throughout the protected area that are connected to a detector. The pipes have small, one-eighth-inch (3.2 mm) holes spaced along the piping network to sample the air in the protected space. Each hole, or port, is considered a detector.

An air-sampling detector contains a fan which pulls air samples into the detector. Older models of air-sampling smoke detectors used a Xenon strobe and a light-obscuration principle. Newer models use a laser to count smoke particles. **See Figure 7-9.**

Air-sampling smoke detectors look somewhat like a control unit. They also have their own primary and secondary power supplies. However, these units are smoke detectors and are not considered control units. A typical arrangement is to send back three or more signals to the control unit: alert, take action, and alarm. An additional signal to actuate suppression systems is available on some models. Additionally, a trouble signal must be sent to the system controls. Relays (dry contacts) inside the air-sampling smoke detector are used to send these signals over initiating device circuits.

Air-sampling smoke detectors are incredibly sensitive, and they are generally not used to provide area protection for

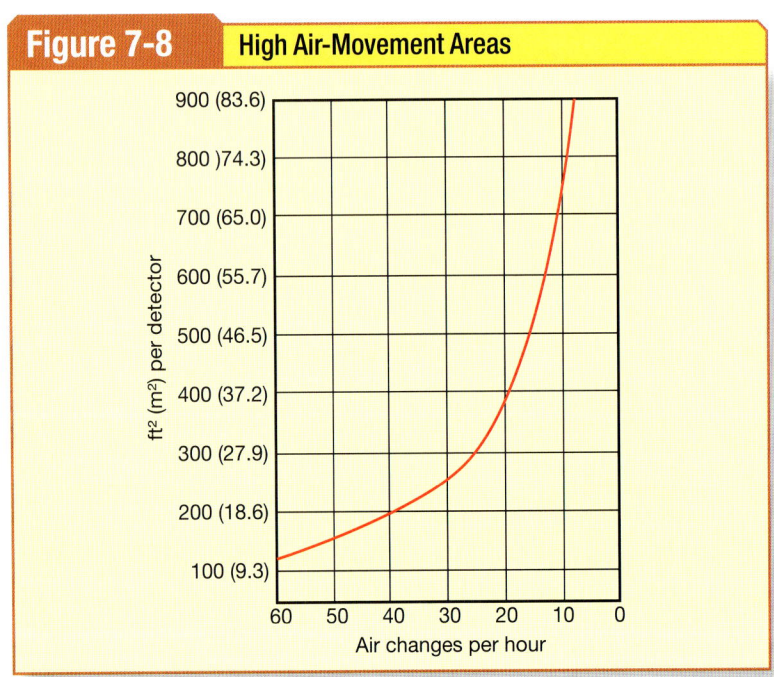

Figure 7-8. The required spacing of spot-type detectors for 34.3 air changes per hour is approximately 230 square feet.

> **Fact**
>
> Air-sampling smoke detectors may be thousands of times more sensitive than spot-type smoke detectors. They are well suited to high air-movement areas, clean rooms, and computer room applications.

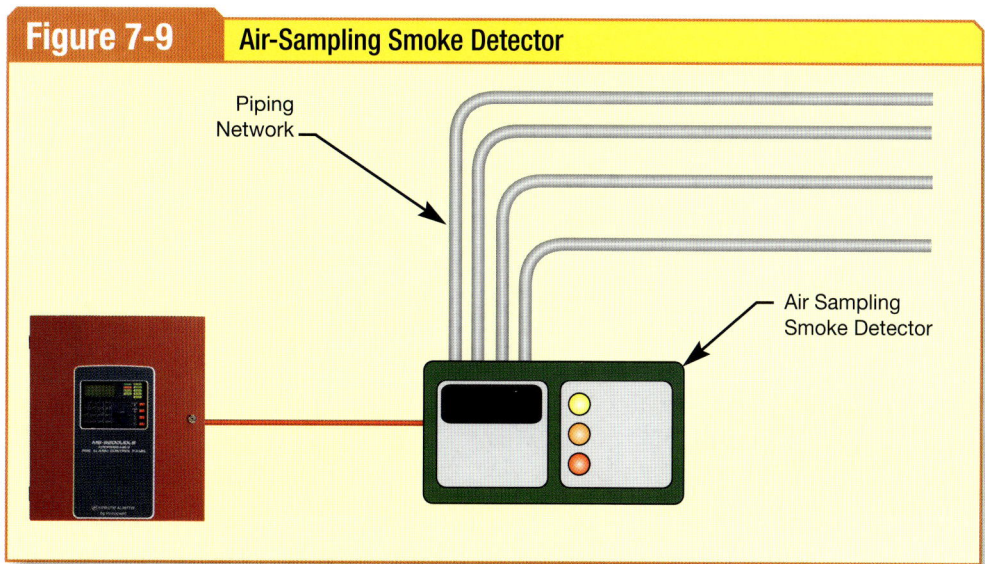

Figure 7-9. The detector is located in or near the protected area, and sampling tubes extend to provide coverage. A fan pulls in samples from the piping network.

For additional information, visit qr.njatcdb.org Item #1015

common areas. However, air-sampling smoke detectors are frequently used to protect high-cost assets, such as computers, data processing centers, or other areas containing sensitive equipment. For the purposes of location and spacing, 17.7.3.6.1 requires each port on the sample-piping network to be treated as a spot-type smoke detector. Most manufacturers of air-sampling smoke detectors provide guidance on the spacing of sampling ports.

Section 17.7.3.6.2 requires the air-sampling smoke detector piping network to be designed so the maximum transport time for air samples does not exceed 120 seconds. Section 17.7.3.6.3 also requires the piping network design to be supported by fluid dynamic principles. The manufacturer's representative or distributor will assist in the computations to ensure these criteria are satisfied.

Some manufacturers use CPVC piping, while others use electrical metallic conduit (EMT). Generally, CPVC piping is the same type used for residential sprinkler systems (orange-colored pipe), and it is approved for use in plenums. Generally, Section 17.7.3.6.7 requires the piping network to be permanently affixed. Additionally, Section 17.7.3.6.8 requires sampling piping to be identified at branches in the piping, changes in direction of a pipe, changes on both sides of penetrated walls or barriers, and at intervals not more than 20 feet (6 m). Labeling must be conspicuous and read as follows:

SMOKE DETECTOR SAMPLING TUBE – DO NOT DISTURB

Labels are available for this purpose.

Air-sampling smoke detectors can sense smoke at levels less than 0.001%/foot obscuration. At this concentration, humans cannot sense smoke. By contrast, most spot-type smoke detectors are designed to detect smoke between 0.5%/foot and 4%/foot obscuration. Air-sampling smoke detectors are well suited to high volume air-movement areas where smoke is diluted by the air-handling system because they actively sample air from the protected area and because they are so sensitive. Spot-type smoke detectors are passive and must wait for smoke to be sufficiently dense to cause an alarm.

Air-sampling smoke detectors are also suited to protection where very early warning is needed to protect high-cost assets like computers or sensitive equipment. Smoke particles and temperatures over 120°F (38°C) can damage computer storage devices. Because air-sampling

Figure 7-10 Projected Beam Smoke Detector

Figure 7-10. *A mirror or reflector is placed on the opposite wall to reflect the beam back to the unit.*

fire continues to develop and grow, the plume will usually reach the detector(s), but system goals may not be satisfied.

A common solution to this problem is to use one or more projected beam smoke detectors. Projected beam detectors use an infrared (invisible) beam between a transmitter and a receiver. Some models use a specially designed reflector to reflect the beam back to a combination transmitter/receiver. **See Figure 7-10.**

The shaded plastic on this photograph is a test screen for detector sensitivity and the reflector is shown in the foreground of the photograph. These detectors operate on an obscuration principle, so that smoke obstructs the light beam and causes an alarm. Total blockages, however, usually result in a trouble signal instead of an alarm. **See Figure 7-11.**

Most projected beam detectors cover a 30-foot (10 m) wide by 330-foot (110 m) long area. However, some models cover 60-foot (30 m) wide areas of the same length. Section 17.7.3.7.1 of *NFPA 72* requires the installation of projected beam detectors to be in conformance with the manufacturer's instructions. Most manufacturers require beams to be installed parallel to the ceiling and within 18 inches (450 mm) of the ceiling. However, very high ceilings may require multiple levels of beams in order to combat stratification. Very large areas or buildings protected with a smoke control system may require specially engineered systems.

Projected beam smoke detectors must be mounted on stable surfaces, as required by Section 17.7.3.7.6. There have been many cases of nuisance alarms caused by transmitters and receivers mounted on surfaces that move, such as a mullion in a large atrium. However, newer technology uses a wider transmitter beam, which is less likely to cause misalignment when moved slightly.

detectors are so sensitive, they can provide multiple levels of warning prior to equipment damage. The early-warning levels are often used to provide an investigation phase and are not generally used to cause a building-wide alarm. Building operations and maintenance staff members must be properly trained to investigate the signals to prevent misunderstandings.

High Ceilings and Projected Beam Detectors

High ceilings can create challenges for smoke-detection systems. High ceilings are often found in atria, warehouses, transportation centers, manufacturing and industrial applications, places of worship, and sporting venues. Smoke can stratify far below the detectors in high-ceiling applications, thereby preventing detection of the signature. If the

 *Fact*

Very large spaces like atria or gymnasiums present significant issues with the design of projected beam detection systems. In some cases, a study of air-movement patterns in the protected area must be conducted.

Fire modeling may be the fastest way to prove the optimal location and quantity of projected beam detectors. Most fire protection engineers are capable of conducting such studies.

Projected beams are generally installed parallel with the ceiling to ensure a larger area of the plume is detected. A vertical orientation of projected beam smoke detector is not usually recommended because a smaller section of the plume will be detected. **See Figure 7-12.**

The detector beam must be kept clear of obstructions like trees, signs, balloons, and other opaque objects to prevent nuisance alarms. Solid objects usually create a trouble signal rather than an alarm signal.

Ceilings over 20 feet (6.1 m) are good candidates for projected beam detectors, but some lower ceiling heights may take advantage of less expensive spot-type smoke detection. Spot-type smoke detectors can be used on ceilings higher than 10 feet (3 m). Unlike the spacing of heat detectors on high ceilings, smoke detector spacings do not require adjustment for high ceilings. However, the spacing of smoke detectors should be adjusted for ceilings over 10 feet (3 m). It is recommended that smoke-detector spacings be adjusted using Table 17.6.3.5.1, for heat detectors. Additionally, spot-type smoke detectors are not recommended on ceilings over 20 feet (6.1 m) in height because of stratification.

Radiant Energy Fire Detectors

Radiant energy fire detectors are designed to react to energy radiated from a fire. Radiant energy fire detectors are designed to react to infrared, ultraviolet, or a combination of these two spectra. These detectors are not off-the-shelf items. They are custom made for specific fuels or spectra. There are two types of radiant energy fire detectors: flame detectors and spark/ember detectors. Spark/ember detectors are generally designed to operate in dark environments, whereas flame detectors are generally used in lit environments. Radiant energy fire detectors may operate in the infrared (IR) range, ultraviolet (UV) range, or both.

Figure A.17.8.2.1 of *NFPA 72* provides an example of a typical spectrum. Every fuel type creates a unique spectrum. Radiant energy fire detectors have extremely fast response times, which make

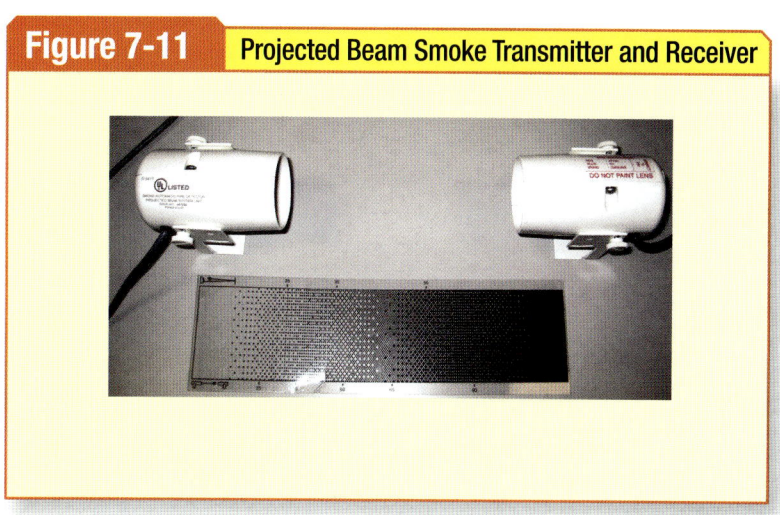

Figure 7-11. *This type of unit requires more wiring than the combination unit.*

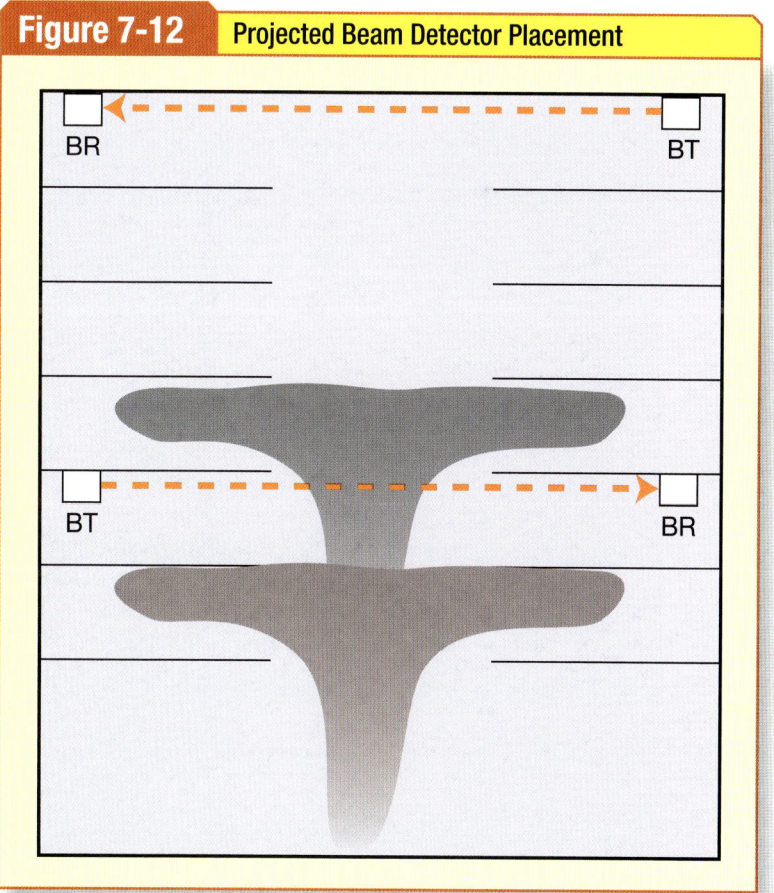

Figure 7-12. *Most manufacturers require beams to be parallel to the ceiling so they pass through more of the plume.*

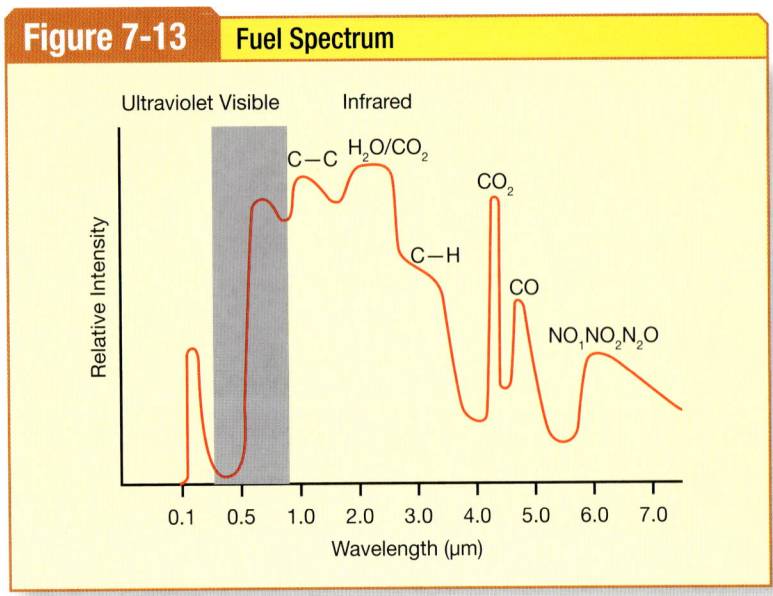

Figure 7-13. Figure A.17.8.2.1 of NFPA 72 provides an example of a typical spectrum. In this case, the spectrum is for free-burning gasoline.

them well suited to suppression actuation. **See Figure 7-13.**

An example of a suppression system used with radiant energy detectors is a spark/ember detector installed in a dust collection tube. **See Figure 7-14.** This type of detector is relatively expensive compared to other types of detection. Other types of detection are more appropriate for less critical applications such as an office building.

Radiant energy fire detectors are commonly used in industrial applications. They will react to any source of energy in the bandwidth they are designed to detect. Other applications of radiant energy fire detectors include tank farms, grain elevators, dust collection systems, conveyors, and paint spray areas. They can also be used in munitions factories, fireworks assembly areas, and other places where a fast detection and application of suppression agent is desired. The detector can be used to trigger a suppression system because it actuates very quickly, usually on the order of a few milliseconds. Smoke detectors and heat detectors are far too slow to be used for this purpose.

The environment plays an important part in the maintenance of the detector. Radiant energy fire detectors can, and often do, react to stimuli from other unintended sources, such as sunlight or welding arcs. Spark/ember detectors are usually installed in a normally dark environment, while flame detectors are normally installed in a lit or dark environment.

Extreme care must be exercised to ensure the detector is aimed at the source without introducing nuisance alarms

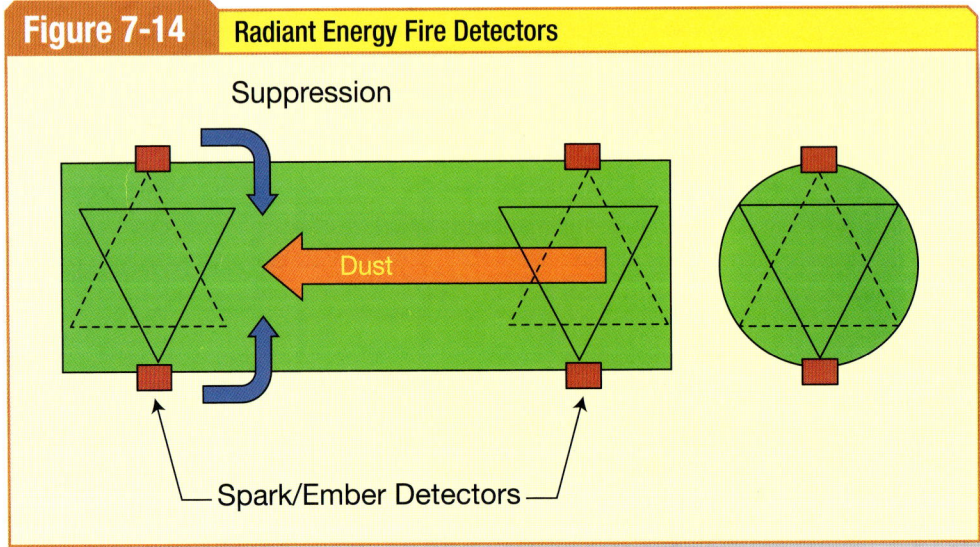

Figure 7-14. Spark/ember detectors are frequently used to detect a spark and cause suppression system actuation downstream, thereby preventing a dust explosion or fire.

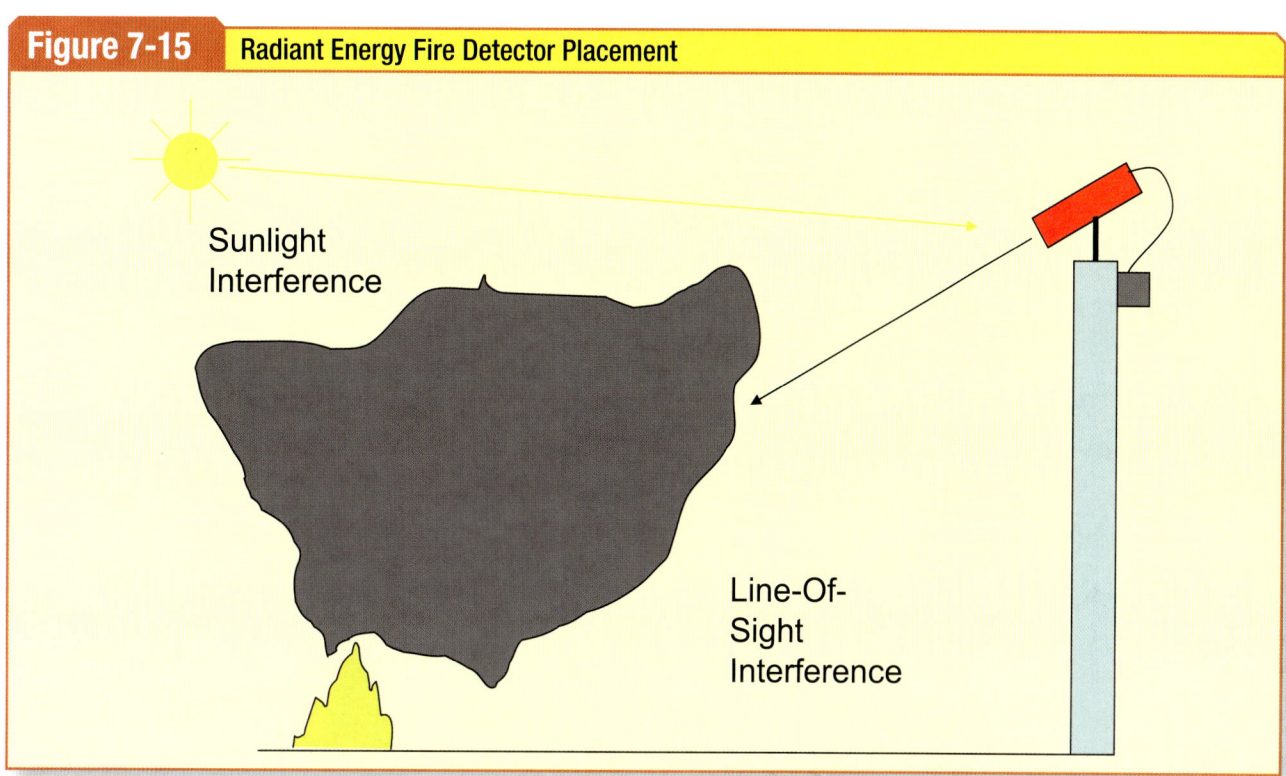

Figure 7-15. *Flame detectors must be able to "see" the fire and must be installed to prevent nuisance alarms from other sources of light in the same spectra as the fuel.*

from outside sources. Many cases of accidental discharges have occurred because of poor installation. **See Figure 7-15.**

Equally important is the need for the detector to see the signature being detected. Vehicles, doors, trees, dust clouds, vapor clouds, and other objects may block the line of sight and prevent detection. Multiple detectors may be required to prevent such an occurrence. For example, lens-cleansing air jets may be required in dusty areas to ensure the lens is kept free of dust, which might block the signal.

Radiant energy fire detectors are often installed in or near hazardous (classified) locations. Every effort should be made to place the detector(s) outside the hazardous area in order to minimize the complexity (and cost) of the wiring system. Where the detectors cannot be located outside the hazardous location, the wiring must conform to *NFPA 70, National Electrical Code,* for the particular hazardous location.

Finally, radiant energy fire-sensing detectors must be located within their area of coverage. The flame or spark must not be located so far away that the detector cannot see the source. Light intensity decreases according to the inverse square law, which means it drops off rapidly. Most flame detectors are designed to detect a one-foot (300 mm) pan fire at 100 feet (30 m).

Video Image Smoke Detection

Video image smoke detection is a relatively new technology involving video cameras, controls, and software used to detect smoke. The requirements contained in Section 17.7.7 cover this technology. Video image smoke detection anticipates the use of video cameras and controls that are able to detect smoke plumes on the raster of the image. The equipment requires the use of sophisticated software that can discriminate between normal images and smoke. Therefore, Section 17.7.7.1 requires all equipment and software to be specifically listed for the purpose of smoke detection. An added benefit to these systems is that they can also be used for video security surveillance.

Video image smoke detection systems must be designed using performance-based criteria. This is because there are

Fact

Video image smoke detection is a fledgling technology which uses custom software. The video cameras and controllers must be listed for fire alarm use, but can be used for security. Video image smoke detection is used in the Ted Williams Tunnel in Boston, Massachusetts.

very few prescriptive requirements that require a qualified design professional to design the system. Section 17.7.7.3 permits the images generated by the video image smoke detection system to be used for other purposes, such as security. These systems are already used in very special applications, such as tunnels. The Ted Williams Tunnel in Boston, Massachusetts, uses this type of system. Video image smoke detection will become more prevalent as the technology develops. **See Figure 7-16.**

SMOKE CONTROL APPLICATIONS

Smoke is toxic to building occupants and causes significant damage to furnishings and equipment. Therefore, every effort is made to control smoke in a fire and prevent its re-circulation. Most building codes and *NFPA 101, Life Safety Code,* require some form of smoke control. The fire alarm system plays a key role in controlling smoke spread. Smoke control is generally accomplished by:

1. Fan pressurization
2. Fan shutdown
3. Exhaust fan control
4. Door releasing service

Smoke detection is used to initiate signals to air-handling systems, smoke door controls, and stair pressurization fans. In complex buildings, like covered malls, transportation centers, or other large buildings, engineered smoke control systems may be used. Both area smoke detectors and duct-mounted smoke detectors are often used for control of smoke spread as referenced in Section 17.7.5.1.

Smoke control systems often use fans to pressurize non-fire areas and prevent smoke from entering those areas. Pressurization of exit stairs is a good example of this principle. Smoke detector actuations can be used to energize injection fans in the stairs or to control dampers to

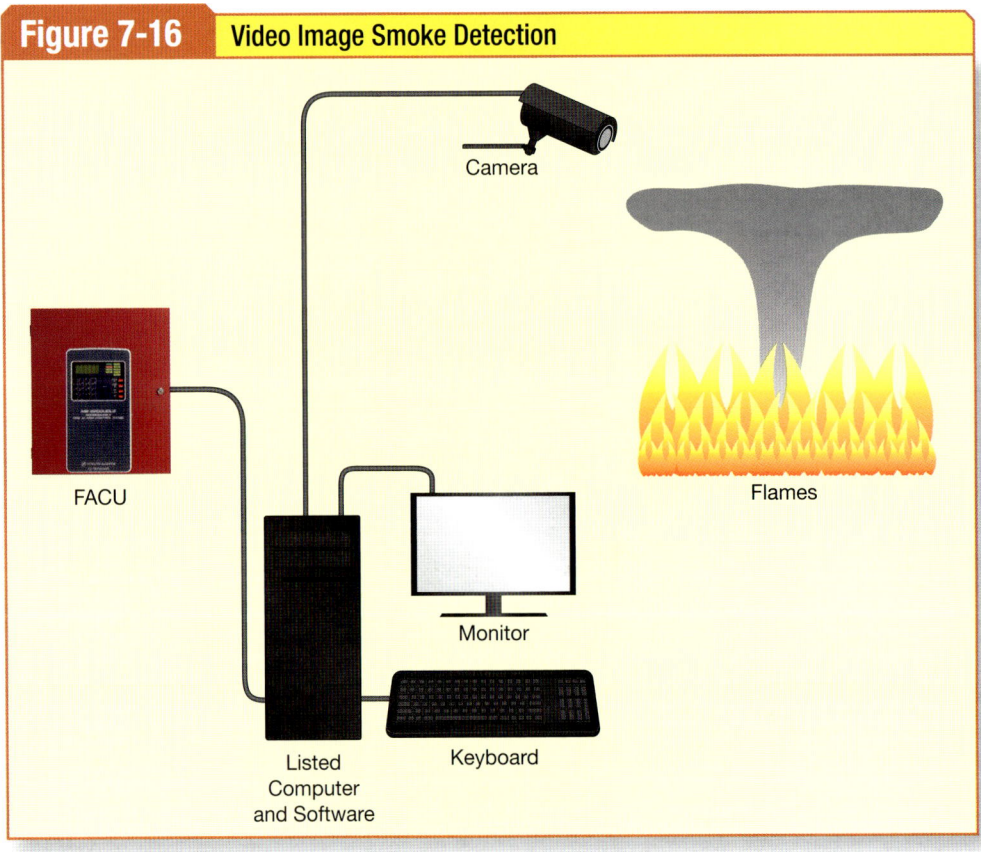

Figure 7-16. All equipment used to meet the requirements of the Code must be listed for such use. The images from this system can be used for other purposes, such as security monitoring.

control air pressure. *NFPA 92A, Standard for Smoke-Control Systems Utilizing Barriers and Pressure Differences*, applies to these applications. This standard is referenced in the *International Building Code*, *NFPA 101*, and other codes.

Covered malls, health care facilities, detention centers, assembly occupancies, high-rise buildings, towers, and underground or windowless structures are often required to have smoke control systems. *NFPA 92B, Standard for Smoke Management in Malls, Atria, and Large Spaces*, provides many more requirements for these types of occupancies. This standard is referenced in the *International Building Code*, *NFPA 101*, and other codes.

Heating, Ventilating, and Air-Conditioning Systems

Section 17.7.5 of *NFPA 72* provides the requirements for the location and installation of smoke detectors used for smoke control when they are required by other codes and standards. Smoke detectors are frequently used in HVAC systems to control smoke spread. They usually control the spread of smoke by energizing or de-energizing certain fans upon actuation. *NFPA 90A, Standard for the Installation of Air-Conditioning and Ventilating Systems*, also requires smoke detectors to control smoke dampers in HVAC systems.

Section 17.7.5.3.1 of *NFPA 72* references *NFPA 90A* for HVAC smoke control. Chapter 6 of *NFPA 90A* requires supply-side smoke detection for HVAC systems over 2,000 CFM (944 L/second). Detectors are generally placed downstream of the fans/filters and ahead of connections. Return-side smoke detection is required on systems over 15,000 CFM (7,080 L/minute) and serving more than one story. Systems of this magnitude that serve one story do not require return-side smoke detection. Return-side detection is located before connection to the common plenum on every story before the smoke is mixed with fresh air. **See Figures 7-17 and 7-18**.

Fact

Many authorities having jurisdiction incorrectly require duct smoke detectors in fresh air intakes because they want to detect smoke entering the building through the fresh air intake. Duct smoke detectors should never be installed in a fresh air intake because humidity, dust, and other influences will cause nuisance alarms.

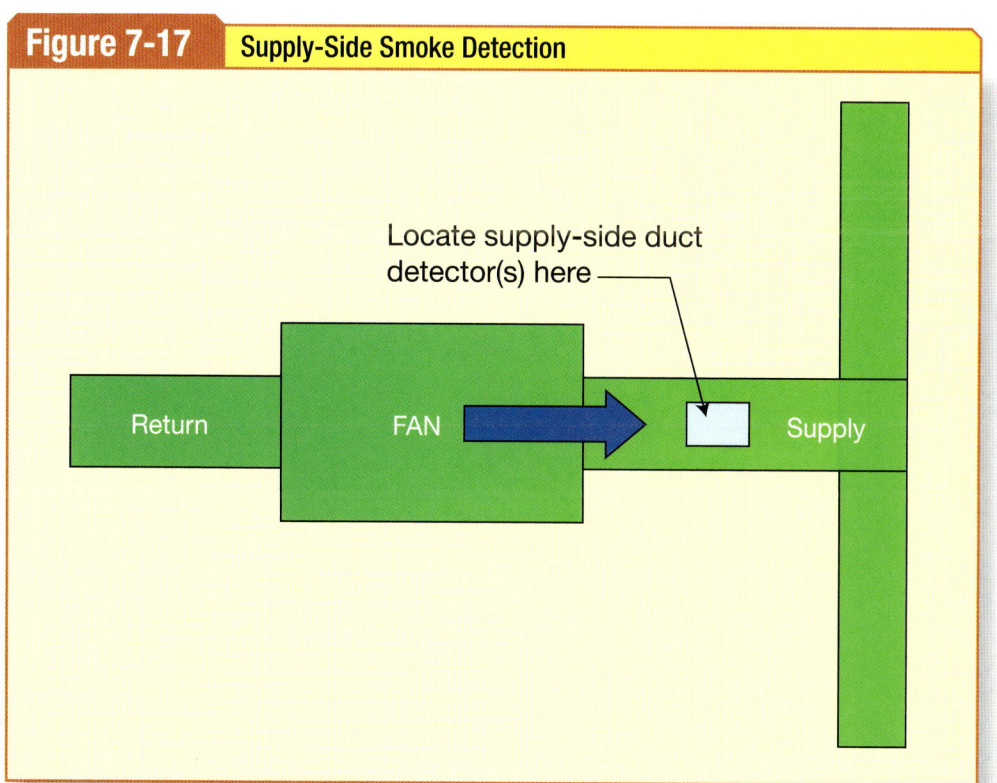

Figure 7-17. NFPA 90A *and the* International Mechanical Code *dictate where supply-side duct smoke detection is required. Duct smoke detectors should be placed downstream of the fan, filters, heaters, and pre-heaters, but before any branches in the line.*

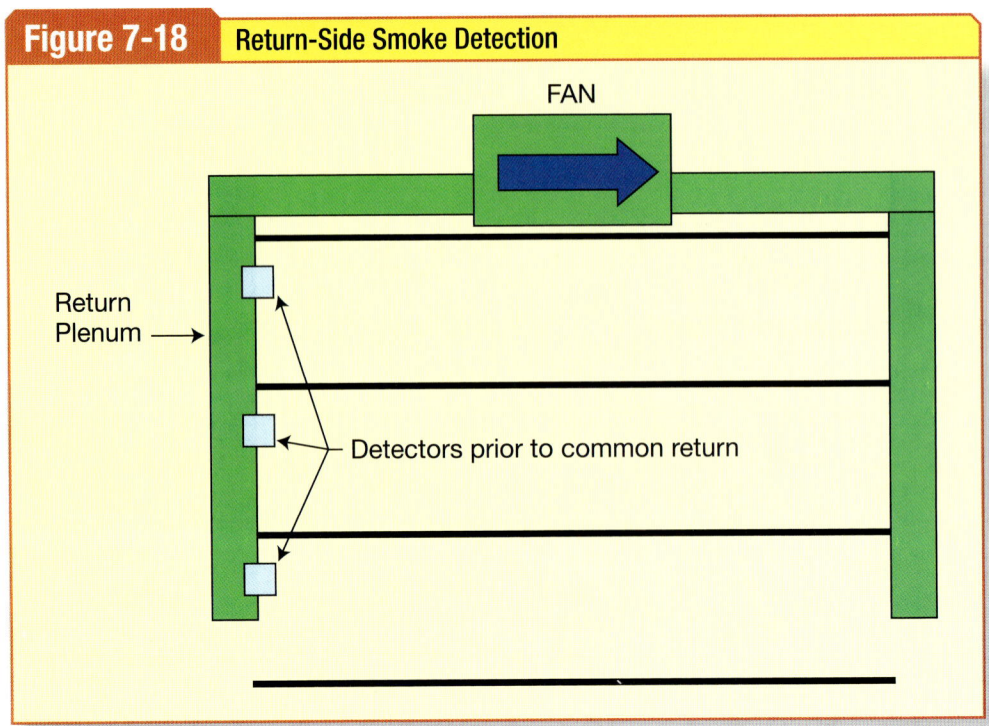

Figure 7-18. NFPA 90A *and the* International Mechanical Code *dictate where return-side duct smoke detection is required. Duct smoke detectors should be placed ahead of all connections, prior to a common return.*

The *International Mechanical Code* is the opposite of the requirements contained in *NFPA 90A*. Supply-side smoke detection is required on systems over 15,000 CFM (6,750 L/minute), and return-side detection is required on systems over 2,000 CFM (944 L/second) serving more than one story.

Return-side smoke detection may be accomplished by smoke detectors in HVAC ducts or by area smoke detectors. Where area smoke detectors are used for this purpose, they must be installed throughout all spaces served by the HVAC system. However, Section 17.7.5.2.1 does not permit duct smoke detectors to be used as a substitute for area detectors. Duct smoke detectors that sample less than 1/2% of the air traveling through the duct are not effective as area detectors. Therefore, they cannot be used as open area detectors. **See Figure 7-19**.

Section 17.7.5.5.2 requires smoke detectors to be installed in such a way as to obtain a sample representative of the air passing through the duct. There are several ways of mounting detectors, including:

1. Rigid (pendant) mounting in the duct, or rigid mounting to the wall of the duct
2. Installation on the outside of the duct with sampling tubes protruding into the duct
3. Projected beam smoke detectors installed so the beam passes through the duct

Pendant-mounted smoke detectors must be independently supported. Only rigid conduit approved for the purpose may be used as a support for pendant-mounted smoke detectors. Most conduit types, including electrical metallic tubing, cannot be used as the sole means of support for the detector. Additionally, Section 17.7.5.5.3 requires access panels to provide maintenance for duct smoke detectors mounted inside ducts. See Figures A.17.5.5.2(a) and A.17.7.5.5.2(b) in *NFPA 72* for pendant mounting and sampling tube installation recommendations. However, smoke detector units on the

Figure 7-19 — Smoke Detection Requirements

	NFPA 90A	International Mechanical Code
Supply Side	>2000 CFM (944 L/s)	>15,000 CFM (7,080 L/s) and serving more than one story
Return Side	>15,000 CFM (7,080 L/s) and serving more than one story	>2000 CFM (944 L/s)

Figure 7-19. The supply-side and return-side requirements for NFPA 90A *are the opposite of those requirements for the* International Mechanical Code.

market today have obviated the need for this arrangement and are much easier to install.

Large openings may also require multiple detectors. Figure A.17.7.5.4.2.2(a) provides guidance on the proper spacing of pendant-mounted detectors in large openings. Typically, one pendant-mounted detector covers a 36-inch (1 m)-by-36- inch (1 m) area. Other types of detection, such as air sampling or projected beam smoke detectors, may be used instead of pendant-mounted detectors.

Although smoke detectors can be pendant mounted inside the duct, most duct detectors are mounted in a housing on the outside of the duct to facilitate maintenance. Sampling tubes extend into the duct from the housing, which creates a pressure differential across the detector in the housing. Sampling tubes are available in standard lengths and must be cut to fit the duct. The manufacturer's installation requirements generally require support for sampling tubes more than two feet (600 mm) in length.

In a sampling tube arrangement, the pressure difference pulls smoke into the sensing chamber, where it is detected. **See Figure 7-20.**

Care must be exercised to ensure the tube openings face into the air stream. Section 17.7.5.5.7 requires all penetrations in return air ducts to be sealed because leakage near detectors can allow dilution of smoke.

Section 17.7.5.5.6 requires smoke detectors to be listed for the environment to prevent nuisance alarms or failure to alarm. Heated and cooled air is often

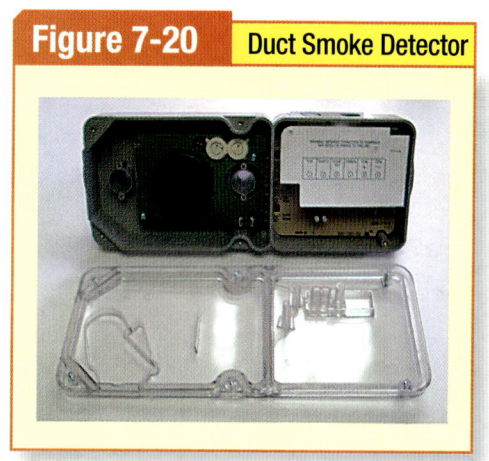

Figure 7-20. Sampling tubes are usually ordered separately for specific duct widths. Most manufacturers require tubes more than 24 inches long to be supported on both ends.

much lower or higher in temperature than the room temperature and it frequently travels at high velocities. Care should be exercised to ensure that the detectors are listed for the environment in which they are installed. Smoke detectors mounted on rooftop units are commonly subject to nuisance alarms because they are installed in a harsh environment. New types of duct smoke detector housings may help prevent problems by providing a climate conditioned space for the detector.

NFPA 72 recommends duct smoke detector and sampling tube placement to ensure the air is well mixed. In some cases, the smoke from a fire may hug one side of a duct and not be detected. Placing the sampling tubes further from openings and bends ensures a well-mixed sample.

For additional information, visit qr.njatcdb.org Item #1016

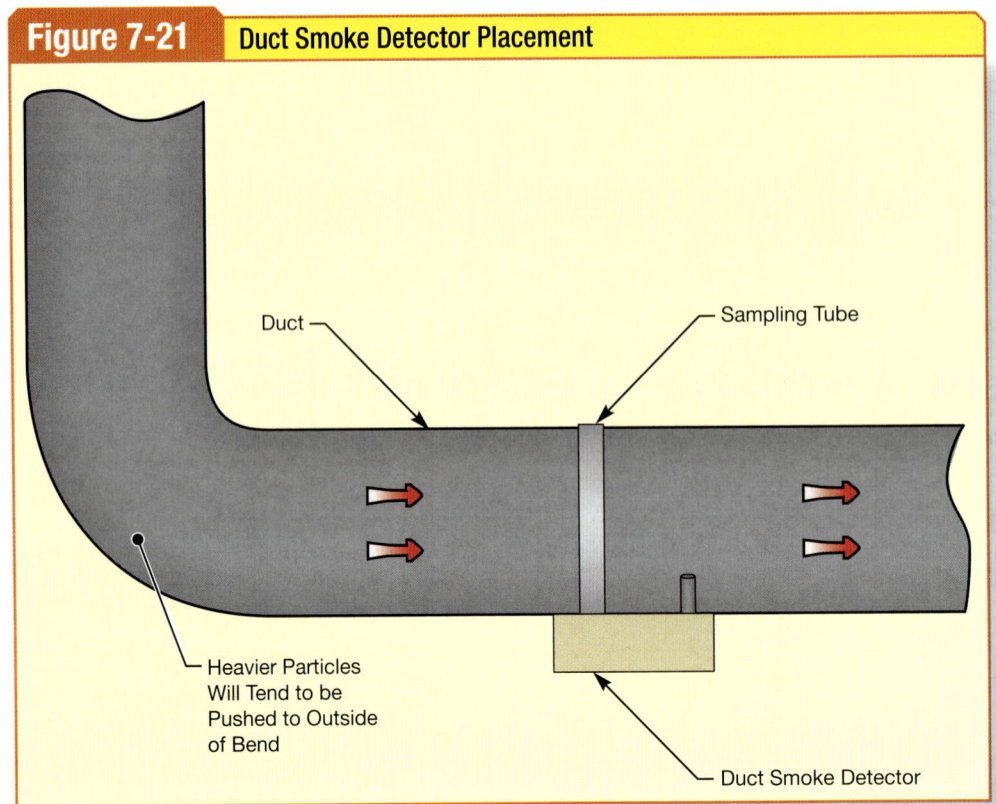

Figure 7-21. Smoke will be forced to the outside of the bend. Placing the sampling tubes perpendicular to the flow in this manner will help ensure any smoke is detected.

New research suggests that smoke particles may be forced to the outside corner of bends because they are heavier than the air in the duct. This research indicates that placing the sampling tubes perpendicular to the air flow may improve chances of detection of smoke that may hug one side of the duct. **See Figure 7-21.**

Duct smoke detectors are not permitted for open area detection. However, open area smoke detection is permitted for the purposes of HVAC shutdown. Where open area detection is used for HVAC shutdown, Section 17.7.5.4.2.2(B) requires smoke detection in all areas served by the HVAC system. This is because the detectors are closer to the fire, and they would tend to react more quickly in most fire scenarios.

Section 17.4.7 requires concealed smoke detectors mounted more than 10 feet (3 m) above the floor, or where the detector's indicator is not visible, to be provided with a remote indicator installed in a location acceptable to the authority having jurisdiction (AHJ). Many installers place a key test switch on the wall or ceiling in the area of the concealed detector. The remote test switch must be labeled to indicate the area and unit served, as required by Section 17.4.8. Unless the specific detector alarm or supervisory signal is indicated at the control unit (and on the drawings with its specific location and functions), Section 17.4.9 requires remote alarm or supervisory indicators to be installed in an accessible location and be clearly labeled to indicate both their function and any device or equipment associated with each detector.

Section 21.7.2 of *NFPA 72* requires smoke detectors connected to the fire alarm system that are used for fan or damper control to be monitored for integrity. Many smoke detectors used for HVAC control are 120 volts alternating

Smoke detectors are often used to cause operation of smoke doors. These doors are controlled by the area smoke detectors on the ceiling.

current (VAC) powered and are not powered by the fire alarm system. Auxiliary contacts can be used to connect to the fire alarm system for trouble monitoring. Section 21.7.4 also permits duct smoke detectors to cause a supervisory signal, where the signal is annunciated in a constantly attended location, such as a guard's desk or supervising station.

SMOKE DETECTORS FOR DOOR RELEASE SERVICE

Smoke detectors are frequently used for door releasing service. They are designed to detect smoke and release their controlled smoke doors, thereby preventing the spread of smoke through a building. Electrically operated magnets are commonly used to hold the doors open until the smoke detector senses smoke and releases the power to the magnets. Integral door closers and detectors can be used as well.

Section 17.7.5.6 permits either stand-alone or system-powered area detectors to be used for door releasing service. Stand-alone smoke detectors (120 VAC) not powered by the fire alarm system are sometimes used, as well as smoke detectors integral to the doorframe. System-powered area smoke detectors can also be used, but the areas on both sides of the door(s) must be protected according to the smoke detector spacing requirements of Section 17.7. Where stand-alone detectors are used, the requirements of Section 17.7.5.6.2 apply. Section 17.7.5.6.5.1 contains a reference to Figure 17.7.5.6.5.1(A), which provides requirements for the location of stand-alone smoke detectors used for door release service.

Section 17.7.5.6.5.1 requires the detector(s) to be placed on the door header or ceiling, depending on the door header and ceiling heights. Figure 17.7.5.6.5.1(A) in the *Code* provides illustrations of detector placement. It should be noted that these rules do not apply when using area detection (on both

> **Fact**
>
> Either stand-alone smoke detectors or open area detectors can be used to accomplish smoke door release. Where open area (system powered) detectors are used, it is not necessary to follow Figure 17.7.5.6.5.1(A) of *NFPA 72*.

Reprinted with permission from NFPA 72®-2013, *National Fire Alarm and Signaling Code*, Copyright © 2012, National Fire Protection Association, Quincy, MA. The information in this figure is intended to be used in conjunction with the requirements of this code. It is not the complete and official position of the NFPA on the referenced subject, which is represented only by the standard in its entirety.

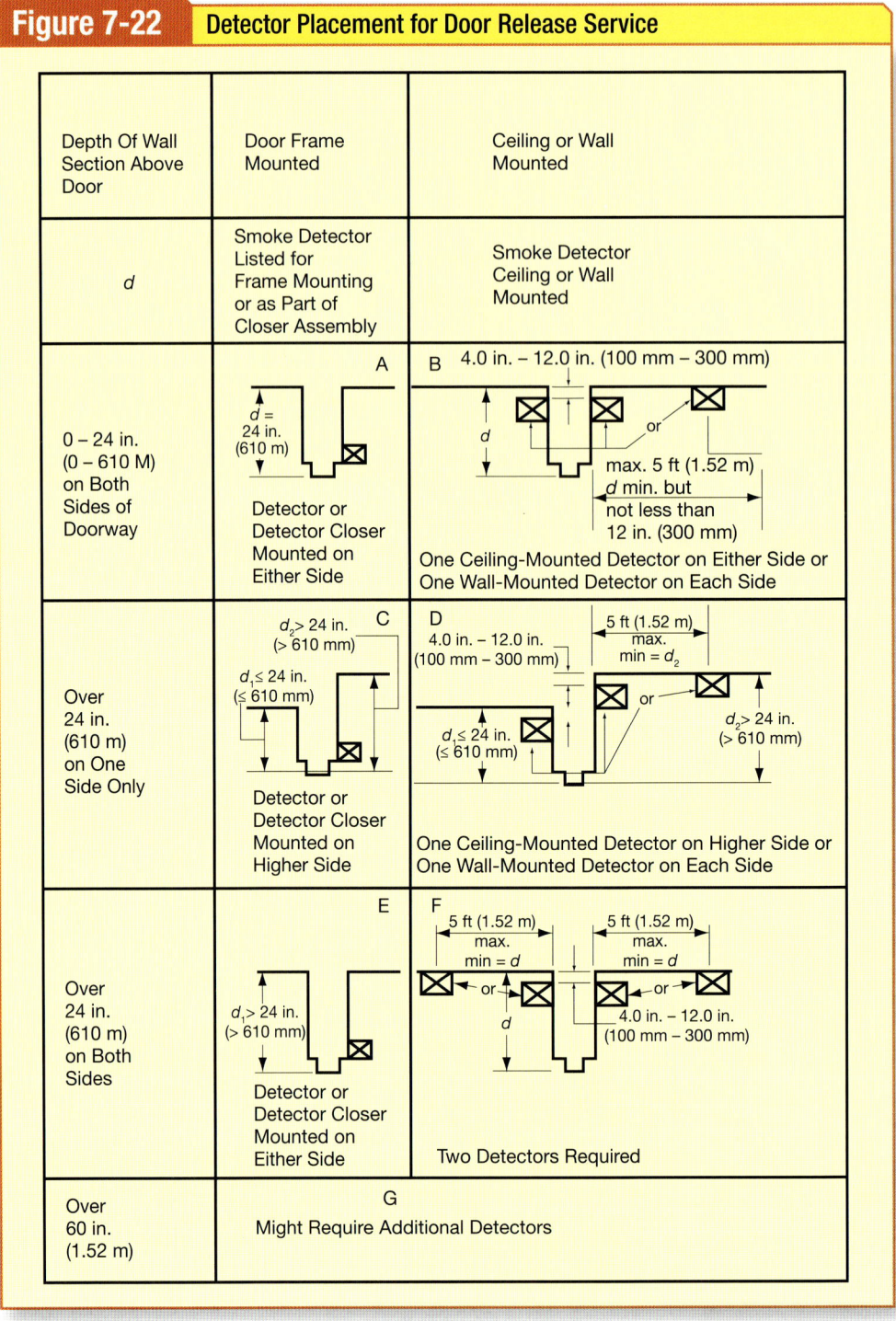

Figure 7-22. Figure 17.7.5.6.5.1(A) in the Code *provides illustrations of detector placement.*

sides of the door) in lieu of stand-alone detection. **See Figure 7-22.**

Ceilings with differing ceiling heights on each side of the header create a unique situation. Detectors must be placed on the high ceiling since smoke rises. Header depths over five feet (1.5 m) usually require additional detectors.

Summary

The study of fire dynamics is a relatively new field. Not all concepts are fully understood at this time, but more research is being conducted with every passing year. The committees responsible for development of the *Code* are constantly incorporating research data into the *Code*. New technologies, such as video image detection, will become more prevalent as manufacturers and designers learn how to detect different products of combustion. Understanding how fires behave will greatly assist in designing and installing systems that operate early, with resistance to nuisance alarms.

Review Questions

1. Which of the following is a product of combustion?
   a. Carbon monoxide
   b. Light radiation
   c. Water vapor
   d. All of the above

2. Which of the following statements is true?
   a. Fire is a self-sustaining, endothermic chemical reaction.
   b. Fire is a self-sustaining, exothermic chemical reaction.
   c. Fire is a self-sustaining, hypothermic chemical reaction.
   d. Fire is a self-sustaining, indothermic chemical reaction.

3. Which of the following is NOT required for combustion to occur?
   a. Carbon monoxide
   b. Fuel
   c. Ignition
   d. Oxygen

4. All fires exist in the flaming stage.
   a. True
   b. False

5. UL268 requires smoke detectors to respond to the kerosene test between __?__ per foot and __?__ per foot obscuration.
   a. 0.5% / 4.4%
   b. 0.5% / 12.9%
   c. 1% / 12.9%
   d. 1.5% / 12.9%

Review Questions

6. UL268 requires smoke detectors to respond to the gasoline test within __?__.
   a. 1 minute
   b. 2 minutes
   c. 3 minutes
   d. 4 minutes

7. Stratification is most likely to be a factor in which of the following locations?
   a. Atrium
   b. Elevator lobby
   c. Office area
   d. Storage room with 12' ceilings

8. Most fires tend to double in size about every __?__.
   a. 1 minute
   b. 2 minutes
   c. 3 minutes
   d. 4 minutes

9. The maximum humidity levels permitted for the installation of smoke detectors is __?__.
   a. 63 RH
   b. 73 RH
   c. 83 RH
   d. 93 RH

10. What is the number of air changes per hour of a space with the following characteristics?

    Room Volume = 21,000 cubic feet
    Diffuser Flow = 7,000 cubic feet per minute
    a. 10
    b. 20
    c. 100
    d. 200

11. Concealed duct smoke detectors located more than __?__ above the floor must have a remote indicator.
    a. 5'
    b. 10'
    c. 15'
    d. 20'

12. Where using smoke detectors for door release service, headers with a depth over __?__ may require additional detectors.
    a. 2'
    b. 3'
    c. 4'
    d. 5'

Emergency Communications Systems (ECS)

Emergency communications systems (ECS) were developed to provide critical communications to large numbers of people. Although fire emergency voice/alarm communications systems have been used for many years, they were not typically used for purposes beyond fire emergencies. Several high-profile bombings at overseas Department of Defense (DoD) and Department of State (DoS) facilities and the terrorist attacks on September 11, 2001, led to the development of Annex E in the 2007 edition of *NFPA 72*. Annex E later became a new Chapter 24 in the 2010 edition of *NFPA 72*. When the material was relocated to Chapter 24, *NFPA 72* was changed to its current title and scope.

Some of the more important changes to Chapter 24 in the 2013 edition of *NFPA 72* include enhanced risk analysis and new documentation requirements. A new Chapter 7 was added to the 2013 edition of *NFPA 72* to introduce requirements for system documentation. Section 24.4.2.8.1 requires new systems employing relocation or partial evacuation to be documented in accordance with Sections 7.3, 7.4, and 7.5.

Emergency communications systems can be used for fire emergencies, terror alerts, paging (emergency and non-emergency), weather alerts, traffic delays, closings, Amber alerts, HAZMAT events, and many other events. Emergency communications systems are primarily used by government agencies. In the wake of the Virginia Tech shootings, many universities began using ECS, and they are likely to become more widely used in the years ahead. ECS can be stand-alone or may be integrated with fire systems

Objectives

- » Describe the various types of ECS
- » Understand power supply requirements for ECS
- » Describe the purpose of a fire command center
- » Recognize the signal types
- » Explain the factors affecting intelligibility
- » Describe circuit survivability requirements

Chapter 8

Table of Contents

Types of Emergency Communications Systems 186
    One-Way Emergency Communications Systems 186
    Two-Way Emergency Communications Systems 189

Power Supplies for Emergency Communications Systems 191

Fire Command Centers 191

Types of Evacuation Signals 192
    Temporal 3 Signals 192
    Coded Versus Non-Coded Signals 193

Intelligibility 194
    Factors Affecting Intelligibility 194
    Intelligibility Standards 195
    Measuring Intelligibility 195

Circuits 196
    Survivability 196
    Class A and Class X Circuit Separation 200

Summary 202

Review Questions 202

TYPES OF EMERGENCY COMMUNICATIONS SYSTEMS

There are two types of emergency communications systems: one-way and two-way systems.

One-Way Emergency Communications Systems

One-way emergency communications systems include the following systems:
- In-building fire emergency voice/alarm communications systems (EVACS)
- In-building mass notification systems (MNS)
- Wide-area mass notification systems (giant voice)
- Distributed recipient mass notification systems (DRMNS)

One-way ECS are primarily used on military facilities, and it is anticipated they will become more widely used as the war on terror continues. These systems all anticipate some kind of instruction from a centralized source, whether recorded or live. EVACS have been used for many years by the fire alarm industry, especially in high-rise buildings, large assembly locations, transportation centers, and the like. Fire emergency voice/alarm communications systems are now called "In-Building Fire Emergency Voice/Alarm Communications Systems." All in-building EVACS and ECS must meet the requirements of Chapter 24. Combination EVACS/ECS have special requirements to avoid confusion. These requirements include special strobes or signage that makes the type of emergency clear to occupants.

In-building ECS are being used more by federal entities to provide warning or instruction in case of a terrorist attack. These systems may be stand-alone or integrated with the fire alarm system. In-building systems can take advantage of strobes, speakers, and visible signs (fixed and variable). All equipment used to meet ECS requirements in Chapter 24 must now be listed for such use.

Wide-area ECS are used extensively on military facilities to provide a variety of alerts or non-emergency information. These systems are often called "giant voice" because they use high-powered speaker arrays (HPSA) to distribute the signal over a wide outdoor area.

Finally, Distributed Recipient Mass Notification Systems (DRMNS) are used in many settings, both private and public. Reverse 911 paging/texting, e-mail notification, and radio broadcasts can be arranged for college students in case of an emergency on campus. These same methods are often used by large organizations for the same reasons. Other pagers and messages, such as variable message highway signs, provide information for drivers on interstate highways. Traffic delays and Amber alerts are just two of the uses for this technology.

In-Building Fire Emergency Voice/Alarm Communications Systems. Large and tall buildings, assembly occupancies, and transportation centers all have special evacuation plans and requirements. The tragic events of September 11, 2001, and other recent tragedies have reinforced the need for good communication as a critical part of any life-safety plan. The ability to clearly convey verbal instructions to occupants during an emergency can save many lives. This was evidenced by the heroic efforts of responders who used the fire alarm system to evacuate the World Trade Center (WTC) buildings following the first crash.

Total evacuation of a high-rise building or health care facility may not be practical or necessary in all cases. For example, a fire on the 79th floor of a building will generally not require the 50th floor to evacuate. In fact, most evacuation schemes for high-rise buildings involve evacuation of the fire floor, the floor below, and the floor above. Some jurisdictions may require evacuation of two floors above the fire. **See Figure 8-1.**

It may be impossible to quickly evacuate a building during a fire or other emergency. Stairways may not be sufficiently wide to accommodate such a large occupant egress. This was proven in the first attacks on the WTC in 1993, where it took several hours to completely evacuate both buildings. Stairways are not designed to evacuate the entire building at once. They can usually only accommo-

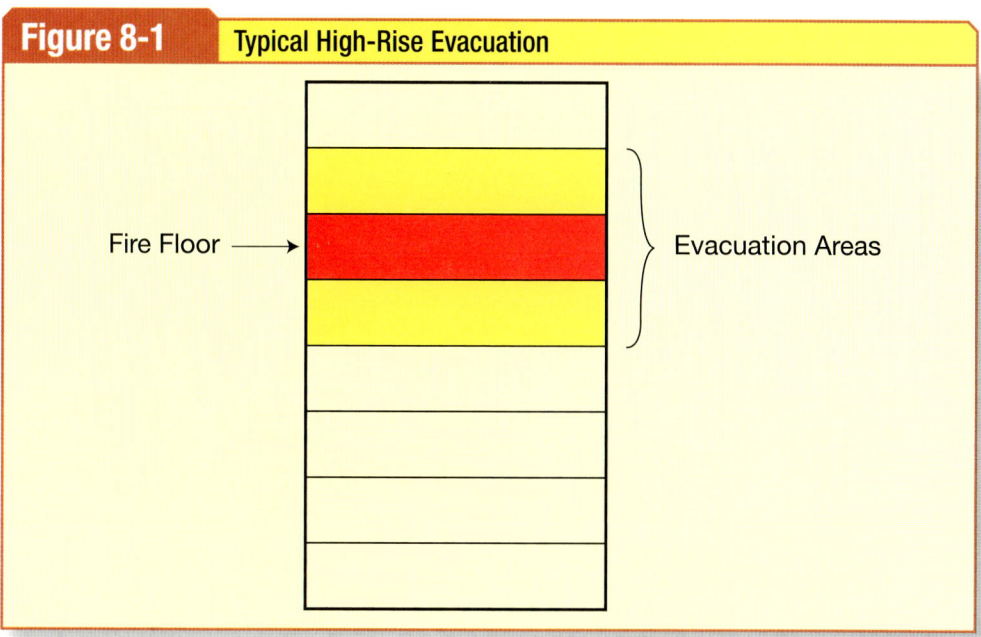

Figure 8-1. Total evacuation of high-rise buildings is not possible or practical. A typical evacuation scheme requires evacuation of the fire floor, floor below, and floor above. Some plans require evacuation of two floors above the fire.

date several floors evacuating at one time, while simultaneously allowing fire department entry.

The need for good emergency communication is not limited to high-rise buildings. Many hospitals, large footprint buildings, assembly occupancies, places of worship, transportation centers, and secure facilities also employ a system of partial evacuation or occupant relocation using an EVAC system. *NFPA 101, Life Safety Code*, and many local building codes require EVACS for these occupancies or wherever it is impractical or impossible to provide total evacuation. Smaller buildings, such as a mercantile occupancy, low-rise office building or a clinic, typically do not require EVACS.

EVACS are a one-way communications system to deliver information or direction to occupants. They may be stand-alone systems, or they may be fully integrated with the fire alarm system controls. The system may be used to provide instructions on a selective basis or may broadcast throughout the entire building.

EVACS are generally used in high-rise buildings, which are defined (by *NFPA 101* and most building codes) as being over seven stories above grade or 75 feet (23 m) in height from the highest level of occupancy to the lowest level of fire department access. EVACS are capable of providing either live or recorded voice instructions to occupants. Recorded messages are usually contained in a memory chip developed by the manufacturer. The voice messages may warn occupants to evacuate or that there is an emergency elsewhere in the building. EVACS can deliver voice messages in almost any language and in either a male or female voice.

In addition to recorded messages, responders can provide live instructions through a microphone when more information is needed. Live or recorded instructions have much greater flexibility than horns or bells. As a rule, occupants will respond significantly faster to voice instructions than to single-stroke

 **Fact**

Emergency communications systems (ECS) have been in use for many years. The Department of Defense (DoD), other government agencies, and universities are embracing these technologies to provide critical information.

In-building fire emergency voice/alarm communications systems (EVACS) were formerly known as emergency voice/alarm communications systems. EVACS have been used in high-rise and assembly occupancies for fire emergencies for decades.

notification appliances, such as bells and horns. Recent studies indicate occupants react almost immediately to voice instructions because they have more information.

EVACS require the use of speakers rather than bells, horns, chimes, or other single-impact audible notification appliances. Speakers must be of sufficient quantity, power output, and spacing to provide adequate sound pressure levels throughout the protected area. The audibility requirements of Chapter 18 will apply to the audibility of the system. **See Figure 8-2.**

For additional information, visit qr.njatcdb.org Item #1018

Figure 8-3. Pre-recorded or live instructions can be provided to any single paging zone or all zones from this location. In this photograph, the paging zones are indicated by the red buttons at the bottom. EVACS controls are usually located in the fire command center.

For additional information, visit qr.njatcdb.org Item #1017

Figure 8-2. Speakers must be used to distribute voice instructions and should not be tapped at high settings to avoid distortion.

EVACS system controls contain amplifiers, which may be centrally located at the fire command center or distributed throughout the building. Some manufacturers require twisted shielded pairs, while some permit non-shielded conductors to be used for speaker wire. Shielding is usually grounded only at the panel end of the circuit to minimize circulation of electrical currents that interfere with the signals. Independently powered speakers can also be used for high-output applications or where electrical isolation from the main fire alarm system is required.

EVACS are capable of providing selected paging to single or multiple zones. Each zone can be selected individually to provide live instructions. EVACS controls also have an "all-call" switch to provide paging to the entire space served by the system. A microphone can be used to provide live voice instructions, if needed. **See Figure 8-3.**

ECS voice instructions must be preceded by an alert tone lasting between one and three seconds. This tone is steady, followed by recorded or live instructions. The alert tone and message are repeated at least three times. Recorded messages are acceptable for most emergencies, but live voice instructions are sometimes necessary when standard evacuation plans fail, such as when an exit stairway is compromised. Live signals always override recorded signals as required by Section 24.4.3.2.8.

Where provided, Section 24.4.2.8.4 requires that each stairwell be a separate paging zone. This allows the fire department responders to broadcast separate instructions to occupants in each stairwell to aid in evacuation.

Section 24.4.2.9.1 does not permit undivided fire areas to be separated into separate zones. The architect and fire protection engineer usually determine fire boundaries based upon the construction of the building. Evacuation zones are generally bounded by smoke and/or fire barriers. These boundaries are shown on life-safety drawings. Section

23.8.6.3.1 requires notification zones to be consistent with evacuation plans for the premises. Additionally, Section 23.8.6.3.2 requires notification zones to be consistent with smoke and fire barriers, outer walls, and fire subdivisions.

Section 24.4.2.9.2 requires all appliance circuits within a single fire area to be arranged to activate or deactivate at the same time. This reinforces the requirements found in Sections 23.8.6.3.1 and 23.8.6.3.2.

In-Building Mass Notification Systems. In-building mass notification systems are intended to provide emergency voice instructions to building occupants in the event of a fire, terrorist attack, earthquake, tornado, flash flood, gas leak, or other life-threatening emergency. Events at Virginia Tech and Columbine High School have illustrated the need for better communication in this type of emergency. The U.S. Department of Defense realized that standardization was needed for these systems following the tragic events of September 11, 2001. Today, nearly all DoD facilities are required by federal law to have mass notification alerting systems. The requirements for the installation of in-building MNS are found in Section 24.4.3 of *NFPA 72*.

Wide-Area (Outdoor) Mass Notification Systems. Mass notification systems are not limited to indoor use. They may be used indoors and outdoors at such places as military facilities, industrial plants, and school campuses. These systems may include distributed recipient MNS, such as Internet-based systems, pagers, and reverse 911 systems. Other outdoor systems include "giant voice" systems, which are used on military facilities. The requirements for their installation and use are contained in Section 24.4.4 of *NFPA 72*. The requirement to install any MNS will be driven by building codes, *NFPA 101*, or organizational policy.

Distributed Recipient Mass Notification Systems. Distributed recipient mass notification systems include reverse 911 systems, pagers, Internet/Intranet messaging, radios, and variable message signs. These systems can be used to provide emergency and non-emergency information to very large numbers of people. Requirements for these systems are found in Section 24.4.5 of *NFPA 72*.

These systems are sometimes slow. Some text messages on reverse 911 systems have been known to take as long as 24 hours to be received. Section 24.4.5 requires backup systems for stand-alone systems and multiple delivery methods for messages. For this reason, layering multiple types of ECS should be done.

Two-Way Emergency Communications Systems

In addition to the one-way emergency communications systems, there are several types of two-way systems.

Firefighters' Two-Way Telephones. Two-way, in-building wired emergency communications systems were formerly called "Firefighters' Telephones." This system is a two-way means of communication for first responders. Handsets connected to a wired system are used to allow responders an easy method of communicating in the event of poor radio communication or when radio traffic is excessive. Fire wardens may also be permitted to use firefighters' telephones during emergencies. Firefighters' telephones allow communications between the fire command center and points throughout the building. Section 24.5.1.2 requires the firefighters' telephone receiving station to be collocated with the in-building fire emergency voice/alarm communications system equipment. In fact, firefighters' telephone systems are usually integral with the EVACS equipment.

There are two basic types of firefighters' telephones in use today: telephone stations and telephone jacks. Telephone stations are self-contained units containing a handset and cradle mounted in a cabinet. Communication is established when the handset is lifted from its cradle or when a button is pushed on the handset.

Fact

Mass notification systems (MNS) are not limited to speaker broadcasts. Systems using telephone systems, cellular telephones, the internet and other media which reach wide audiences can be used with AHJ permission.

Telephone jacks require the user to bring a handset with him or her. In this arrangement, telephone handsets are kept at the fire command station so firefighters have access to them during a fire. Communication is established when a handset is plugged into a jack. A private mode signal sounds at the fire alarm controls, which prompts the operator at the fire command center to answer the call. Telephone stations and jacks are located throughout a protected area at exits and lobbies.

Firefighters' telephones are usually designed for a party line or conference call arrangement. The caller is connected to the line when his or her handset is connected. Section 24.5.1.6 requires the system to support no fewer than five handsets under simultaneous use.

Section 24.5.1.11 of NFPA 72, *National Fire Alarm and Signaling Code*, requires telephone stations or jacks in the following locations:

1. Each floor level
2. Each notification zone
3. Each elevator cab
4. Elevator lobbies
5. Elevator machine room(s)
6. Emergency and standby power room(s)
7. Fire pump room(s)
8. Area(s) of refuge
9. Each floor level inside an enclosed exit stair(s)
10. Other room(s) or area(s) as required by the authority having jurisdiction (AHJ)

A handset is usually integrated with the operating controls in the fire command center. **See Figure 8-4.**

These stations or jacks are usually mounted in or near stairways and elevator lobbies on every floor. Additionally, Section 24.5.1.11 of *NFPA 72* and many AHJs require an additional jack or telephone station at the fire pump.

Some firefighters' two-way telephones indicate the station location; however, most do not. If the two-way telephone system does not indicate the location of the caller, Section 24.5.1.14 requires the location of the telephone station or jack to be permanently labeled for the caller to identify his or her location. All firefighters' telephone circuit wiring must be protected to provide Level 2 or Level 3 survivability as required by Section 24.3.6.7.

Two-Way Radio Communications Enhancement Systems. Fire departments are increasingly allowing the elimination of two-way, in-building wired emergency communications systems (firefighters' telephones), provided that their radio communication is adequate throughout the building. Where radio communication is inadequate, the AHJ may permit a radio repeater to boost the signals. This is covered by Section 24.5.2 and is referred to as two-way radio communications enhancement systems or linear amplifiers. However, there have been cases where excessive radio traffic or building steel prevents good radio communication. The 1988 fire at the Interstate Bank in Los Angeles is a good example of an event where fire department radio traffic overwhelmed the system and firefighters' telephones were very helpful.

Section 24.5.2.1.2 requires AHJ approval for this type of system. A renewable permit is required at the successful completion of testing. Section 24.5.2.2.1 requires 99% of critical areas to be provided with radio coverage. These areas specifically include exit stairs, fire pump room, elevator lobbies, standpipe cabinets, sprinkler valve locations, and other areas identified by the AHJ.

Section 24.5.2.6.1 requires the system to be monitored for integrity, including

Figure 8-4. *These stations or jacks are located in critical areas for first responder use.*

antenna malfunctions and signal booster failure. Section 24.5.2.6.2 requires a dedicated monitoring panel located at the emergency command center to annunciate the status of all booster locations. The following must be provided and labeled:

1. Normal AC power
2. Signal booster trouble
3. Loss of normal AC power
4. Failure of battery charger
5. Low-battery capacity

POWER SUPPLIES FOR EMERGENCY COMMUNICATIONS SYSTEMS

In-building fire emergency voice/alarm communications systems must be capable of operating for a period of not less than 24 hours in standby followed by two hours of operation. It is difficult to know exactly how much power will be required for two hours of operation. Therefore, Section 10.6.7.2.1(2) of the *NFPA 72, National Fire Alarm and Signaling Code,* permits a substitution of 15 minutes of maximum connected load to be used for this calculation. An all-connected load includes both audible and visible notification appliances, such as strobes and speakers.

Many EVACS are integrated with automatic detection and alarm systems, which may require the entire system to operate for 15 minutes of all-connected load. The same rules apply for other systems integrated with EVACS, such as access control, because there is no way to isolate them during a power outage. Section 10.6.7.2.1(1) requires a 20% safety factor on all battery calculations for fire alarm systems.

In-building mass notification systems are subject to the same requirements for secondary power, as found in Section 10.6.7.2.1(7).

Section 24.4.4.4.2.2 also requires seven days of standby followed by 60 minutes of operation for high-power speaker arrays (HPSA) used on outdoor wide-area emergency communications (giant voice) systems.

FIRE COMMAND CENTERS

NFPA 72, National Fire Alarm and Signaling Code, does not require a fire command center, but *NFPA 101, Life Safety Code,* and local building codes do require one in some occupancy types, such as high-rises, assembly occupancies, and where ECS or EVACS are used. The fire command center must be designed and built so the fire department can maintain communication with occupants during a fire. Additionally, Section 911 of the *International Building Code* requires the fire command center to be of a minimum one-hour construction rating.

The fire command center may also contain controls or indicators for other systems, such as elevators, HVAC, and smoke control.

The fire command center must contain system controls and displays for the detection, alarm, and communications systems. Operating controls must be clearly marked. Generally, all controls, including firefighters' telephones, for fire department use are located in the fire command center.

The fire command center is where firefighters begin their response during a fire emergency. Where required by another code, Section 10.18.1.1.1 requires the location of an actuated initiating device to be visibly indicated by zone, floor, or other means, by visible means, or printout at the fire command center. A liquid crystal display (LCD) or printed or graphic annunciator is often used to assist firefighters in a response. **See Figure 8-5.**

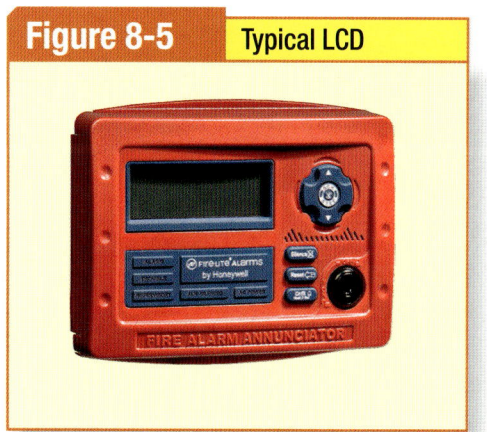

For additional information, visit qr.njatcdb.org
Item #1019

Figure 8-5. LCDs are used to provide system information for the responder and user. Most modern addressable systems use LCDs.

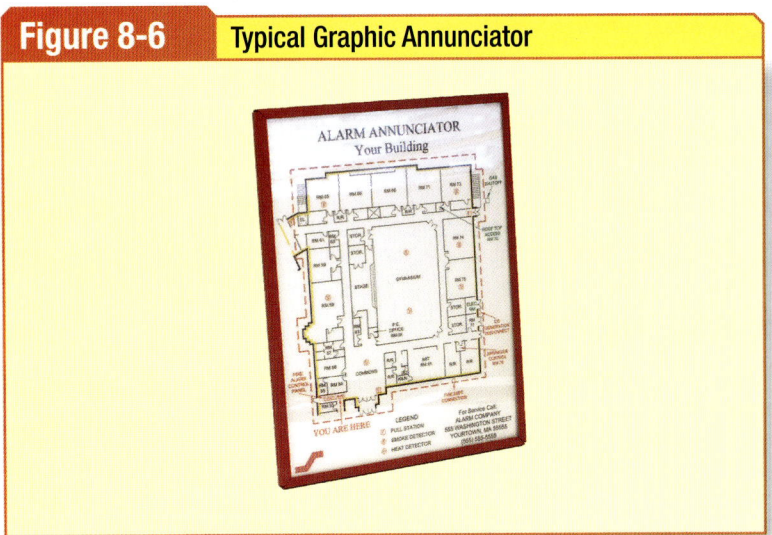

Figure 8-6. Graphic annunciators are particularly helpful to first responders. Graphic annunciators are frequently used for large footprint or complex structures and may be required by some local codes.

Graphic annunciators provide an outline of the building floor plan with indicating light-emitting diodes to provide the location of actuated devices. **See Figure 8-6.**

Graphic annunciators work well in large or complex buildings and campus arrangements. Many jurisdictions require a graphic annunciator to assist with the emergency response. Building codes often provide zoning requirements, but Section 10.18.6 requires each floor of the building and each separate building to be annunciated separately.

The fire command center is required by local building codes to have a minimum one-hour fire-resistance rating.

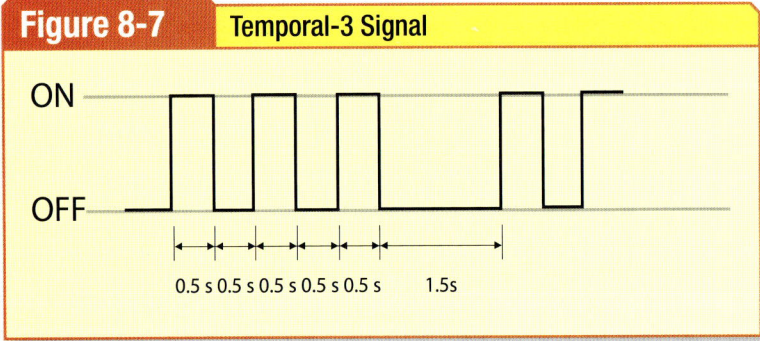

Figure 8-7. This signal is used internationally for evacuation of an area and has been a requirement of NFPA 72 since July 1996.

This requirement corresponds with the requirement for the EVAC system to operate for at least one hour during a fire emergency. The fire command center must also have an outside entrance so firefighters do not have to enter a burning building to reach it.

Many existing fire command centers are located inside the lobby or interior of a building. These fire command centers were built before these requirements were placed in the *Code*. Fire alarm system retrofits and upgrades might require the fire command center to be relocated, so one should be sure to check with the AHJ before assuming the current location of the fire command center is acceptable. The fire command center must also contain a minimum level of equipment, including a telephone and adequate lighting.

TYPES OF EVACUATION SIGNALS

There are three types of evacuation signals:
1. Standard (Temporal 3)
2. Coded
3. Non-coded

Temporal 3 Signals

Many buildings follow a total evacuation plan during a fire emergency. Low-rise buildings with low occupant loads or small area can be quickly evacuated, generally within five minutes or less. Where total evacuation is planned, Section 18.4.2 of *NFPA 72, National Fire Alarm and Signaling Code*, requires the fire alarm system to produce an audible Temporal-3 signal, in accordance with ANSI Standard S3.41, *American National Standard Audible Emergency Evacuation Signal*. This signal is one-half second ON, followed by one-half second OFF, repeated three times, followed by one and one-half seconds OFF. However, visible signals are not required to flash in this pattern. **See Figure 8-7.**

Several notification appliances operating out of sequence will negatively impact the audible signal in atria or other large open spaces. Section 18.4.2.4 requires the Temporal-3 signal to be synchronized within evacuation notification zones. This is very important because two notification appliances producing Temporal-3 signals out of phase will simply generate noise, rather than a recognizable evacuation signal.

In addition to the Temporal-3 pattern, Section 10.10.2 requires alarm signals to be distinct or different from other signals used in the area. Alarm signals may not be the same as burglary or hold-up signals, tornado alarms, and other signals. Similar requirements exist in Sections 10.10.5 and 10.10.6 for supervisory and trouble signals, respectively.

This requirement does not preclude the use of different types of audible appliances, such as bells and horns, in the same area. However, every effort should be made to use the same type of appliance within a single building or facility to minimize occupant confusion.

Coded Versus Non-Coded Signals

Coded signals are designed to be understandable only by trained staff. The purpose of coded signals is to allow an investigation or response to the signal by trained staff without alerting untrained occupants evacuating the building(s). Coded signals may be either voice or non-voice. Coded signals were commonly used on older systems before the advent of addressable systems and were designed to assist the responders in determining the area of the fire. It should be noted that the Temporal-3 pattern is a non-coded signal.

On coded systems, standard notification appliances such as bells were frequently used to produce a signal. Coded manual stations were designed with a spring-wound mechanism that, when actuated, would turn a notched wheel, which tapped a unique signal on the circuit by closing electrical contacts at specific intervals. The panel would then broadcast this coded signal throughout the zone or protected premises. After the manual station was actuated, the spring required rewinding.

For example, a manual station on the east wing of the second floor in a building might have a code wheel that creates a 2-1 pattern signal. **See Figure 8-8.**

In this arrangement, a single pulse follows two pulses. This type of coded signal can be made using almost any notification appliance, such as a bell or horn.

This signal is recognizable to trained individuals as a signal meaning there is a fire on the second floor, east wing. Each zone has manual stations with code wheels designed to produce a signal unique to the floor and zone.

There are still many coded systems installed in older buildings, but manufacturers no longer produce this technology. Some new addressable systems use coded signals during walk test mode to make testing easier. The coded walk test signal is broadcast throughout the premises or zone to alert the service personnel what initiating device is actuated. The signal denotes the zone or loop number (signaling line circuit number) and device address. However, the signal is generated by software in the fire alarm controls, rather than by a code wheel in the initiating device.

In-building fire emergency voice/alarm communications systems can be used to alert occupants in either a coded or non-coded format. However, EVACS use audible textual (voice) signals to deliver the messages. A coded EVACS signal uses instructions that only trained staff can understand. For example, hospitals use coded textual signals like "Paging Dr. Firestone; Third Floor, East Wing." This type of signal is used to avoid inducing stress in patients. The medical and security staff are trained to understand the meaning of the signals, while patients and visitors are generally not alerted to their meaning. A non-coded version of this signal might be, "May I have your attention please? A

Fact
Tone coded signals are no longer widely used. Coded voice signals are now used on voice systems, especially where the occupants cannot or do not need to take action, such as in a hospital.

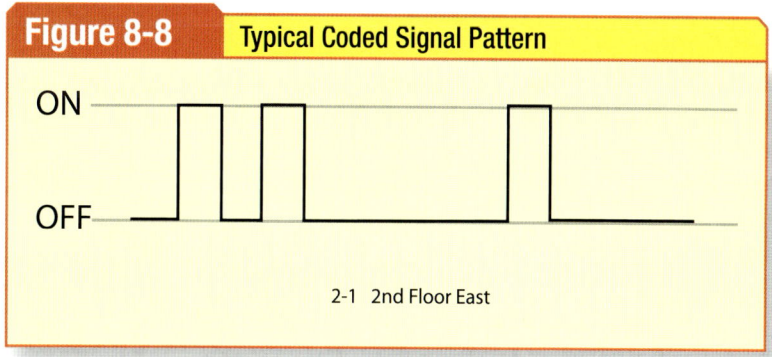

Figure 8-8. This pattern is a 2-1 pattern which could, for example, be used to indicate the 2nd Floor, East Wing of a building. Coded signals are rarely used, but a significant population of legacy systems are still installed.

fire has been reported on the Third Floor, East Wing. Please evacuate the building using the nearest exit."

It is imperative that EVACS remain in service for longer periods during the fire in a building where total evacuation is not planned than during a total evacuation. In such situations, because not all occupants are required to evacuate during the emergency, continued communications will be needed. The secondary power supply must be capable of operating the system for a period not less than two hours. Additionally, the system must be resistant to attack by fire. It must be capable of delivering communication to occupants who do not immediately evacuate. Therefore, the system must be more resistant to attack by fire than other fire alarm systems. Resistance to attack by fire is called "survivability."

Being able to clearly hear and understand voice instruction during an emergency is critical to occupant safety. For this reason *NFPA 72* requires all ECS to produce signals with intelligibility.

INTELLIGIBILITY

Clear verbal announcements are essential in transmitting emergency instructions to the occupants of public and private structures whose lives depend on their comprehension of the message. The consequences of unintelligible emergency messages were made tragically clear on January 6, 1995, when six people died in a high-rise apartment complex fire in North York, Ontario. A major factor in the loss of life was the occupants' inability to understand or hear the instructions over the EVACS.

Factors Affecting Intelligibility

Voice intelligibility is the measure of a listener's ability to understand a spoken message. Among the many variables affecting intelligibility is distortion, which results when significant portions of an audible signal are chopped from the output, so it sounds more like noise than speech.

Another variable affecting intelligibility is echo, which occurs when the signal bounces off a surface before reaching the listeners, causing the words to run into each other. Echoes occur in spaces with large amounts of acoustically hard surfaces such as tile, concrete, untreated gypsum board, and hardwood paneling.

Acoustically hard surfaces and a shotgun approach to distributing signals are also responsible for excessive reverberation. Reverberation, which is similar to echo, can blur words and make them difficult to understand.

Another factor affecting intelligibility is the announcement's signal-to-noise ratio or the relationship of a signal's sound pressure level to that of the background noise. If a signal is below the background, or ambient SPL, it will not be heard or understood. One way to think of this is to compare it to using a flashlight outside during the day. The signal to noise ratio is so low that the signal from the flashlight is not as noticeable as when used in the dark.

Low signal-to-noise ratios contribute to a large number of unintelligible systems, particularly in places of worship, sports arenas, large atria, airports, covered malls, exhibition areas, and other places where there is a high level of ambient noise. Most designers, installers, and authorities having jurisdiction understand the importance of the appropriate SPL. However, a system that provides an appropriate SPL still may not provide adequate intelligibility.

Section 18.4.10 of *NFPA 72, National Fire Alarm and Signaling Code,* requires EVACS to be capable of the reproduction of prerecorded, synthesized, or live messages with intelligibility within acoustically distinguishable spaces (ADS). The ADS is determined by the system designer and is defined as follows:

Acoustically Distinguishable Space (ADS). An emergency communications system notification zone, or subdivision thereof, that might be an enclosed or otherwise physically defined space, or that might be distinguished from other spaces because of different acoustical, environmental, or use characteristics, such as reverberation time and ambient sound pressure level."

Section 18.4.10.2 requires each ADS as requiring or not requiring intelligibility. Not all areas will be designated as an ADS.

Fact

Speech intelligibility is affected by wall and floor coverings, furniture, sound pressure level, distortion, reverberation, and echoes. Prudent designers will disperse the signal using more low power appliances, rather than a few high-powered appliances.

The designer must evaluate if intelligibility will be required or not through analysis. This analysis must be submitted for review by the AHJ. Areas not designated as ADS are places where occupants will not spend great amounts of time. Note that Section 24.4.2.2.2.2 requires audibility in all areas, as required by Chapter 18 of the *Code*.

Intelligibility Standards

To achieve intelligibility, D.2.4.1 recommends a minimum measured score of 0.65 on the Common Intelligibility Scale (CIS), as defined by International Electrotechnical Commission (IEC) Standard 60849, *Sound Systems for Emergency Purposes*. The minimum average score must be at least 0.70. This score must be achieved in 90% of measurement locations in each ADS, as defined by the designer. This method of testing requires trained listeners in the area being tested. A value of 0.70 on the CIS is roughly equivalent to an 80% word-comprehension rate and a 95% sentence-comprehension rate, which are sufficient to transmit accurately an emergency message. It must, however, be noted that intelligibility requirements are not enforceable because they appear in Annex D of *NFPA 72*, rather than in the body of the document.

The Speech Transmission Index (STI) method is the intelligibility test method that allows a true quantitative measurement without the use of human subjects. An STI measurement instrument is available to measure intelligibility using the STI method in a timelier, less labor-intensive manner. The instrument requires a special signal generator connected to the fire alarm system panel amplifier, which transmits a test signal to the evacuation zones being tested. A hand-held device takes measurements at various points in the protected area, while simultaneously measuring sound pressure levels to deliver both an SPL measurement in dBA and an intelligibility measurement in STI. Thus, a technician can easily take several measurements within a few minutes. **See Figure 8-9.**

A minimum STI score of 0.5 is approximately equal to 0.7 on the CIS and is required for the communication to be considered intelligible.

> **18.4.10.2.1*** Unless specifically required by other governing laws, codes, or standards, or by other parts of this Code, intelligibility shall not be required in all ADSs.
>
> **18.4.10.3*** Where required by the enforcing authority; governing laws, codes, or standards; or other parts of this Code, ADS assignments shall be submitted for review and approval.
>
> *(Excerpt from NFPA 72.)*

Measuring Intelligibility

Chapters 18 and 24 of *NFPA 72* require voice signals to be intelligible throughout all ADS areas. A message must be as clear in occupancies traditionally plagued by low intelligibility, such as an atrium or an arena, as it is in a small meeting room in an office building. The designer will determine what areas are considered as ADS.

Although *NFPA 72* does not contain requirements for intelligibility test methods or test frequencies, Annex D does recommend such methods, most of which involve using trained talkers and listeners to test a system once it has been installed in an occupancy. Generally, the talker repeats words from a list, and the listener writes down what he or she has heard.

These human-based methods of measurement are expensive and time consuming, and some tests are not consistently repeatable. Others require large numbers of

Figure 8-9 STI Meter

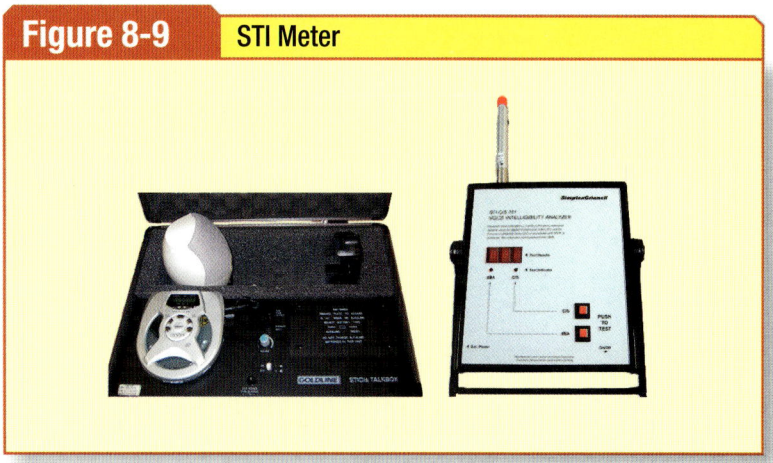

Figure 8-9. *A quantitative Speech Transmission Index (STI) meter (right) can be used to measure the intelligibility of an ECS using the STI scale. Annex D of* NFPA 72 *provides guidance in this area.*

test words and listeners, and they all have the potential for bias, intentional or not. For these reasons, human-based measurement methods are generally used only to settle disputes among the parties involved in designing and installing a system.

Remedies for low intelligibility must address the root of the problem causing the low intelligibility. For distortion, more speakers may need to be added at lower power settings. For echoes and reverberation, it may mean adding acoustical dampening materials or aiming a signal directly at the audience, or both. For low signal-to-noise ratios, remedies may include increased signal strength.

Even if a system meets intelligibility requirements when it is first installed, renovations or changes in furnishings may alter its performance. There is no *Code* requirement for re-testing a system's intelligibility when changes such as these have occurred.

CIRCUITS

Article 760 of *NFPA 70, National Electrical Code*, contains the requirements for wiring of fire alarm circuits. However, *NFPA 72-2010, National Fire Alarm and Signaling Code*, contains special performance requirements for survivability of fire alarm systems, which supplement *NEC* wiring requirements.

Survivability

Section 24.6.3.4.1 of *NFPA 72* clearly indicates the section applies only to systems used for partial evacuation or relocation of occupants. Section 24.6.3.4.2 clearly states that systems requiring total evacuation of occupants are not required to meet survivability requirements because occupants are required to immediately evacuate. The system would not be needed to provide further instructions to occupants who have already evacuated. Therefore, survivability requirements usually apply only to ECS.

Only those circuits providing occupant notification are required to have survivability. These provisions specifically apply to audible and visible notification appliance circuits, and any circuits that are necessary to provide communications. A network riser in a distributed system must, therefore, be survivable. Initiating (input side) device circuits and control circuits are exempt from these requirements.

The first survivability requirements are found in Section 24.6.3.4.1. This section requires in-building fire emergency voice/alarm communications systems to have a Level 2 or Level 3 survivability. The intent of this section is to prevent a fire in one notification zone from causing loss of

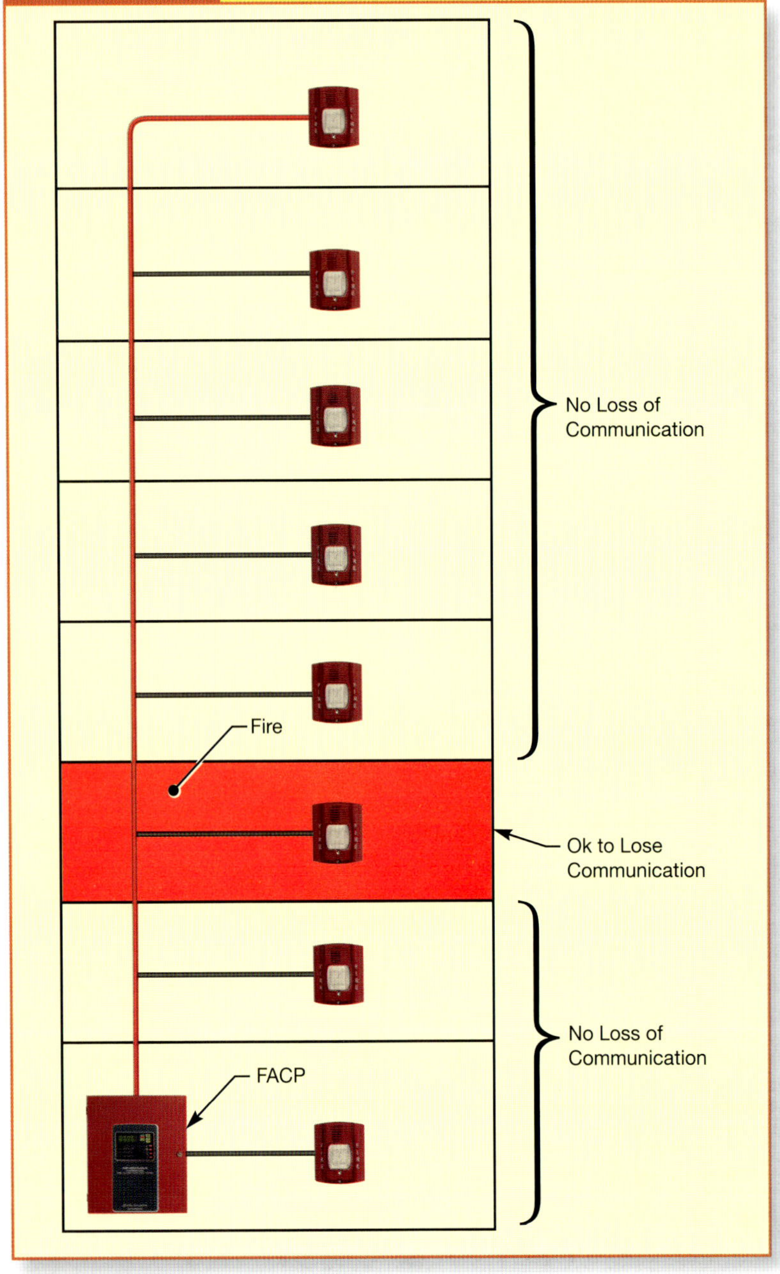

Figure 8-10. Survivability is intended to imply that a fire in one area cannot cause loss of communications to another area for a prescribed amount of time.

communication to another zone by destroying the circuit. **See Figure 8-10.**

The level of survivability for ECS circuits is defined by Chapter 12 of the *Code*. Section 12.4 defines the four levels of survivability.

This section was added to clearly establish levels of protection for systems installed to the requirements of Chapter 24.

In a high-rise building, a notification zone is usually a single floor, as required by 10.18.5.1 of *NFPA 72*. *NFPA 101*, *Life Safety Code*, and most building codes will not permit fire zones to consist of more than one floor, with very few exceptions. Therefore, notification appliance circuits must serve only the fire zone for which they provide annunciation.

12.4 Pathway Survivability. All pathways shall comply with *NFPA 70, National Electrical Code*.

12.4.1 Pathway Survivability Level 0. Level 0 pathways shall not be required to have any provisions for pathway survivability.

12.4.2 Pathway Survivability Level 1. Pathway survivability Level 1 shall consist of pathways in buildings that are fully protected by an automatic sprinkler system in accordance with NFPA 13, *Standard for the Installation of Sprinkler Systems*, with any interconnecting conductors, cables, or other physical pathways installed in metal raceways.

12.4.3 Pathway Survivability Level 2. Pathway survivability Level 2 shall consist of one or more of the following:

(1) 2-hour fire-rated circuit integrity (CI) cable
(2) 2-hour fire-rated cable system [electrical circuit protective system(s)]
(3) 2-hour fire-rated enclosure or protected area
(4) 2-hour performance alternatives approved by the authority having jurisdiction

12.4.4 Pathway Survivability Level 3. Pathway survivability Level 3 shall consist of pathways in buildings that are fully protected by an automatic sprinkler system in accordance with NFPA 13, *Standard for the Installation of Sprinkler Systems*, and one or more of the following:

(1) 2-hour fire rated circuit integrity (CI) cable
(2) 2-hour fire rated cable system (electrical circuit protective system(s))
(3) 2-hour fire rated enclosure or protected area
(4) 2-hour performance alternatives approved by the authority having jurisdiction

(Excerpt from NFPA 72.)

24.3.6 Pathway Survivability.

24.3.6.1 Pathway survivability levels shall be as described in Section 12.4.

24.3.6.2 Other component survivability shall comply with the provisions of 24.4.2.8.5.6.

24.3.6.3* The pathway survivability requirements in 24.3.6.4 through 24.3.6.12 shall apply to notification and communications circuits and other circuits necessary to ensure the continued operation of the emergency communications system.

24.3.6.4 In-building fire emergency voice/alarm communications systems shall comply with 24.3.6.4.1 or 24.3.6.4.2.

24.3.6.4.1 For systems employing relocation or partial evacuation, a Level 2 or Level 3 pathway survivability shall be required.

24.3.6.4.2 For systems that do not employ relocation or partial evacuation, a Level 0, Level 1, Level 2, or Level 3 pathway survivability shall be required.

24.3.6.4.3 Refer to Annex F for previous nomenclature and cross reference.

24.3.6.5 Pathway survivability levels for in-building mass notification systems shall be determined by the risk analysis.

24.3.6.6 Pathways survivability levels for wide area mass notification systems shall be determined by the risk analysis.

24.3.6.7 Two-way in-building wired emergency communications systems shall have a pathway survivability of Level 2 or Level 3.

(Excerpt from NFPA 72.)

Figure 8-11 — Level 0, Level 1, Level 2, and Level 3 Survivability

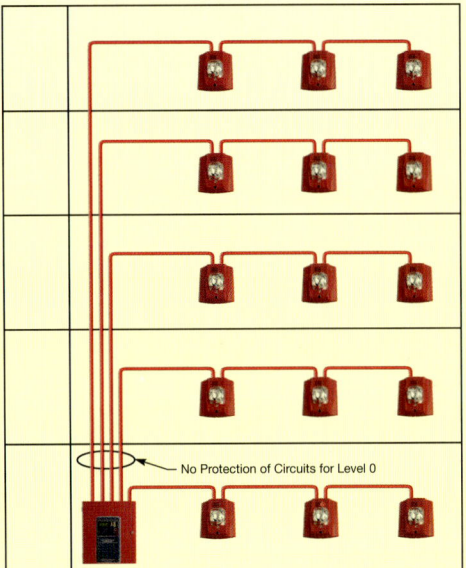

Figure 8-11(a). Pathway Level 0 survivability has no requirements for any survivability.

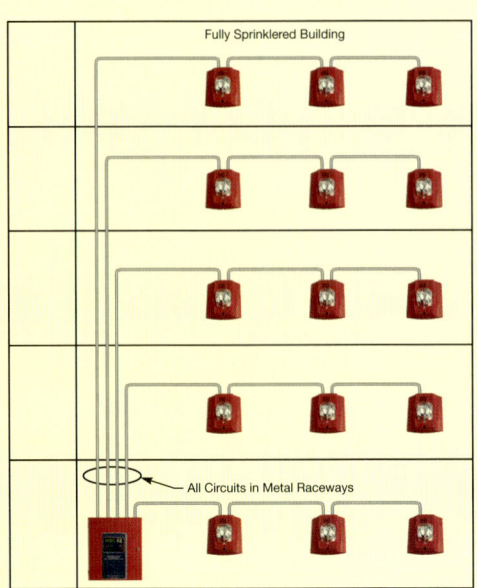

Figure 8-11(b). Pathway Level 1 survivability requires the entire building to be sprinklered per NFPA 13 *and all ECS wiring installed in metal raceways.*

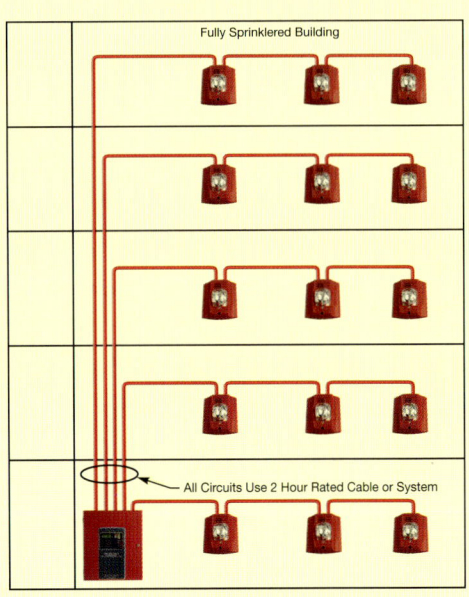

NOTE: Level 3 Survivability Option A is the same as Level 2, except that the building must be fully sprinklered.

Figure 8-11(c). Pathway Level 2 survivability requires all ECS wiring to be survivable for not less than two hours.

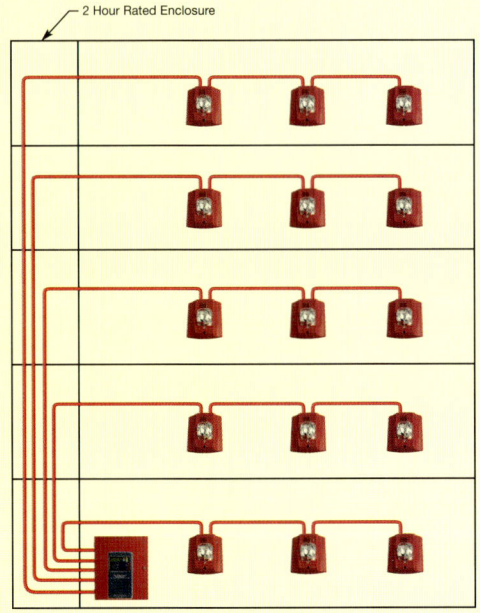

NOTE: Level 3 Survivability Option B is the same as Level 2, except that the building must be fully sprinklered.

Figure 8-11(d). Pathway Level 3 survivability requires all ECS wiring to be survivable for not less than two hours. Level 3 survivability also requires the entire building to be fully sprinklered, per NFPA 13.

Figure 8-11. Section 12.4 of NFPA 72 *defines four levels of survivability for pathways.*

The requirement for the equipment designed to resist attack by fire essentially requires the equipment to be evaluated and listed to ensure that the system will not fail if one circuit develops a fault during a fire. Using equipment that is listed by an independent testing laboratory will help ensure this requirement is satisfied. Unlike the *NEC*, Section 10.3 of *NFPA 72* requires all equipment to be listed for the purpose for which it is used. This means that the equipment will be listed for fire alarm signaling service.

Survivability basically requires the conductors of affected circuits to be protected between the fire command center or control equipment and the zone in which they serve. In other words, a fire in one zone cannot cause loss of communications to another zone. It is, however, assumed that a fire in a zone will cause loss of communications in that zone. This is deemed acceptable because the occupants will not stay there during a fire. Therefore, circuits do not require survivability in the paging zones they serve. **See Figure 8-11.**

Level 0 survivability requires no protection. Level 1 survivability requires building sprinklers and metal raceways. Level 2 survivability requires a minimum of two-hour circuit protection. Level 3 survivability is the same as Level 2, except that the entire building must be fully sprinklered, per *NFPA 13*.

Survivability is defined in Chapter 12 and is required by Chapter 24. At the moment, survivability is only required for systems using live voice instructions. It is not required to provide survivability for systems using a total evacuation scheme.

Survivability requirements apply to the operation of notification appliance circuits and to other circuits that are necessary for the transmission of communications signals. This specifically includes a signaling line circuit (data riser) providing a network connection between the main control unit and satellite control units containing a distributed amplification system. **See Figure 8-12.**

A two-hour rated shaft or enclosure can serve as a means of routing conductors from the fire command center to the notification zone they serve. However, a two-hour rated shaft or enclosure can be expensive to construct and will take up valuable floor space. This requirement applies to the entire conductor run between the control equipment and the notification zone. A two-hour rated enclosure or shaft will generally be constructed of masonry walls or steel studs with double five-eighths-inch layers of gypsum board on each side. One should consult the Underwriters Laboratories' *Fire Resistance Directory* or a fire protection engineer when exercising this option. The use of two-hour rated cables may eliminate the need to use a two-hour rated shaft, thereby saving money and usable floor area.

There are two types of cables that are two-hour rated. Mineral insulated (MI) cable and circuit integrity (CI) cable are both designed to maintain integrity for at least two hours under direct attack by fire. MI cable is certainly able to meet this requirement and will also provide good resistance to mechanical damage. However, it requires special termination fittings and can be difficult to work with, especially for those accustomed to low-voltage systems.

CI cable is a relatively new cable that is

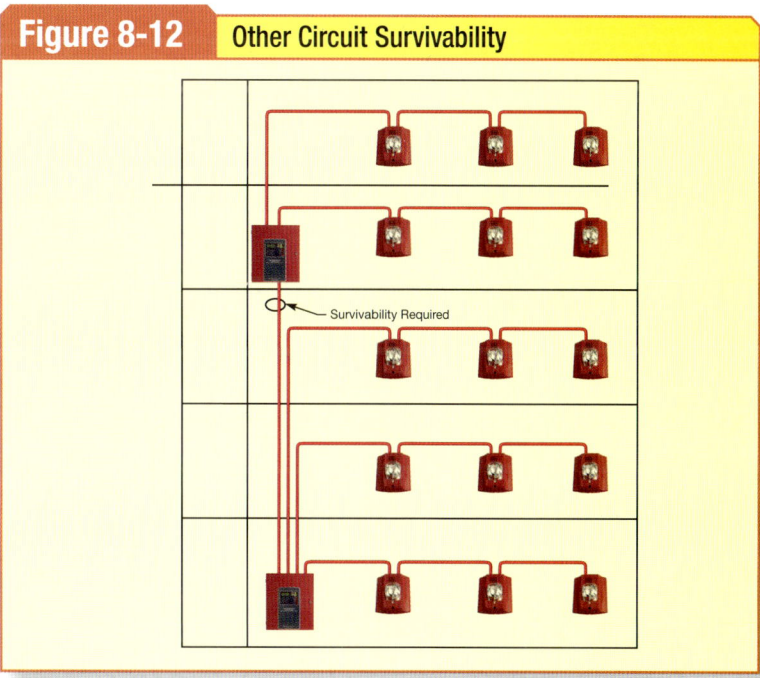

Figure 8-12. In this case, survivability is required for circuits which are essential for operation of communications between control units.

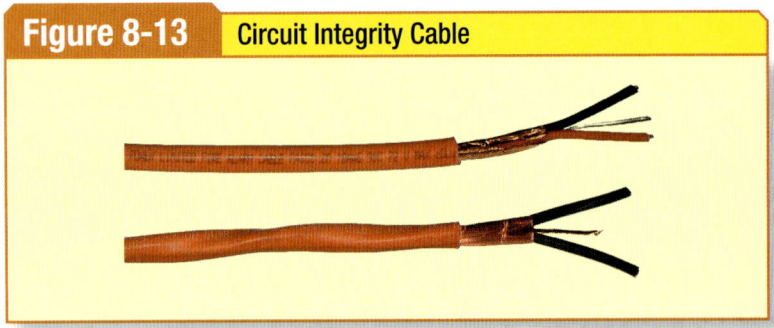

Figure 8-13. Depending on the brand used, it may be necessary to sleeve CI cable in a metal raceway in order to achieve the full 2-hour rating.

designed to provide fire alarm circuit integrity under attack by fire. The appearance of CI cable is similar to power-limited cables such as FPL, and most installers find it easy to install. **See Figure 8-13.**

CI cable can be identified by a "-CI" designation following the cable type on the outer jacket (for example, Type FPLP-CI). The plastic jacket and insulation is designed to crystallize and solidify when burned, thereby providing good insulation resistance. One drawback of CI cable is that the cable is relatively vulnerable to mechanical damage after burning. Running CI cable in a metallic conduit can provide resistance to mechanical damage, but it does not increase the fire rating of the cable.

An additional concern is that the cable must not be spliced between the fire command center or control equipment and the fire zone it serves because there are no listed fittings or terminations that have the two-hour rating. Splices in the cable will not have a two-hour rating, and they would compromise the cable's integrity when burned. By the same token, CI cable requires no special fittings and is installed just like any other power-limited cable.

Other wiring methods permitted by Chapter 3 of the *NEC* are not two-hour rated and cannot be used to satisfy this requirement. Additionally, using a metal conduit does not automatically provide a two-hour rating for the conductors inside. However, there is at least one CI cable on the market requiring installation in metallic conduit in order to meet a two-hour rating. The CI cable manufacturer also requires the metallic conduit to be securely mounted on a two-hour surface, such as concrete.

Sometimes, the central controls for the system are located in a different location from the fire command center. This is because the fire command center is not large enough to house all of the system controls. If the fire command center is remote from central control equipment, special requirements apply to the interconnected wiring between them. Section 24.4.2.8.5.4 requires the wiring between the fire command center and any central control equipment to be protected using Level 2 or Level 3 survivability.

Class A and Class X Circuit Separation

Class A and Class X circuits are not required by *NFPA 72*, nor are they required for emergency voice/alarm communications systems. However, where Class A and Class X circuits are installed, Section 12.3.7 of *NFPA 72, National Fire Alarm and Signaling Code*, requires the outgoing and return circuits to be routed separately from each other. This means the outgoing and return circuits must not be in the same cable, conduit, or raceway. The purpose of this requirement is to ensure that a fire in one space does not completely disable communications within that space. Class A and Class X circuit separation should not be confused with survivability.

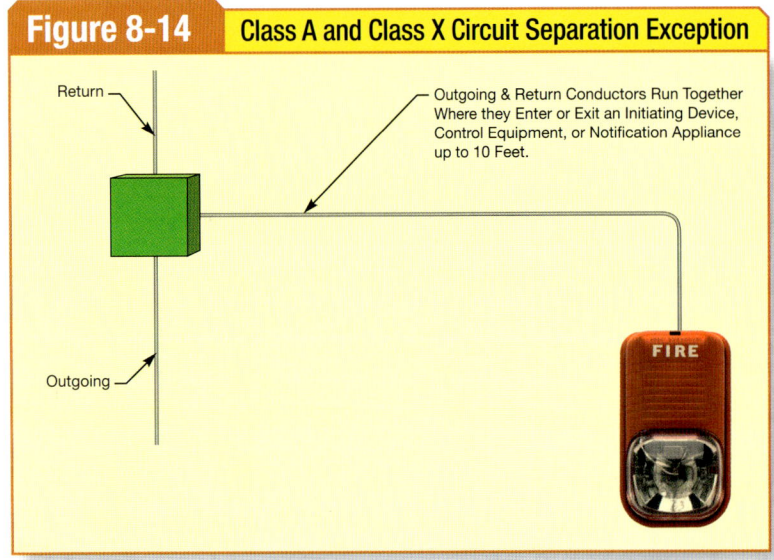

Figure 8-14. In this case, outgoing and return conductors can be routed together for up to 10 feet (3.0 m) where they enter or exit the initiating device, notification appliance, or control unit enclosure.

Section A.12.3.7 provides guidance on the Class A and Class X circuit separation requirement. This section recommends a separation of 12 inches (305 mm) where conductors are installed vertically, and 48 inches (1.22 m) where installed horizontally. However, this section is not enforceable as *Code* because it is located in the Annex. Some federal agencies, such as the Federal Aviation Administration (FAA), amend the *Code* to adopt Section 12.3.7 as mandatory text. In any event, the outgoing and return conductors should be installed as far apart as practicable.

Section 12.3.7 permits three limited cases where the outgoing and return conductors are in the same cable, conduit, or raceway. The first case is where outgoing and return conductors enter or exit an initiating device, notification appliance, or control enclosure. This "exception" permits conductors to be run in the same cable, conduit, or raceway for a distance of 10 feet (3 m). **See Figure 8-14.**

The second case is where the outgoing and return conductors may be in the same conduit or raceway drop to a single device or appliance. This exception does not contain a distance limitation, but it does require the use of raceway. **See Figure 8-15.**

The third exception applies to a single conduit or raceway drop to multiple devices or appliances installed in a single room not exceeding 1,000 square feet (92.9 square meters) in area. Again, the use of cable wiring methods is not permitted under this exception. **See Figure 8-16.**

The decision to use a Class A or Class X circuit is made by the designer. However, there are jurisdictions, such as the State of Rhode Island and the Nuclear Regulatory Commission (NRC), which require all alarm system wiring to be Class A circuits. Where the designer has a choice, the decision to use Class A or Class X circuits is based upon criteria, such as:
- Transmission media (copper or fiber)
- Length of the circuits
- Area of system coverage
- Hazards involved
- Condition of the occupants

Class A and Class X circuits are used on both survivable circuits and non-survivable systems. However, Class A and Class X circuit separation has no bearing on survivability issues.

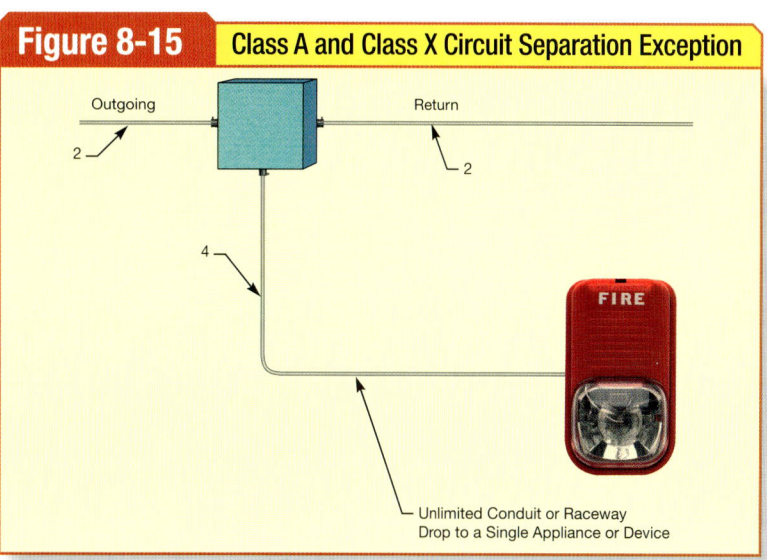

Figure 8-15. In this case, outgoing and return conductors can be routed together to a single initiating device or notification appliance using an unlimited length of conductors in raceway.

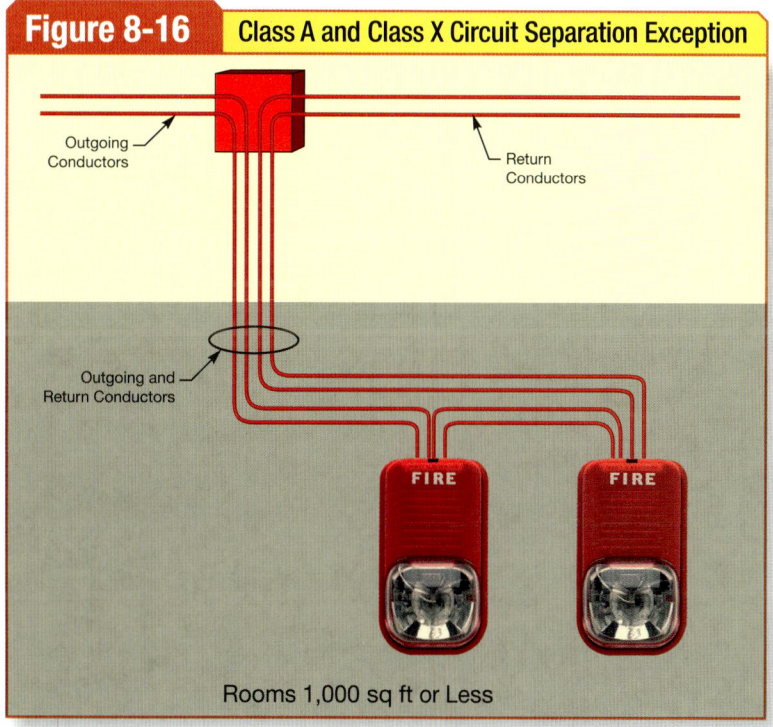

Figure 8-16. In this case, outgoing and return conductors can be routed together to multiple initiating devices or notification appliances using an unlimited length of conductors in raceway, to a room not exceeding 1,000 square feet (93 m²) in area.

Summary

In the post 9/11 world, many organizations need a better method of communicating with large numbers of people. Emergency communications systems achieve this goal and will become much more commonplace in the coming years. The fire protection community has been using voice/alarm communications systems for many years with good success. These systems specifically include one-way in-building fire emergency voice/alarm communications systems and two-way in-building wired emergency services communications systems. Other emergency services are adopting the same systems. Chapter 24 of *NFPA 72* was added to the *Code* to address such systems. With the addition of Chapter 24 came the need to address circuits and pathways. This promulgated the need for a new Chapter 12, which includes circuit survivability and circuit class.

Review Questions

1. An emergency voice/alarm communications system is most likely used where __?__.
    a. evacuation consists of the Temporal-3 signal
    b. evacuation is conducted in stages
    c. evacuation takes under 5 minutes
    d. none of the above

2. A high-rise is defined as a structure or building where the highest occupied level is more than __?__ above the lowest level of fire department access.
    a. 55'
    b. 65'
    c. 75'
    d. 85'

3. An emergency voice/alarm communications alert tone must sound at least __?__ times.
    a. 2
    b. 3
    c. 4
    d. 5

4. Smoke and fire barriers are usually located on these drawings.
    a. Architectural plans
    b. Floor plans
    c. Life-safety plans
    d. Sprinkler/fire alarm plans

Review Questions

5. **NFPA 72, National Fire Alarm and Signaling Code, requires a fire command center (FCC).**
   a. True
   b. False

6. **Emergency voice/alarm communications systems (EVACS) are required to have a secondary power supply capacity of __?__.**
   a. 1 hour
   b. 2 hours
   c. 3 hours
   d. 4 hours

7. **NFPA 72 requires signals to have an intelligibility level not less than __?__ on the STI.**
   a. 0.7
   b. 0.8
   c. 0.9
   d. 1.0

8. **Survivability requirements apply to which of the following systems or circuits?**
   a. Emergency voice/alarm communications systems
   b. Initiating device circuits
   c. Power supply branch circuits
   d. All of the above

9. **Which of the following might be used to satisfy survivability requirements?**
   a. Metal raceways in fully-sprinklered buildings
   b. 2-hour rated CI cable
   c. 2-hour rated enclosure
   d. All of the above

10. **Class A or Class X circuits __?__ be separated by a distance of not less than __?__ where run vertically.**
    a. Shall/1'
    b. Should/1'
    c. Shall/3'
    d. Should/3'

11. **Class A or Class X circuit conductors may be run together to a single appliance in a room not exceeding __?__.**
    a. 500 ft^2
    b. 750 ft^2
    c. 1,000 ft^2
    d. 1,500 ft^2

Public Emergency Systems and Supervising Stations

Fire alarm systems limited to alerting occupants of a building are called "protected premises" or "local" systems. Other kinds of systems are intended to alert people outside the building and to initiate some kind of response. This requires connection to a supervising station. Supervising stations have existed since Moses Farmer and William Channing built the first telegraph system in Boston in 1852. Technically, this system was a public fire emergency reporting system because only signals from street boxes were transmitted to the dispatcher. Later versions of public fire reporting systems transmitted alarm signals from inside buildings through contacts inside the street box. This arrangement is known as a master box. Today, these systems are called "public emergency alarm reporting systems." This chapter discusses off-premises reporting systems, including supervising stations and public emergency reporting systems.

Objectives

» Explain key events in the history of public alarm systems
» Describe the operating principles of the street box
» Identify the types of supervising stations and the *Code* requirements that apply
» Describe the communication methods used by supervising stations, including:
  › Multiplex
  › Radio
  › DACS
  › DARS
  › Direct-connect
  › Radio frequency

Chapter 9

Table of Contents

Public Emergency Alarm Reporting Systems 206

    Operation of Street Boxes and Circuits.... 206

    Code Requirements 207

    Master Boxes 208

Supervising Stations 209

    Central Station Alarm Systems and Central Station Service 210

Proprietary Supervising Station Alarm Systems 215

Remote Supervising Station Alarm Systems 217

Communications Methods for Supervising Stations 218

Summary 224

Review Questions 224

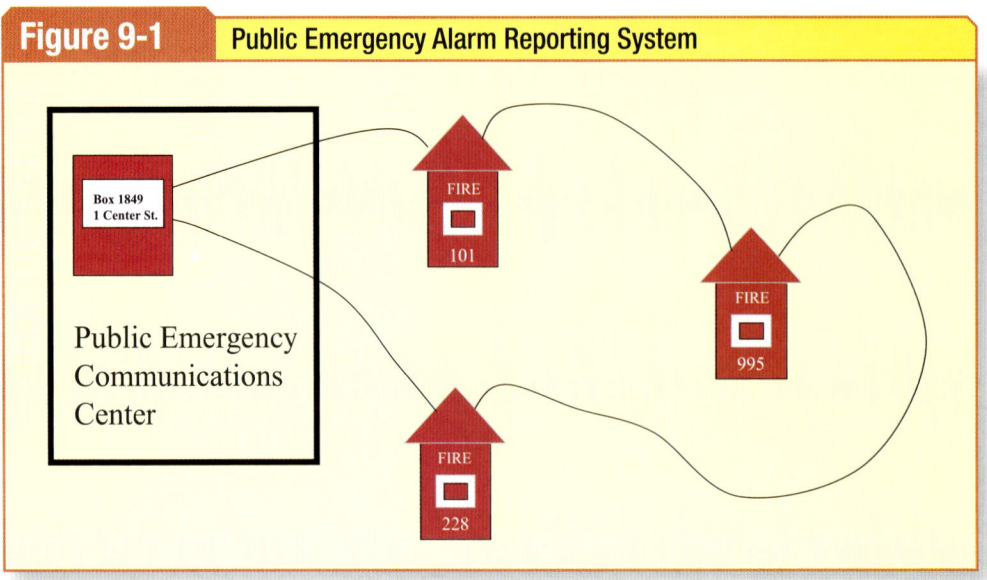

Figure 9-1. Each alarm box is distributed throughout the protected area and has a unique number. When the box is actuated, the unique number is sent to the public emergency communications center, the location where dispatchers will respond.

PUBLIC EMERGENCY ALARM REPORTING SYSTEMS

At noon on April 28, 1852, in Boston, Massachusetts, Moses Farmer and William Channing established the first public fire reporting system. Many of these systems are still in use today.

Public emergency alarm reporting systems were formerly called public fire reporting systems and are usually operated and maintained by local jurisdictions. There are many public emergency alarm reporting systems installed in the northeastern United States and on government facilities across the United States. A typical public emergency alarm reporting system uses street boxes, placed at specific intervals and key locations, which allow the public to actuate the box and report an emergency. **See Figure 9-1.**

Operation of Street Boxes and Circuits

Street boxes are connected to the public emergency communications dispatch center by a pair of conductors, sometimes referred to as a McCulloh loop or a Gamewell loop. McCulloh loops and Gamewell loops consist of a pair of conductors extending from the public fire dispatch center throughout neighborhoods in the protected area. Street boxes are connected to the loop and transmit a unique number in a fashion very similar to a telegraph. **See Figure 9-2.**

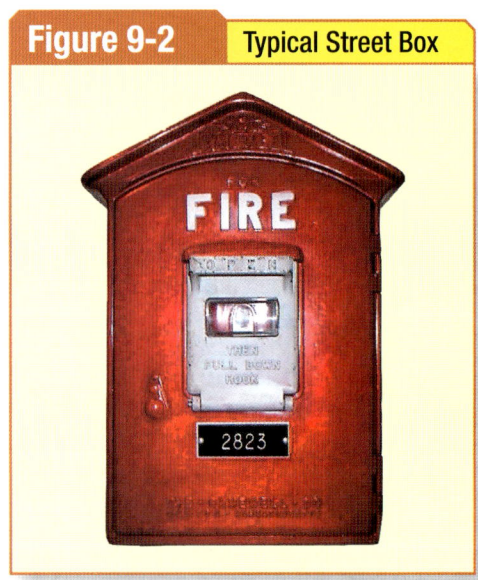

Figure 9-2. Each box has a unique number that, when actuated, is tapped out by a notched wheel inside the box. The box is spring wound and must be reset after each use.

Copper conductors are not the only means of transmitting signals to the public emergency communications dispatcher. Newer technologies, such as radio frequency (RF) and optical fiber cable, are also being used on public emergency alarm reporting systems. Unlike copper-wired public emergency alarm reporting systems, these systems can also transmit alarm, supervisory, and trouble signals.

Public emergency alarm reporting system loops are very reliable because they can operate on a single conductor using the earth as a return.

Code Requirements

Requirements for public emergency alarm reporting systems are found in Chapter 27 of *NFPA 72, National Fire Alarm and Signaling Code*. Public emergency alarm reporting systems are not considered supervising station alarm systems, but they are discussed here because of their similarities with supervising station alarm systems. In fact, Chapter 26 of *NFPA 72* recognizes the transmission medium (McCulloh loops) and in previous editions, permitted it to be used for communication between protected premises and a public emergency communications dispatch center. A public emergency communications dispatch center, however, is not a supervising station and is covered by *NFPA 1221: Standard for the Installation, Maintenance, and Use of Emergency Services Communications Systems*.

Public emergency alarm reporting systems were developed before the widespread use of telephone service and were the best technology at the time. One disadvantage to this arrangement is that only alarm signals are transmitted. Trouble and supervisory signals cannot be transmitted using this technology.

Each box on a system has a unique number. The street box shown is Box #2823. When the box is actuated, a spring-wound mechanism rotates a notched code wheel three times, which taps out the box number three times. Transmission of the number three times is intended to prevent misreading a garbled number. The number is transmitted to the emergency services communications center dispatcher over conductors maintained by the municipality, using a technology similar to telegraphs. The notched wheel operates a set of contacts, which send signals to the public emergency services communications center.

Most modern circuits use a Gamewell proprietary circuit technology similar to a McCulloh loop. See **Figure 9-3**. Box #245, for example, would tap out the signal three times.

Street boxes must be rewound following actuation in order to ensure they have sufficient energy to transmit another signal in the future.

Early public emergency alarm reporting systems required an operator to listen for the signal. Later systems used paper tapes, which punched the box number on the paper. An operator would then record the number and cross reference it to a street address from a list of boxes on the system. Newer systems often use

Fact

Protected premises (local) fire alarm systems will detect fires or actuation of a suppression system. However, nobody outside the building will know the system is actuated until it is connected to a supervising station.

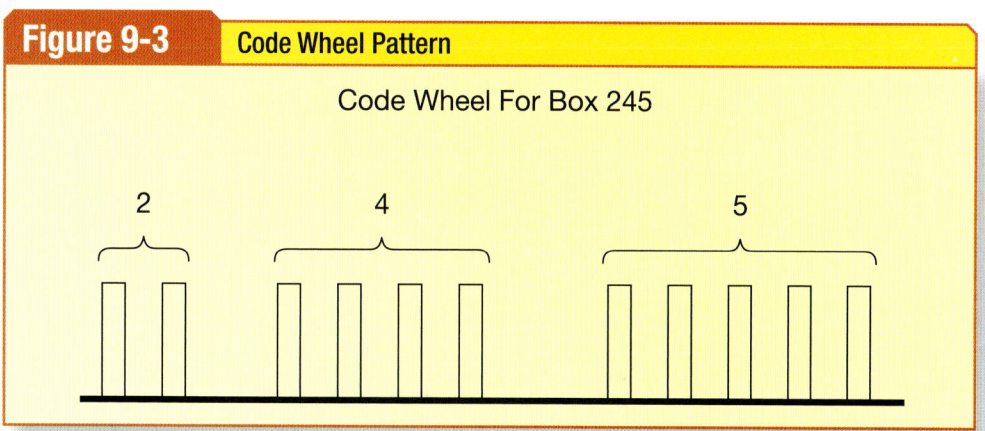

Figure 9-3. This signal is transmitted three times when the box is actuated. The only way to change the signal is to change the code wheel inside the box.

electronic converters (sometimes referred to by the trade name Digitize), which record the box number and convert it to a display with the street address of the box. This greatly shortens response time for the dispatcher. Street boxes are still, however, required to transmit each box number not less than three times. Following a box actuation, maintenance staff must rewind the box to ensure it has sufficient spring tension to operate the next time it is tripped.

Signals from public emergency alarm reporting systems are usually transmitted to the public fire dispatcher because these systems are owned and operated by local municipalities. However, there are some industrial and military facilities that use public emergency alarm reporting systems to protect their premises.

Public emergency alarm reporting systems usually operate on a 100 milliamperes direct-current loop. Public fire alarm reporting systems are incredibly resistant to open circuits and ground faults. Most of these systems can operate on a single conductor using the earth as a return path for the current. Since these systems operate on direct current, the secondary power supply is usually a large lead-acid battery located at the public emergency dispatch center.

Type A systems are required by Section 27.5.21.2 of *NFPA 72* where the total number of alarms on the system exceeds 2,500 per year. Type A systems are permitted to use automatic retransmission, provided that there is automatic receipt, storage, and retrieval of signals and the operator has the capability to revert to manual retransmission. Type B systems are used where there are fewer than 2,500 alarms per year. In all cases, there must be a device at the receiving center to produce a permanent recording of all signals.

27.6.3.2.2.1 of *NFPA 72* provides requirements for local energy and shunt-type systems. Shunt-type and local energy public emergency alarm reporting systems actually permit the city loop to be routed through buildings. Contact closure-type initiating devices, such as heat detectors and manual fire alarm boxes, were connected to the circuit inside the building. Fires or mechanical damage inside the building sometimes resulted in loss of the loop or a portion of the loop. Current *Code* requirements permit the city loop to be routed through buildings only where isolation disconnects are installed outside the building, where the master box is as close to the entrance as possible, and where the conductors are installed in rigid conduit or EMT. It should be noted that these systems are no longer widely installed but are still permitted by *NFPA 72*.

Master Boxes

A master box arrangement is designed so actuation of a fire alarm control unit inside the protected premises trips a relay located inside the master box. The actuation of the master box relay has the same effect as pulling the hook on the box, thereby transmitting an alarm to the public fire dispatcher. Trouble and supervisory signals are not transmitted on Gamewell or McCulloh loops because the loop is capable of transmitting only alarm signals. Newer systems using RF or optical fiber transmission may transmit all three signals.

Master boxes must be isolated from the system when the protected premises system is being tested. Isolation of the box is commonly referred to as "plugging out" the box. Plugging out the box is accomplished through the use of a shorting plug located inside the box, which, when inserted into the master box relay, will prevent transmission of alarm signals to the public fire dispatcher. The box will still, however, transmit alarms if manually pulled. In most jurisdictions, only the fire department signaling shop is permitted to plug out a box.

There are few, if any, new public emergency alarm reporting systems being installed today. The widespread use of cellular telephones and the high maintenance costs of plant equipment have caused a decline in the installation of public fire reporting systems. Some jurisdictions have actually removed existing public emergency alarm reporting systems and some have replaced them with wireless (RF) systems. Many wireless public emergency alarm system boxes look like

Fact

Many public fire reporting systems require a significant amount of maintenance and are maintained by the local fire department.

standard street boxes used on wired public emergency alarm reporting systems and operate the same way.

Public emergency alarm reporting systems look and act much like supervising station systems. But for the purposes of *Code* requirements, they are a different technology. Chapter 27 of *NFPA 72, National Fire Alarm and Signaling Code,* provides requirements for the installation and operation of public fire reporting systems.

SUPERVISING STATIONS

Supervising station alarm systems are capable of monitoring fire alarm systems from distant locations, either across town or across the country. Signals are transmitted from the protected premises to a supervising station, where the signals are retransmitted to the public fire dispatcher. Chapter 26 of *NFPA 72* covers supervising station alarm systems and their associated communications methods.

Supervising stations may be stand-alone facilities, but they may also be operated in parallel with other supervising stations known as subsidiary supervising stations. Connections between the protected premises and the supervising station are known as "legs." Communication pathways between the supervising station and subsidiary supervising station are known as "trunks." **See Figure 9-4.**

There are three types of supervising stations permitted by *NFPA 72*:
1. Central station alarm systems
2. Proprietary supervising station alarm systems
3. Remote supervising station alarm systems

Supervising station connections may be required by an insurance carrier, *NFPA 101, Life Safety Code, International Building Code,* local codes, or by an authority having jurisdiction (AHJ). Many local codes do not require any type of supervising station connection for most occupancy types. *NFPA 101* and local building codes usually require schools and hospitals to have some type of supervising station connection, called "emergency forces notification."

According to Section 26.3.7.2, the operation of central station alarm protective service must be the primary function of the operators. Although not specifically stated in *NFPA 72,* supervising stations cannot operate other businesses at the same location as the protective

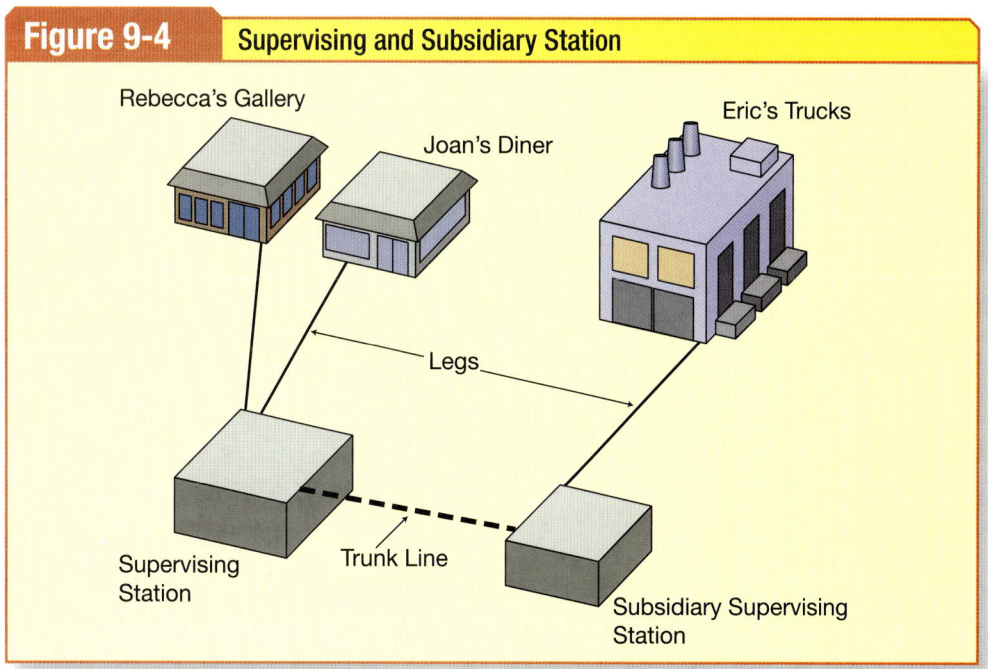

Figure 9-4. The supervising station may or may not have a subsidiary supervising station. Protected premises are monitored from centralized locations, where trained staff are on duty 24/7.

service. However, burglary/intrusion detection systems are often monitored from the same location because fire alarms and burglar alarms are closely related and may be integrated as a single system.

The scope of supervising station systems includes the control equipment and transmitters at the protected premises. Therefore, fire alarm controls at the protected premises must be listed for the type of connection or service provided. Controls used for central station service must be listed for central station use. The specification sheets must indicate that the equipment is listed for the type of service to be provided. It is possible that the equipment may be listed for one or more of the following: central station, remote station, and proprietary supervising station connections.

Chapter 26 of *NFPA 72* is divided into several major parts. Section 26.3 applies to central stations, Section 26.4 applies to proprietary supervising stations, and Section 26.5 covers remote supervising stations. Section 26.6 provides the requirements for the transmission methods between the supervising station and the protected premises. Although there are many transmission methods, the most common method is the digital alarm communicator system (DACS). RF transmission is becoming popular in urban areas. Additionally, newer methods, such as internet based supervision, are being implemented.

Central Station Alarm Systems and Central Station Service

Central stations are usually remotely located receiving points for alarm, trouble, and supervisory signals from protected premises. In addition to central stations, subsidiary supervising stations are sometimes used to expand areas of coverage and to provide redundancy. Subsidiary supervising stations are built to the same requirements as central stations and can act as the central station in the event of a failure at the central station. Subsidiary supervising stations may or may not be staffed.

Central stations offer the highest level of service that can be provided among the three types of supervising stations.

An independent testing laboratory must evaluate central stations to national standards. If the facilities and operations meet these standards, the central station is published in a listing, just like any other listed product or service. Underwriters Laboratories is the only nationally recognized testing laboratory that evaluates and lists central stations.

Listing Requirements. A central station connection includes more than monitoring. It is a contracted service and must include periodic inspections, testing, and maintenance as required by Chapter 14. Central station service includes the supervising station, receivers, transmitters, and the entire system installed at the protected premises.

Central station service must be provided under contract and Section 26.3.2 requires the following six elements:
1. Installation of the transmitters
2. Signal monitoring
3. Retransmission
4. Record keeping and reporting
5. Testing and maintenance
6. Runner service

These elements are provided as part of central station service and must be provided under contract with the subscriber. Section 26.3.3 of *NFPA 72* provides the methods by which the central station contract is provided. **See Figure 9-5.**

There are four basic methods by which a contract may be provided:
1. The listed central station provides all six elements.
2. The listed central station provides the monitoring, retransmission, and record keeping and subcontracts all or part of the installation, testing and maintenance, and runner service. Any part of the installation, testing, maintenance, and runner service may be subcontracted to another party.
3. A listed alarm service–local company provides the installation, testing, and maintenance and subcontracts the monitoring, retransmission, record keeping, and reporting to a listed central station. Runner service can be provided by either the listed

Fact

Central station service is the highest level of service available to the consumer. It requires a contract between the owner (subscriber) and the service provider (prime contractor). Only listed prime contractors may provide central station service.

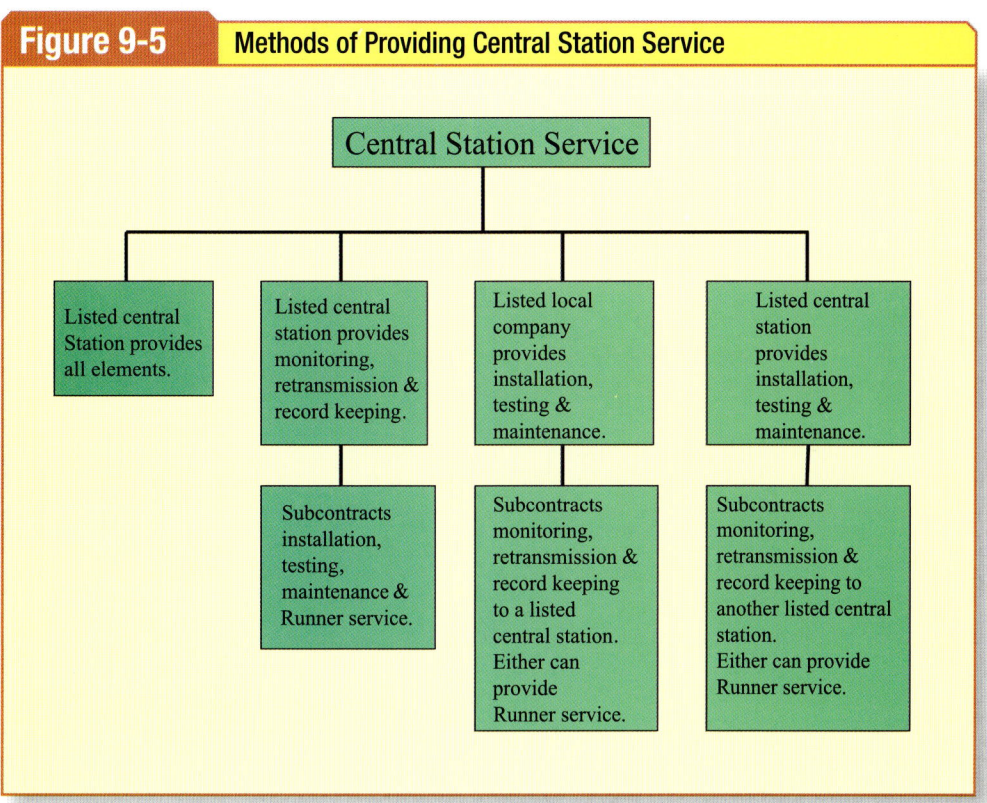

Figure 9-5. In all cases, the prime contractor must be listed and the contract for central station service must be between the prime contractor and subscriber.

alarm service–local company or the listed central station.
4. A listed central station provides the installation, testing, and maintenance and subcontracts the monitoring, retransmission, and record keeping to another listed central station. Either listed central station can provide the runner service.

The prime contractor must always be either a listed central station or a listed local fire alarm service company. There are hundreds of listed central stations and approximately 100 listed local fire alarm service companies in the United States. Section 3.3.202 of *NFPA 72* defines the term "prime contractor" as follows:

Prime Contractor. The one company contractually responsible for providing central station services to a subscriber as required by this Code. The prime contractor can be either a listed central station or a listed alarm service–local company.

Central stations and central station service can be listed just like any other listed product or service. An initial investigation of the product or service is conducted to ensure compliance. If the product or service complies with the applicable standards, the listing authority gives the manufacturer or service provider permission to apply a listing mark to products, which can include fire alarm system installations. In the case of central station service, the system is certified by hanging a certificate or placard at the protected premises.

Underwriters Laboratories maintains a directory of listed central stations and local fire alarm service companies that appear in the UL Fire Protection Equipment Directory, under Category UUJS. UL also lists remote supervising stations, sprinkler supervisory service systems, and local systems. These and all UL listings are also available online at the UL website, www.ul.com.

The UL listing process involves a modest initial investment and long-term commitment on the part of the prime contractor. Annual follow-up inspections are required as part of this process. The listed central station or local fire alarm service company must be completely committed to strictly adhering to *Code* requirements and to high quality. UL works with the prime contractors to ensure they follow appropriate UL standards, NFPA codes/standards, specifications, and drawings during an investigation period and provides follow-up inspections on a periodic basis.

For central stations, UL also assesses the effectiveness of the staff monitoring the receiving equipment. For example, alarm signals must be retransmitted to the public fire dispatcher in the appropriate amount of time, without error. The facilities must also meet UL827, *Standard for Central-Station Alarm Services*. This standard requires facilities receiving signals to be of fire-resistant construction, be located outside flood plains, have a sally port for security, have a minimum 24-hour secondary power supply, and have other safety features. Subsidiary stations must meet these same requirements because they may be staffed in emergencies. **See Figure 9-6.**

Central stations or local fire alarm service companies must be listed by a nationally recognized testing laboratory. Again, the only listing authority for prime contractors is Underwriters Laboratories. Section 26.3.4 of NFPA 72 requires an indication of compliance for all central station systems. Compliance of a central station system must be identified by a document that provides a minimum amount of information, as required by Section 26.3.4.2, including the following:

1. Name of the prime contractor involved with the ongoing *Code* compliance of the central station service
2. Full description of the alarm system as installed
3. Issue and expiration dates of the documentation
4. Name, address, and contact information of the organization issuing the document

Figure 9-6. *Trained staff members are on duty 24/7 to receive signals from the protected premises. Alarms are immediately dispatched to the local fire department.*

5. Identification of the authority(ies) having jurisdiction for the central station service installation

The requirements of Section 26.3.4 do not preclude other listing organizations. Factory Mutual Global (FM Global) requires a placard, containing the same information as a UL certificate.

When a system installation is near completion, the prime contractor applies for the compliance documentation by sending the project drawings, specifications, cut sheets, and other documentation to the listing agency for approval. If all documentation appears to satisfy *Code* requirements, the documentation is issued. **See Figure 9-7.**

Some entities, including FM Global, may have similar requirements for certificates or placards indicating compliance with applicable codes and standards. *NFPA 72* only requires an indication of compliance. The only entity requiring a certificate is Underwriters Laboratories.

Once the system is installed, the prime contractor must indicate compliance with the *Code* by placing the documentation within three feet (1 m) of the main controls, as required by Section 26.3.4.3.

The documentation required by Section 26.3.4.1 does not guarantee that the system complies with applicable codes and standards. It is a declaration by the listed central station or listed alarm service–local company that the system

Fact

Central station service is generally more expensive than other supervising station services because it includes testing and maintenance. Additionally, there is a certain level of overhead for the listing, which is intended to ensure minimal *Code* compliance.

complies with applicable codes and standards. The listed central station or listed alarm service–local company will generally not risk revocation of its listing by doing non-compliant work. The installation is often inspected by the AHJ upon completion, but this is not considered adequate by UL.

Listing agencies do not inspect every installed system for compliance. However, the same can be said for listed products of any kind. Most listing agencies do, however, conduct follow-up service where a sampling of installations (products) is field inspected by their staff. For central stations, inspectors conduct surprise inspections in conjunction with the local fire department. During these inspections, the operations are tested to ensure signals are properly handled.

According to the Central Station Alarm Association, nearly 60% of all monitored fire alarm systems are connected to listed central stations. However, only about 3% to 5% of these systems actually comply with Section 26.3. The other connected systems more likely fall into the remote supervising station category, even though many people still refer to them as central stations. Again, only a system meeting all the requirements of Section 26.3 is considered a central station.

Code Requirements. Central station facility requirements are covered by the requirements of Section 26.3.5 of *NFPA 72*. Section 26.3.5.1 requires that the building or buildings be resistant to intrusion and have fire protection features, emergency lighting, and standby power according to UL827, *Standard for Central-Station Alarm Services*. Central stations are often connected to subsidiary supervising stations to expand their areas of coverage. Subsidiary stations must also conform to UL827 and must be monitored by the central station.

Section 26.3.6 provides requirements for central station facilities and personnel. Central stations must receive trouble, supervisory, guard's tour, and alarm signals. Guard's tour signals are not frequently used because of high labor costs. However, when used, a

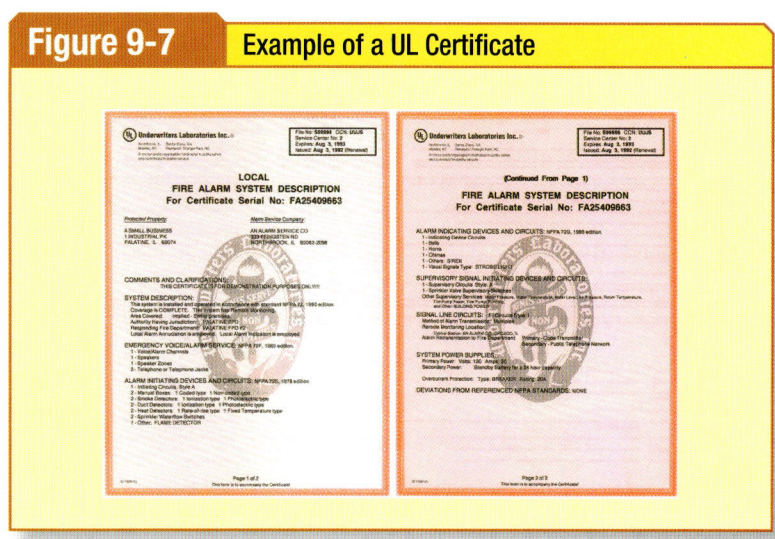

Figure 9-7. *This certificate indicates compliance of the system to UL and Code requirements and must be posted within three feet (1 m) of the main system controls at the protected premises.*

roving guard must initiate switch stations around the premises in a prescribed order and in a prescribed amount of time. Failure to initiate the signals in the proper time and order results in a guard's tour delinquency signal.

All signals received at the central station must be recorded on a permanent means, such as a paper printer. Following a fire, these records may be used by the AHJ or insurance interests to reconstruct a sequence of events leading up to the fire. There have been cases where these records have helped to convict arsonists.

Section 26.3.6.3 requires central station equipment and operating software to be listed for the application. Software or equipment listed only for burglary use is not appropriate for central station fire alarm use. Section 23.3.6.4 requires central station power supplies to comply with Chapter 4 of the *NEC*. In other words, primary power may be supplied by a utility, and at least 24 hours of secondary power is required.

Section 26.3.6.6 requires at least two independent means of retransmission to be provided. Retransmission is the relaying of the alarm signals to the public fire dispatcher. "Two independent means of retransmission" implies at least two separate telephone instruments. Most central stations have many operators, and at least one telephone instrument must be provided for each operator.

Central station operators cannot call 911 to retransmit alarm signals because the fire department dispatcher would receive the address for the central station. Therefore, Section 23.3.6.6.1 does not permit this practice. Central stations must have a list of direct-dial fire department dispatcher numbers for each of the protected premises. These numbers are usually stored in the software database for the premises. The central station personnel must also record the time and date of each retransmission. Additionally, each means of retransmission must be tested at least daily.

Section 26.3.7.1 requires central stations to have at least two operators on duty at all times. This staffing level helps ensure that the station can meet the time limits for retransmission. Central station operators are not permitted to conduct other functions that would preclude them from providing the protective service. However, central stations frequently monitor burglary systems or remote supervising station signals. This practice is permitted by UL with the understanding that fire alarm and life-safety signals have priority.

Section 26.3.8 provides the requirements for the disposition of signals by operators. Section 26.3.8.1.2 requires alarm signals to be immediately retransmitted to the communications center. Retransmission is accomplished by the use of a telephone instrument. The word "immediately" is intended to imply a period of time not exceeding 90 seconds, as noted in A.26.3.8.1.2(1).

A verification call to the premises (before retransmission) is not permitted in any case. Section 29.7.9.2 only permits a verification call for dwellings that use remote station systems, provided that it does not delay retransmission by more than 90 seconds. With that said, very few, if any, dwellings have central station service.

Following successful retransmission of an alarm signal, the operator must dispatch a technician or runner to arrive within two hours when equipment requires resetting. The runner or technician will be required to reset the equipment or investigate alarm signals.

Guard's tour delinquency signals must be processed quickly, as the guard may be incapacitated. The operator must communicate with the guard at the protected premises. But, failure to communicate with the guard must result in the dispatch of a runner within 30 minutes. There have been documented cases where the delinquency signal has resulted in action that saved the life of a guard who was having a heart attack.

Supervisory signals do not have the urgency of fire alarms and guard's tour signals. For this reason, central station operators must immediately communicate with the subscriber or the designated representative. The operator must also dispatch a runner to investigate within two hours. The operator may be required by the AHJ to report all supervisory signals and must inform the AHJ when a suppression system has been out of service for more than eight hours.

Trouble signals are processed by operators in a similar fashion to supervisory signals. The operator must immediately communicate with the subscriber or the subscriber's designated representative. However, a runner must be sent to investigate within a four-hour period. Fire alarm system impairments lasting more than eight hours require the operator to notify the AHJ.

Section 26.3.8.5.1 requires the central station to record all test signals along with the time and date. This includes test signals sent to the central station by the subscriber's equipment. Of great importance to installer/maintainers are Sections 26.3.8.5.5 and 26.3.8.5.6. These sections require all prime contractors to issue unique passwords to service personnel and for service personnel to produce a unique password when placing the system into test. Unless the central station operator receives the correct password, the system is not placed in the test mode. This helps prevent arsonists from delaying a response to alarms. Central stations must retain all records for a

Fact

Proprietary supervising stations are permitted to monitor only accounts under ownership of the proprietary supervising station owner. The properties monitored are not, however, required to be located on the same property as the supervising station.

period of not less than one year, as required by Section 26.3.9.1. The AHJ may request these records at any time, for any reason. There have been cases where supervising station records were used to re-create a timeline for the fire event during an investigation.

Proprietary Supervising Station Alarm Systems

As the name implies, proprietary supervising stations are owned and operated by the owner of the protected premises. By definition, a proprietary supervising station cannot monitor properties owned by others. The supervising station is required to be located on property owned by the owner of the protected premises. This arrangement offers control over hazards that can adversely affect the ability to monitor the premises.

Many proprietary supervising stations are located in a separate building on the same property as the protected premises, such as a college campus. It is not necessary, however, for the supervising station to be located on the same property as is the protected premises' system(s). Proprietary supervising stations may also be on a separate property, as would be the case for a supervising station that monitors a chain of grocery stores. In any event, the owner of the protected premises must own the property where the supervising station is located.

Proprietary supervising stations often look much like central stations in terms of equipment. They may contain computers and computer displays running specially designed software providing operators with information. They may also use smaller systems with LCD displays located on the main controls and may look more like a protected premises system. **See Figure 9-8.**

Section 26.4 of *NFPA 72* provides requirements for proprietary supervising station facilities. Sections 26.4.5.4 and 26.4.5.5 do not permit the operators to engage in activities that could preclude

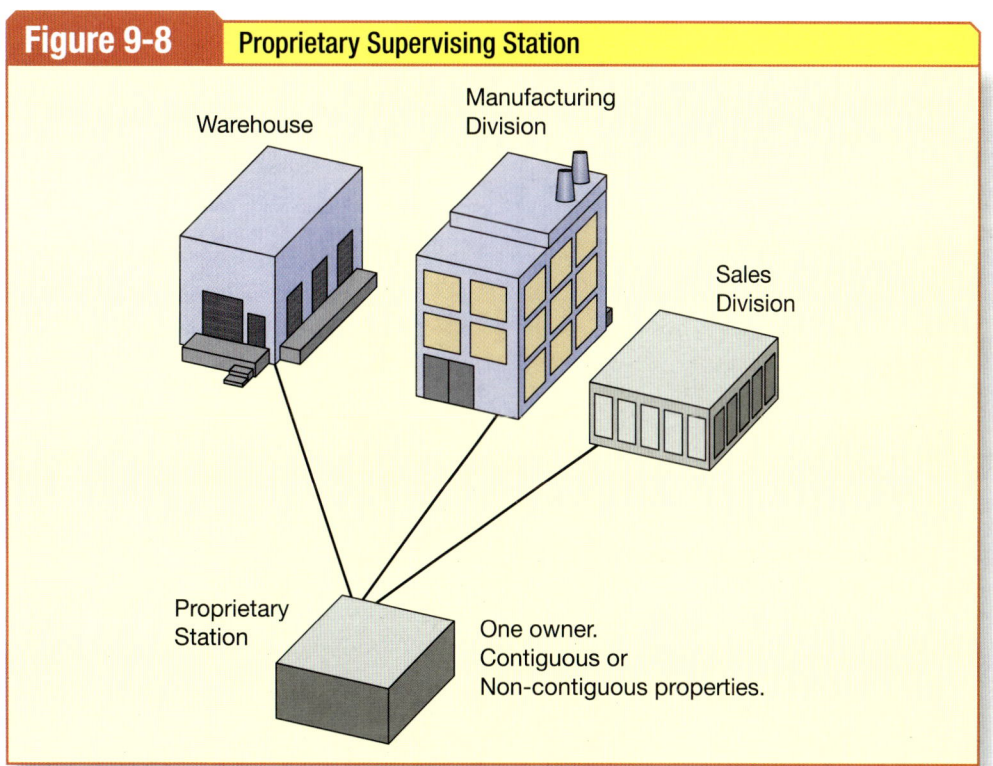

Figure 9-8. Proprietary supervising stations may only monitor properties owned by the owner of the supervising station. The supervising station may be located on the same property as the protected premises, or may be located elsewhere, but must be protected from hazards.

them from performing their duties. This is similar to the requirements for central station service.

As with central stations, security of the facility is of great concern. Access to a proprietary supervising station must be restricted to persons concerned with operation of the service, as required by Section 26.4.3.2. This prevents the operations area from becoming a gathering place for security guards on break, which may distract the operators.

Section 26.4.3.1 requires the proprietary supervising station to be located in a fire-resistive cut-off room or detached building that is not exposed to other hazards on the property. For example, it would not be permitted to install a proprietary supervising station next to a bulk storage facility for propane gas. A safe location would mean operating the protective service from a building located a safe distance away from the hazard.

Section 26.4.3.3 requires portable fire extinguishers to be provided in proprietary supervising stations, but there is no requirement to otherwise provide automatic fire sprinklers. Additionally, Section 26.4.3.4 requires the proprietary supervising station to be provided with an emergency lighting system in case of a primary power failure. This system must provide lighting for a period of not less than 26 hours.

Proprietary supervising stations must have equipment that is capable of indicating the building where the signal originates. Section 26.4.4.1.3 also requires either the protected premises system or the proprietary supervising station system to indicate the floor, zone, or subdivision where the signal originates, unless otherwise permitted by the AHJ. Signaling in the proprietary supervising station is considered as private mode signaling because operators are supposed to pay attention to the equipment.

Section 26.4.4.2 contains requirements for signal-alerting equipment. Section 26.4.4.2.1 requires at least two different means for alerting operators, in addition to a permanent recording device such as a printer. At least one of the alerting means must be an audible signal, which sounds until the signal is acknowledged. The audible signal must sound for trouble, supervisory, and alarm signals. Additionally, the same audible signal may be used to alert operators of all signals.

The staffing of proprietary supervising stations depends on the number of signals received. Section 26.4.5.1 requires a minimum of two operators on duty at all times. However, the exception to Section 26.4.5.1 permits a single operator if signals are automatically retransmitted to the fire department. The intent of this section is to prevent delays if an operator is incapacitated or taking a bathroom break.

As with central stations, runners are an integral part of the operation. Runners are sent to investigate signals and reset equipment. Section 26.4.5.2 requires the runner to establish two-way communication with the station at 15-minute intervals when he or she is not in attendance at the station. Section 26.4.5.3 was revised in the 2013 edition of *NFPA 72* to address the case where a runner is responsible for a single protected premises. This section permits telephone contact with a runner physically in attendance at a noncontiguous protected premises and immediately available, if that runner is not responsible for another protected premises. This section, however, requires two operators in the proprietary station at all times.

The primary duty of the operators is to operate the protective service. Sections 26.4.5.4 and 26.4.5.5 do not permit operators to take on additional duties that might interfere with the operation of the supervising station.

Section 26.4.6.6 provides the requirements for the disposition of signals by operators. The requirements for proprietary supervising station signal processing mirror those found in Section 26.3 for central stations. Section 26.4.6.6.1 requires alarm signals to be immediately retransmitted to the public fire dispatcher, emergency response team, or other party as required by the AHJ. Retransmission is accomplished by use of a telephone instrument. The word "immediately" is intended to imply a period of time not exceeding 90 seconds.

Fact

Most proprietary supervising alarm systems are found on campus settings. This may include colleges and universities as well as industrial and military applications.

A verification call to the premises is not permitted in any case. Following successful transmission of the alarm signal, the operator must dispatch a technician or runner to arrive within two hours. The runner or technician will be required to reset the equipment or investigate alarm signals. Following retransmission, the operator must also dispatch a runner to arrive within two hours after receipt of the alarm signal. Only after the signal cause is determined can the operator reset the system.

Section 26.4.6.6.2 provides requirements for disposition of guard's tour delinquency signals. The operator must immediately communicate with the guard at the protected premises by telephone, radio, or other approved means. Failure to communicate with the guard must result in dispatch of a runner within 30 minutes.

Supervisory signals require immediate communication with the subscriber or the designated representative to determine the nature of the problem, as covered by Section 26.4.6.6.3. The operator must also dispatch a runner to investigate within two hours. The operator may be required by the AHJ to report all supervisory signals but must inform the AHJ when a suppression system has been out of service for more than eight hours.

Section 26.4.6.6.4 requires trouble signals to be processed by operators in a fashion similar to supervisory signals. The operator must immediately communicate with the subscriber or the subscriber's designated representative. However, a runner is required to be sent to investigate within a four-hour period. Impairments lasting more than eight hours require the operator to notify the AHJ.

Section 26.4.7 requires the proprietary supervising station to retain all records for a period of not less than one year. The AHJ may request these records at any time, for any reason.

Remote Supervising Station Alarm Systems

Remote supervising stations were first developed in the 1950s. They were often operated by a local fire department dispatcher, but they could also be operated by a private entity. Remote supervising stations were almost always collocated with the public fire dispatcher. However, many remote supervising stations were privatized in the 1970s.

Requirements for remote supervising stations are found in Section 26.5 of *NFPA 72*. Generally, remote station connections are used where central station service is neither chosen nor available. The original intent of this section was to offer a type of supervision for small sprinkler supervisory systems where a central station would not be practical. The architecture of remote supervising stations is often very similar to central stations, but there are far fewer restrictions on this type of connection. Many central stations provide remote station connections even though they far exceed the requirements found in Section 26.5.

Many remote station alarm systems are not properly maintained. Section 26.5.2 requires an indication of compliance for remote station service. Owners of the system must now provide annual documentation to the AHJ that identifies the party responsible for inspection, testing, and maintenance of the system. This documentation must take one of the following forms:

1. An affidavit attesting to the responsibilities and qualifications of the parties performing the inspection, testing, and maintenance of the system signed by a representative of the service provider
2. Documentation indicating *Code* compliance of the remote station alarm system issued by the organization that listed the service provider
3. Other documentation acceptable to the AHJ

Remote stations originally were collocated with the public fire dispatcher, and only alarm signals were transmitted to the operator. Trouble and supervisory signals were usually transmitted to another location and were not required to be transmitted. Modern remote stations are generally no longer located at the public fire dispatcher, and newer equipment is capable of receiving all

Fact

Remote station service is an entry-level supervising station alarm system. It is generally chosen because the owner wishes to have a minimum level of supervision or is required by an insurance carrier to have the system. The *Code* does not require remote station service to include maintenance and testing.

signal types. Additionally, remote supervising stations are permitted to manually record signals, unlike central stations and proprietary supervising stations.

Section 26.5.3.1.1 permits remote supervising station signals to be received at the public emergency service communications center, provided it complies with *NFPA 1221, Standard for the Installation, Maintenance, and Use of Emergency Services Communications Systems*. Section 26.5.3.1.2 permits alarm, supervisory, and trouble signals to be received at the fire station, or governmental agency which has public responsibility for taking prescribed action.

Since most local jurisdictions choose not to receive alarm signals, Section 26.5.3.1.3 gives the AHJ latitude to permit the signals to be sent to acceptable alternate locations. Section 26.5.3.2 also permits trouble signals to be sent to a different location from that where alarm or supervisory signals are sent, provided staffing meets *NFPA 72* requirements. Many smaller remote supervising stations are located in homes or small office suites. Section 26.5.3.3 requires restricted access to the receiving equipment when the remote station signals are not sent to the public fire service communications center. As can be seen, the requirements for remote stations are significantly less restrictive than the ones for central stations.

Section 26.5.4 provides requirements for remote station equipment. Like central and proprietary stations, the receiving equipment for remote stations must indicate receipt of signals by audible and visible means. However, there is no requirement for the automatic recording of signals. Section 26.5.4.4 requires only a single telephone instrument, private fire department radio, or other means acceptable to the AHJ.

Staffing levels are specified by Section 26.5.5.1, which requires a minimum of two trained operators on duty at all times. Like central and proprietary stations, operators must not engage in activities that would distract or prevent them from performing the protective service.

Section 26.5.6.1 requires alarm signals to be immediately retransmitted when the remote station is not collocated with the public fire service communications center. Section 26.5.6.2 requires the operator to immediately notify the subscriber when trouble or supervisory signals are received. However, there is no requirement to dispatch a runner for these conditions.

Section 26.5.7.1 requires the remote supervising station to retain all records for a period of not less than one year. As with central and proprietary supervising stations, the AHJ may request these records at any time.

Communications Methods for Supervising Stations

There have been many significant changes to communications methods in recent editions of *NFPA 72*. Modern communications methods are finding their way into the *Code*, and a few older "legacy" methods were removed from the *Code*. Much of the material for these methods was placed in Annex A, making it non-enforceable. These legacy systems include active multiplex, direct connect non-coded, and private microwave. It is understood that existing systems using legacy systems are still acceptable and need not be replaced.

At this time, the only remaining requirements in Section 26.6 pertain to digital alarm communicator systems, two-way RF multiplex systems, and one-way private radio alarm systems. However, Section 26.6.3.1.1 permits other methods as long as they meet performance requirements of Section 26.6. The key issues are the power supply for the transmitter or portal during a power outage and supervision of fault conditions. Section 26.6.3.1.5 permits a single

Fact

For alarm signals, the subscriber may only be contacted after re-transmission. The subscriber, however, may be contacted first for trouble or supervisory signals.

26.6.3.1.1 Conformance. Communications methods operating on principles different from specific methods covered by this chapter shall be permitted to be installed if they conform to the performance requirements of this section and to all other applicable requirements of this *Code*.

(Excerpt from NFPA 72.)

transmission method (for example, Internet supervision), where communications faults are annunciated at the supervising station within 60 minutes. *NFPA 72* does not have jurisdiction over communications service providers and cannot dictate secondary power requirements.

Most of the major manufacturers of fire alarm controls now offer an Internet Protocol (IP) gateway, which allows a user to see the system from anywhere internet access is available. The gateway permits users to remotely monitor systems through an internet connection. This tool is especially useful for diagnosing problems without conducting a site visit. These and other methods must be listed for such use and will require AHJ approval before use.

Section 26.6.3 provides the requirements for transmission of signals between the protected premises and the supervising station. These methods are permitted for all types of supervising stations. The requirements of Section 26.6.3 apply to:
1. Transmitters at the protected premises
2. Transmission media between the protected premises and the supervising station or subsidiary supervising station
3. Any subsidiary supervising station and its communications channel
4. Signal receiving, process, display, and recording equipment at the supervising station

Once a signal is created at the protected premises, a transmitter sends a signal to the supervising station. There are three methods of communications permitted by Section 26.6.3. Currently, these recognized transmission methods include:
- Digital alarm communicator systems
- One-way private radio alarm systems
- Two-way radio frequency multiplex systems

Other methods not specifically mentioned in the *Code* may include IP supervision. This type of communications method must meet the requirements of Section 26.6.3.1.10 of *NFPA 72*. Key issues include the requirement for a maximum 90-second transmission time for all signals, as well as power supply requirements. Section 26.6.3.1.15 requires all equipment necessary for the transmission of signal to be provided with secondary power that conforms to Section 10.6.7 of the *Code*. This essentially requires 24 hours, plus five minutes of alarm and monitoring for integrity. However, a new exception added to the 2013 edition of *NFPA 72* permits the shared transmission equipment (for example, cable modem) to have at least eight hours of standby power where approved by the AHJ and risk analysis is performed to ensure acceptable availability of signal transmission.

Digital Alarm Communications Systems. Digital alarm communications (DAC) systems are the most commonly used communications means and utilize the public switched telephone network as the means of transmitting data. The digital alarm communicator transmitter (DACT) and digital alarm communicator receiver (DACR) act as a modem to use voice lines that transmit digital data. **See Figure 9-9.**

This is the most common communications method because telephone service is widely available and is relatively inexpensive. Digital alarm communications are passive in nature because they do not transmit data until the fire alarm system controls generate a signal.

For additional information, visit qr.njatcdb.org Item #1020

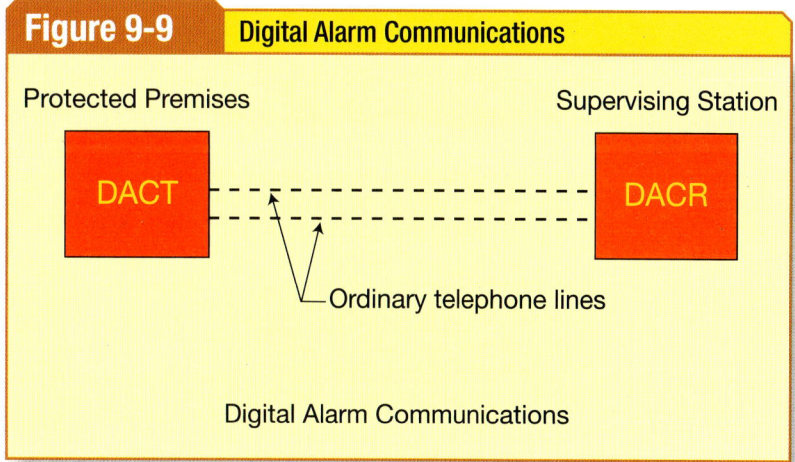

Figure 9-9. DACs are used with loop start telephone lines and send signals over the public switched telephone network.

Some equipment has an optional DACT card available for installation inside the control unit, while others use a stand-alone DACT. Since digital alarm communicator transmitters use the public switched telephone network, the connection must be reliable. Therefore, the telephone line(s) used for the DACT must be under the control of the subscriber at the protected premises. Tenants sometimes have short-term leases, so it would not be recommended or permitted to use telephone lines under their control.

Extensions of a private branch exchange (PBX) are not suitable for use on a DACT because they use ground start lines. Ground start lines produce a dial tone when the tip is temporarily grounded, and there is no voltage present until a dial tone is generated. Therefore, they cannot be monitored for integrity. Loop start lines create a dial tone when tip and ring are temporarily connected. There is always voltage present on a loop start line, making it easy to monitor for integrity. Digital alarm communicator transmitters must use loop start lines so they can monitor for integrity, as required by Section 26.6.3.2.1.1(B)

An interesting problem with DAC systems has arisen with the introduction of optical fiber and cable communications, such as FiOS or broadband communications. Section 26.6.3.2.1.1(B) requires a loop start line for a DACT. Neither an optical fiber telephone connection nor a telephone connection from a cable modem meets the requirement for a loop start telephone line. Many service providers no longer provide copper circuits between subscribers' locations and the telephone office switch. Cable companies also have a similar problem because the lines they provide are not loop start. The issue here is that the power supply to convert electrical signals to a compatible means does not have 24 hours of standby capacity. Section 26.3.1 does contain requirements for communications methods not covered by the *Code* and makes provisions for the lack of supervision. *NFPA 72* does not have scope over utilities, and it is hoped this issue will continue to be addressed in future editions of *NFPA 72*.

Section 26.6.3.2.1.1 requires the DACT to be connected upstream of any other telephone instruments at the premises. In fact, the telephone lines must be connected to RJX-31 jacks directly from the network interface. This permits the DACT to seize the line when a signal must be transmitted. Party lines cannot be used because they cannot be controlled or seized when signals must be transmitted. **See Figure 9-10.**

Additionally, each of the protected premises must have a separate DACT. Sharing of a DACT and associated telephone lines between two or more protected premises (buildings) is not permitted by the *Code*.

Section 26.6.3.2.1.4 requires all DACT to send test signals every six hours. Failure of either channel must be annunciated within four minutes, and a trouble signal must be sent over the other channel. Failure of both means must be annunciated locally.

When a DACT calls its associated receiver, the DACT and DACR must handshake before data can be transmitted. The DACT sends an indication of completion, and the DACT and DACR then "kiss off" at the end of communicating. Section 26.6.3.2.1.3(B) requires transmission of data to take 90 seconds or less per attempt from the time the DACT goes off hook to on hook.

Section 26.6.3.2.1.4(B)(5) requires the DACT to deliver an indication the signal has been received. Other technologies, such as tape dialers, simply repeat the same message without verifying the receipt of data. Because there is no indication the message is complete, they cannot be used for fire alarm systems and are not listed for fire alarm use. Additionally, tape dialers can cause loss of use on the receiver because they tie up the incoming line by endlessly transmitting the same data.

Failure of a DACT to connect and communicate with its associated receiver must result in successive attempts.

Figure 9-10 — Telephone Line Connections to a DACT

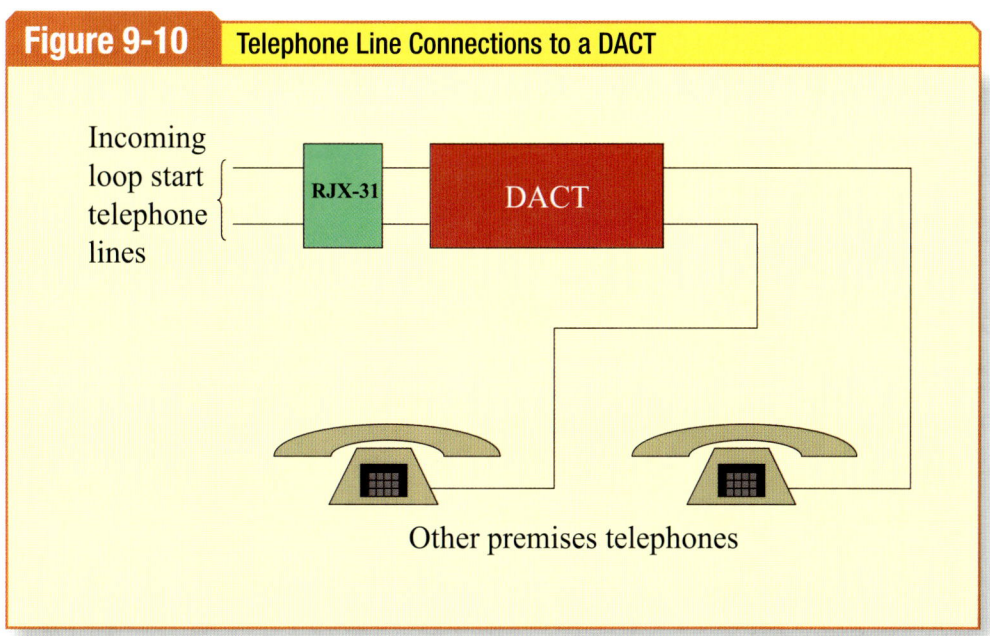

Figure 9-10. The telephone lines must be first routed to the RJX-31, and if desired, to other telephone instruments. This allows the DACT to seize the line to transmit signals.

Section 26.6.3.2.1.3(C) requires a DACT to make at least five, but not more than 10, attempts. Section 26.6.3.2.1.3 (D) requires an indication of failure to communicate at the protected premises (a trouble signal) when the maximum number of attempts has failed. Signal transmission may take as long as 15 minutes from the time the alarm sounds.

Section 26.6.3.2.1.4 requires all digital alarm communicator transmitters to use two means of communication. The primary means must be a loop start telephone line. The second means may be any of the other means permitted by Section 26.6.3.2.1.4(A). These include one-way private radio systems, two-way RF multiplex systems, or performance-based systems covered by Section 26.3.1 (for example, Internet supervision). If a loop start line and one of the technologies mentioned here are not available, the exception to Section 26.3.2.1.4(A) permits a second telephone number to be used. This limits the possibility of failure. If two telephone lines are used, they must be tested at six-hour intervals.

Active Multiplex Transmission Systems. Active multiplex was the most frequently used communications method for proprietary supervising stations. Active multiplex communications involve a two-way flow of information over the circuit or pathway. The requirements for active multiplex systems have been removed from the *Code* and now appear in Annex A. However, existing systems are acceptable. The *Code* also now permits DAC systems to be used for proprietary supervising stations.

Signaling line circuits (SLC) provide active multiplex communications and are frequently used for proprietary supervising stations. Most proprietary supervising stations are located on contiguous properties, such as a college campus, and SLC loops are a good choice for this arrangement. Many campus-style settings have optical fiber networks, which can be used to transmit fire alarm signals. Where optical fiber cables are shared, fire alarms should be on a separate fiber to prevent interference.

Derived local channel (DLC) is a form of active multiplex because it allows constant two-way communications between transmitter and receiver. DLC is a service very similar to digital subscriber line (DSL) technology used for Internet access. DLC allows a telephone line to be simultaneously used for both digital and voice signals. This service has been successfully marketed in both fire and burglary applications as "cut-line service." Because communications are active, the receiver will detect a loss of communications very quickly. This method is no longer widely used and is offered by very few telecommunications service providers.

A premises using DLC must be within 10,000 feet (3,048 m) of the telephone company office serving the area. Additionally, the telephone company must offer this service in the area. Another active multiplex technology uses integrated services digital network (ISDN) lines, but these technologies have become less available and less popular with the advent of faster Internet connections.

Direct Connect Non-coded Systems. Direct connect non-coded connections are common in large cities like Boston and Chicago. Direct connect non-coded circuits are actually notification appliance circuits that extend from the protected premises to the fire department or supervising station. Leased telephone lines are used to provide this extension between protected premises and a supervising station. These circuits can be used to transmit a single bit of information or can transmit digital data, depending on what the jurisdiction can receive.

Direct connect non-coded systems are rare in rural or suburban settings because leased lines are not available. New leased lines in urban settings are also becoming difficult to obtain because communications providers are switching most networks to optical fiber cables. It is no longer permitted practice to replace these lines once they have failed because they are no longer permitted by the *Code* and are costly.

Radio Systems. One-way radio systems, one-way private radio systems, two-way RF multiplex systems, and private microwave systems are ideal for remote locations where telephone service is not available. However, a direct line of sight or repeaters are required to ensure good signals.

Two-way RF systems act much like hard-wired multiplex systems, using multiplex communications. Integrity of the pathway (frequency) must be monitored for integrity (loss of signal) just like any other wired fire alarm circuit. One-way private radio systems obviously do not use active multiplex technology because signal transmission is one way. One-way private radio systems may be shared between subscribers. Supervision of one-way private radio systems must be conducted at the supervising station to ensure a reliable connection. One-way private radio systems only transmit alarms from protected premises to the supervising stations and cannot receive alarms.

Private microwave systems are also monitored for integrity at the supervising station. These systems are very expensive compared to other transmission media and are, therefore, very rare.

There are other methods of communicating between the protected premises and the supervising station. These include two-way RF multiplex systems. Two-way RF systems transmit data in both directions, as they are multiplexed. This makes them more secure and somewhat more reliable.

In the two-way RF multiplexed network, a supervising station uses RF transceivers (repeaters) spread across a large geographic area. The transceivers must be located in the geographic area so that each transmitter can communicate with one or more transceivers. RF systems are ideal for campus-style arrangements and other similar dense urban settings with a high density of protected premises in a relatively small area.

A typical RF arrangement is to install a network of transceivers across the area served, so that there are multiple redundant pathways. Section 26.6.3.3.1 of *NFPA 72* provides requirements

Fact

The DACT is the most common method of transmission of signals to a supervising station. However, it is not listed for proprietary supervising station use because it cannot transmit signals in the prescribed timeframe.

for two-way RF multiplex systems. **See Figure 9-11.**

The reporting times required by *NFPA 72* for RF technologies are similar to other technologies and require alarm signals to be transmitted in 90 seconds or less. There are two basic categories of RF technologies: Type 4 and Type 5. Section 26.6.3.3.1.4(A) contains the requirements for Type 4 systems. Type 4 systems are required to be provided with the following features:
- Each site has an RF receiver connected to the supervising station by a separate channel.
- Each RF receiver at the protected premises can access at least two receiving sites.
- The system has at least two RF transmitters located at one site with the capability of interrogating all RF transmitters/receivers on the premises or two transmitters dispersed with the ability to be interrogated by two different transmitters.
- Each RF transmitter allows immediate use, and every transmitter is operated at least once every eight hours.
- Failure of any RF receiver does not interfere with other receivers and the failure is annunciated at the supervising station.

Type 5 systems have far fewer requirements than Type 4 systems. Specifically, Section 26.6.3.3.1.4(B) contains the two primary requirements for Type 5 systems as follows:
1. There must be one RF receiving site.
2. There must be a minimum of one RF transmitting site.

Type 4 systems have more reliability because of the redundancy of transmitters and receivers. This redundancy is provided as a means of mitigating failures caused by lightning strikes. Because Type 4 systems are more reliable, the maximum loading is significantly higher than for Type 5 systems. Table 26.6.3.3.1.5(B) of *NFPA 72* permits loadings of approximately four times those for Type 5 systems.

A possible drawback with RF systems is that antennas are often struck by lightning and are subject to wind and ice damage. However, land-based transmission methods are subject to the same problems.

Digital Alarm Radio Systems. Digital alarm radio systems (DARS) provide a means of two-way communication but do not require dedicated transceivers located in the geographic area. Instead, the transceivers at each protected premises act as repeaters for other transmitters, and vice versa. This arrangement makes for a very reliable system because signals have multiple pathways back to the supervising station. DARS systems may be slightly more reliable when widespread telephone outages occur because they do not rely on cellular technology, which eventually switches signals to land-based lines. This technology is sometimes referred to as "adaptive network systems" and is often used on burglary systems because of their ability to withstand faults on land lines.

Fact

Depending on the type, radio systems may require a large installed population of transmitters and receivers in order to be effective. Radio systems operate on a line-of-sight principle, so the area of coverage may be limited in some terrain. They work best in populated areas with high densities of systems.

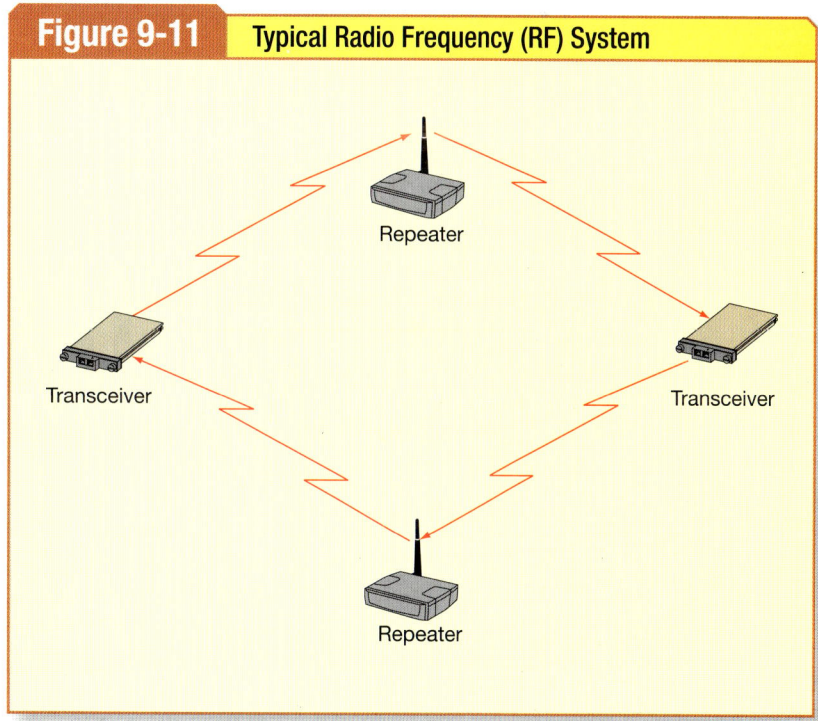

Figure 9-11. More repeaters will provide greater redundancy and reliability.

DARS systems are sometimes used wDARS systems are sometimes used with DAC systems as a way of providing another means of transmission in the event of a DAC system failure. Section 26.6.3.2.3 provides requirements for DARS. Specifically, the DARS must transmit the alarm signal if the DAC system transmission by telephone land line is not successful. In this case, the DACT must continue to attempt to connect with its receiver. The DARS must transmit a signal at least once daily to ensure reliability of the systems. Again, DARS are now considered as legacy methods and are no longer permitted for new installations.

Summary

Supervising station alarm systems have been in existence for more than half a century. Public fire emergency reporting systems have been around for over 150 years. Countless lives have been saved because signals were sent from a protected premises to a supervising station or public emergency dispatch center. The supervising stations themselves have changed somewhat over the past 50 years. The fire alarm community is scrambling to catch up with the fast-paced world of communications. Many new communications methods, such as the Internet, have caused a decline in copper line telephone service. This will most certainly have an impact on how signals are transmitted and how the *Code* addresses them.

Review Questions

1. The first public fire reporting system was located in __?__.
    a. Baltimore
    b. Boston
    c. New York
    d. Providence

2. Public fire reporting systems transmit alarm, trouble, and supervisory signals.
    a. True
    b. False

3. Public fire reporting systems must tap out the signals they send not less than __?__ time(s).
    a. 1
    b. 2
    c. 3
    d. 4

4. A secondary supervising station operating in parallel with a supervising station is called a __?__.
    a. remote station
    b. satellite station
    c. secondary station
    d. subsidiary station

Review Questions

5. Listed central station service requires __?__ elements.
   a. 3
   b. 4
   c. 5
   d. 6

6. Underwriters Laboratories inspects every central station system it certificates.
   a. True
   b. False

7. Central stations must have no fewer than __?__ operators on duty at all times.
   a. 1
   b. 2
   c. 3
   d. 4

8. Supervising stations must dispatch a runner to the protected premises within __?__ of receipt of a supervisory signal.
   a. 1 hour
   b. 2 hours
   c. 3 hours
   d. 4 hours

9. Proprietary supervising stations must provide emergency lighting for a period of not less than __?__ in the event of a power failure.
   a. 22 hours
   b. 24 hours
   c. 26 hours
   d. 28 hours

10. Remote supervising stations must retain records for at least __?__.
    a. 6 months
    b. 9 months
    c. 12 months
    d. 18 months

11. When connecting telephone lines to a DACT, the telephone lines must be connected to the RJ-31X before being run to any other telephone instrument.
    a. True
    b. False

12. Active multiplex is often used for this type of supervising station connection.
    a. Central station
    b. Proprietary supervising station
    c. Remote supervising station
    d. All of the above

Household Fire Alarm Systems

Chapter 29 of *NFPA 72, National Fire Alarm and Signaling Code,* contains the requirements for single- and multiple-station smoke alarms and household fire alarm systems. These requirements are applicable to dwelling units and include one- and two-family dwellings, hotel rooms, guest suites, lodging and rooming houses, residential board and care facilities, dormitory rooms, and apartments. Chapter 29 must be referenced by other codes such as the *International Building Code (IBC), NFPA 101, Life Safety Code,* or *NFPA 5000, Building Construction and Safety Code,* in order for it to be required by law.

Objectives

- » Describe proper smoke detector/alarm location and mounting
- » Describe proper heat detector/alarm location and mounting
- » Describe proper workmanship and wiring
- » Describe appropriate notification systems
- » Describe the power requirements

Chapter 10

Table of Contents

Basic Requirements 228
    Permitted Equipment..........................228

Required Detection......................................231
    Smoke Alarm/Detector Location and Mounting 231
    Heat Alarm/Detector Location and Mounting 234

Workmanship and Wiring 236

Occupant Notification.................................236
    Audible Notification 236
    Visible Signaling 237

Power Supplies..238
    AC Primary Power Source..................239
    Secondary Power Source...................239
    Non-Rechargeable Battery Primary Power Source 240
    Rechargeable Battery Primary Power Source 240

Household Fire Alarm Systems 240
    Non-Supervised Interconnected Wireless Alarms.................................241
    Supervising Station Connections 241

Summary 242

Review Questions.......................................242

BASIC REQUIREMENTS

The scope of Chapter 29 of *NFPA 72* is intended to include one- and two-family dwellings, sleeping rooms of lodging and rooming houses, individual dwelling units of apartment buildings (to include dwelling areas of high-rise buildings), hotel guest rooms and suites, day-care homes, board and care facilities, and dormitory rooms. Chapter 29 of *NFPA 72* applies only where required by another *Code* or standard. However, most local building codes contain references to Chapter 29 of *NFPA 72* for each type of dwelling within their respective chapters on fire protection. Finally, Section 29.1.4 makes it clear that Chapter 29 does not apply to manufactured homes, which are covered by Housing and Urban Development (HUD) standards.

The *Code* states its purpose as follows:

> **29.2* Purpose.** Fire-warning equipment for residential occupancies shall provide a reliable means to notify the occupants of the presence of a threatening fire and the need to escape to a place of safety before such escape might be impeded by untenable conditions in the normal path of egress.

(Excerpt from NFPA 72.)

Chapter 29 of *NFPA 72*, *National Fire Alarm and Signaling Code* contains the requirements for the location, quantity, and performance of single- and multiple-station alarms and dwelling fire alarm systems. Chapters 10 through 27 of *NFPA 72* do not apply to Chapter 29, unless specifically referenced. None of the documentation requirements in Chapter 7 applies to Chapter 29. However, inspection, testing, and maintenance requirements contained in Chapter 14 apply to single- and multiple-station smoke alarms and household fire-warning equipment. The primary purpose of equipment and systems installed in dwellings is to provide life safety, and property protection is not anticipated.

Section 29.3.1 requires all devices, combinations of devices, and equipment to be listed for their intended purpose. Further, Section 29.3.2 requires all equipment to be installed according to the manufacturer's published instructions. This requirement mirrors Section 10.3 because Chapter 10 does not apply specifically to dwellings or household equipment.

The requirements in Chapter 29 of *NFPA 72* apply only to dwelling units and do not apply to tenantless (common) areas of a hotel, high-rise apartment building, or dormitory. The remaining chapters of *NFPA 72* cover the tenantless spaces, such as lobbies, corridors, meeting rooms, and common-use areas for these spaces. **See Figure 10-1.**

Permitted Equipment

Section 29.1.2 requires smoke alarms to be installed where required by applicable laws, codes, or standards. This requirement does not preclude the use of detection in areas otherwise not required to be protected. However, certain areas of dwellings, such as kitchens and bathrooms, are not appropriate for smoke detection. Heat alarms or detectors may be more appropriate for these areas. There are several Annex A sections for Chapter 29 that indicate other ways to reduce fire risks associated with dwellings, such as the use of heat detection in garages or attics. Furthermore, Section 29.1.5 states the primary purpose of Chapter 29 of *NFPA 72* is to provide life safety and does not contemplate property protection.

The vast majority of dwelling fire-warning equipment consists of single- and multiple-station smoke alarms. Although fire alarm systems are permitted by *NFPA 72*, they tend to be more expensive and, therefore, less frequently used than smoke and heat alarms.

Fire-warning equipment is intended to save lives rather than provide property protection. Fire-warning equipment is intended to provide occupant evacuation warnings within the protected dwelling

Fact

The National Institute of Standards and Technology (NIST) estimates that 4% of American homes do not have a smoke alarm and 20% of American homes have non-operational smoke alarms.

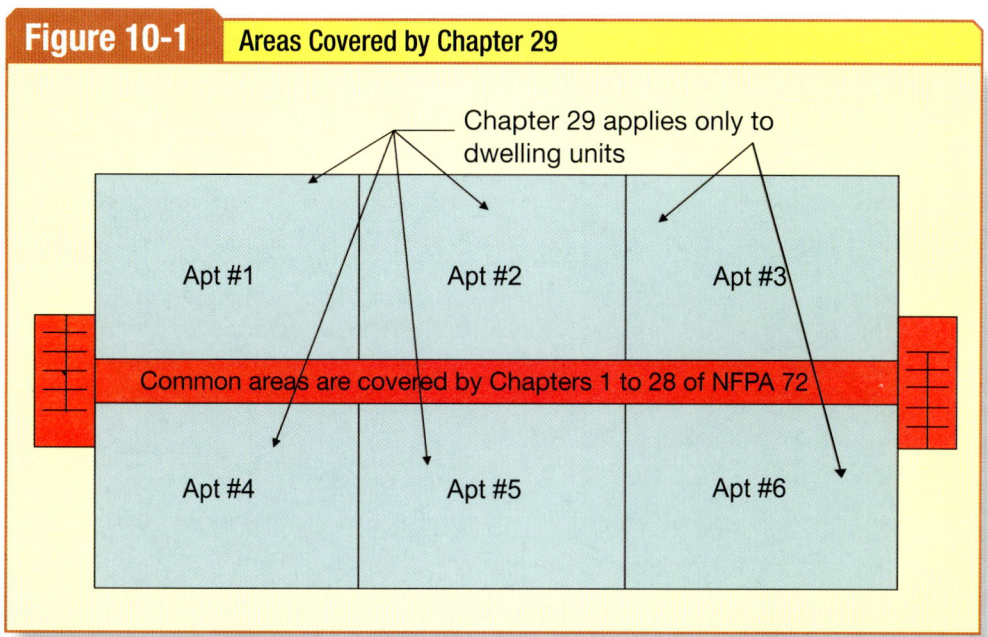

Figure 10-1. Chapter 29 applies to dwelling units, not tenantless areas, which are covered by Chapters 1 through 28.

unit only. It is assumed that occupants will call for help or initiate an alarm on the building system. However, the protected premises (building) fire alarm system must be capable of evacuating the entire building (where provided).

Single- and multiple-station smoke alarms are self-contained units and represent the vast majority of devices installed in dwellings. They contain a sensor, controls, an audible notification appliance, and a primary and/or secondary power supply in a single package. **See Figure 10-2.**

Multiple-station smoke alarms can be used as stand-alone alarms, but they can also be interconnected so the actuation of one alarm will cause all alarms to operate. Single-station alarms cannot be interconnected. In new construction, multiple-station smoke alarms are required to be interconnected.

Multiple-station smoke alarms require a third conductor between all interconnected alarms. This conductor is used to provide the trigger signal between alarms. Note that it is not permitted to interconnect smoke alarms from one dwelling unit to another.

Single- and multiple-station smoke alarms are only listed for use in household applications and cannot be used as part of a protected premises (commercial) system under any circumstances. Section 29.3.1 requires all equipment to be listed for the purpose for which it is used. It is permitted, however, to use protected premises fire alarm systems in a dwelling because the equipment requirements are more restrictive. Section 29.3.3 permits either single- and multiple-station smoke alarms

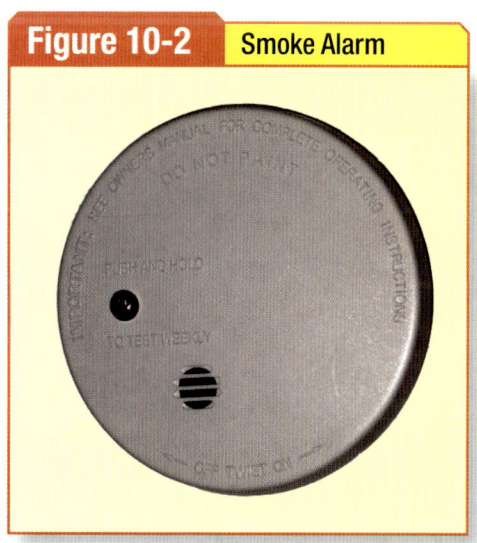

Figure 10-2. Single- and multiple-station smoke alarms are self-contained units containing sensor, notification appliance, and power supply.

or fire alarm systems using system-powered smoke detectors in dwellings. In either event, the quantity and location of the detectors must meet the requirements found in Section A.29.3.3 of *NFPA 72*, which reads as follows:

> **A.29.3.3** This Code establishes minimum standards for the use of fire-warning equipment. The use of additional alarms or detectors over and above the minimum standard is encouraged. The use of additional devices can result in a combination of equipment (e.g., a combination of single- and multiple-station alarms or a combination of smoke alarms or smoke detectors that are part of a security/fire system and existing multiple-station alarms). Though a combination is allowed, one type of equipment must independently meet the requirements of the Code. Compliance with the requirements of the Code cannot rely on the combination of the following fire-warning equipment:
>
> (1) Single-station alarms
> (2) Multiple-station alarms
> (3) Household fire alarm system (includes a security/fire system with smoke alarms or smoke detectors)
>
> It is encouraged that the highest level of protection be used where possible. For example, if multiple-station alarms are added to an occupancy with compliant single-station alarms, the multiple-station alarms should be installed to replace all of the single-station alarms. Similarly, if a monitored household fire alarm system is added to a house that has compliant multiple-station alarms, monitored smoke alarms or smoke detectors should be installed to replace the multiple-station alarms or be installed to provide the same required coverage.
>
> *(Excerpt from NFPA 72.)*

Several fire safety organizations recommend that consumers utilize both ionization and photoelectric technologies in their home smoke alarm systems to permit the longest potential escape. This is not intended to preclude the development of new technology with equivalent performance. The responsiveness of ionization and photoelectric smoke alarms depends on a number of factors, including the type of fire (smoldering, flaming), the chemistry of materials involved in the fire, and the properties of the resulting smoke.

Dwelling fire alarm systems typically use control units similar to those found in protected premises systems, but household fire alarm control units may or may not have integral operator interfaces. However, keypads with liquid crystal displays or light-emitting diode displays provide a user interface and are frequently used for aesthetic purposes. The control unit is commonly hidden in a closet or mechanical room, especially if it is a combination fire/burglary system. **See Figure 10-3.**

Unlike single- and multiple-station smoke alarms, system-powered smoke or heat detectors do not usually contain notification appliances. Therefore, the system will use separate notification appliances like mini-horns. Alternatively, system-powered smoke detectors can also be used to distribute the alarm signal instead of mini-horns where they are outfitted with sound-producing bases. These arrangements are referred to as "sounder bases." Either is acceptable, and the decision will likely be based upon cost.

A dwelling owner sometimes has both system-powered detectors and single- and multiple-station smoke alarms installed. This practice is permitted by Section 29.3.3, provided that one of the two arrangements (smoke and heat alarms or a system) completely provides all of the required detection. Two separate and incomplete systems of smoke alarms or detectors may result in delayed or insufficient occupant notification and are not permitted by Chapter 29 of *NFPA 72*.

A common arrangement in dwellings is to use a combination burglary/fire alarm system. Section 29.7.7 permits supplementary functions of this nature, provided that the supplementary functions do not interfere with the performance of the fire alarm system. Many homeowners opt to install combination burglary/fire alarm systems. Unfortunately, many combination burglary/fire alarm systems do not meet *Code* requirements because an inadequate number of smoke detectors are installed by the contractor. Again, Section 29.3.3 permits either smoke alarms or a fire alarm system, provided that one of the two arrangements independently meets all *Code* requirements in their entirety.

Section 29.4 covers assumptions made by the technical committee. Since dwelling fire alarm equipment can only warn occupants of the fire, it is up to the occupants to have an escape plan, free access to a means of egress, and the ability to evacuate. Sections A.29.2 and A.29.4.2 discuss and recommend a family escape plan. Section 29.4.3 assumes the equipment will be properly installed, tested, and maintained according to manufacturer's instructions and the *Code*. Finally, *NFPA 72* assumes the use of smoke detection as the first means of detection. This is because measurable quantities of smoke are almost always produced in dwelling fires. However, the *Code* permits heat detection where smoke detection is inappropriate, such as in kitchens or attics, which may be subject to nuisance alarms where smoke detectors are used.

REQUIRED DETECTION

Detection devices may be smoke or heat detectors. The proper selection of detection devices will produce alarms quickly without nuisance alarm signals. This will prevent the "cry wolf" syndrome.

Smoke Alarm/Detector Location and Mounting

Section 29.5.1.1 provides the requirements for detection in dwelling units where they are required by other codes. Smoke detection is required in:

- All sleeping rooms and guest rooms
- Outside each separate dwelling unit sleeping area, within 21 feet (6.4 m) of any door to a sleeping room, measured in the path of travel
- On each level of the dwelling unit, including basements
- On every level of the residential board and care occupancy (small facility), including basements, but excluding crawl spaces and unfinished attics
- In the living areas of a guest suite
- In the living areas of a residential board and care occupancy (small facility)

If the sleeping area(s) and living area(s) are separated by a door, Section 29.5.1.2 requires additional smoke detectors or alarms outside the door to the living area. This is in addition to the smoke alarms or detectors outside the sleeping area and is intended to protect the path of egress. It should be noted, again, that Section 29.1.4 clearly states that Chapter 29 of *NFPA 72* does not apply to manufactured homes, which are covered by Housing and Urban Development Standards. See **Figure 10-4**.

Where an area of a dwelling (excluding garage area) exceeds 1,000 square feet (93 m²), Section 29.5.1.3 requires additional smoke alarms or detectors. Section

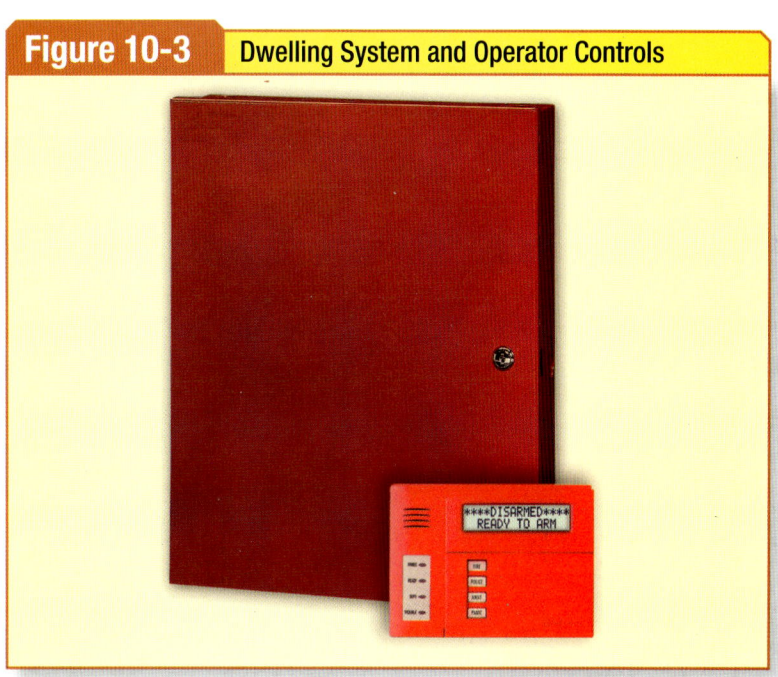

Figure 10-3. *Some control units have sophisticated control pads and displays.*

For additional information, visit qr.njatcdb.org Item #1021

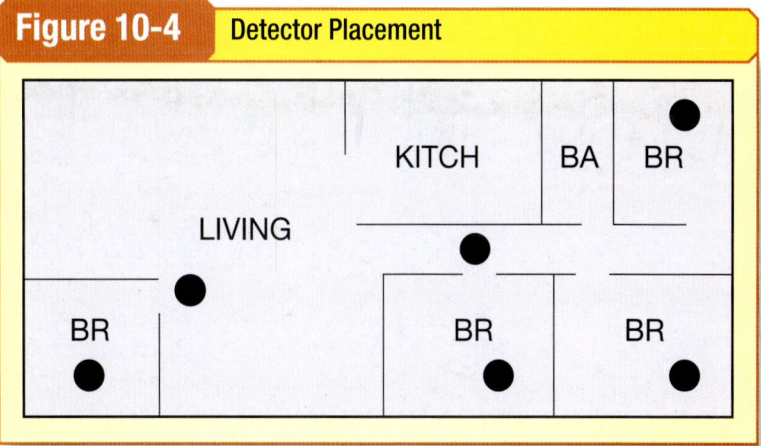

Figure 10-4. *Smoke alarms or detectors must be spaced every 30 feet in the path of travel and for every 500 square feet of area.*

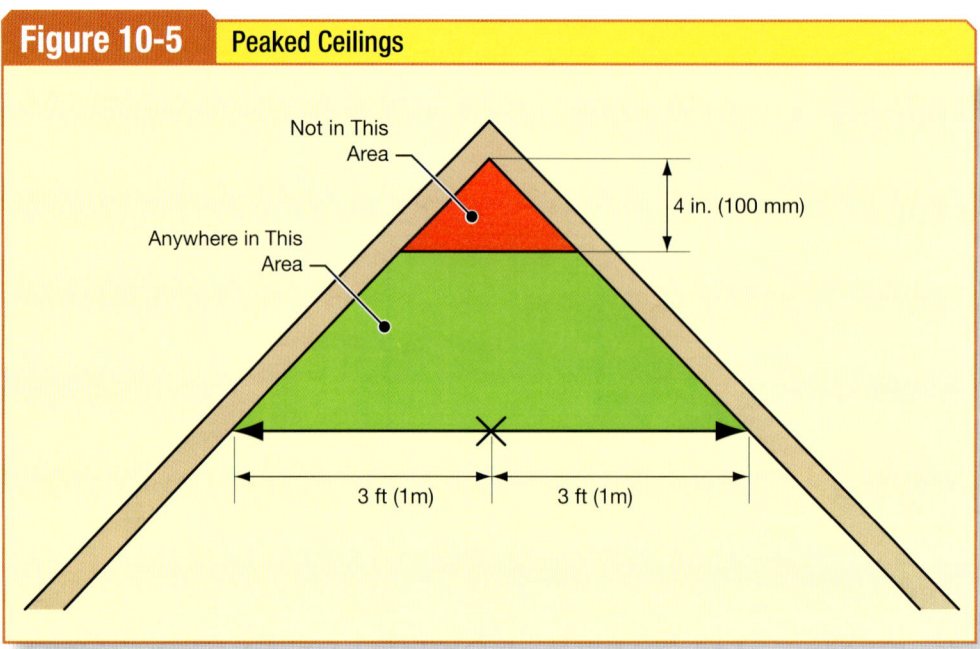

Figure 10-5. Smoke detectors must be positioned within three feet of the peak in a peaked ceiling, but not closer than four inches measured vertically.

29.5.1.3.1 requires smoke alarms or detectors every 30 feet (9.1 m) of travel distance or spaced in such a way that there is one smoke alarm or detector for every 500 square feet (46.5 m²) of floor area. Larger dwellings may have great rooms that extend over one or more floors of the dwelling. In this case, Section 29.5.1.3.2 permits smoke detectors or alarms installed on upper levels to serve for the floor below. Finally, if there is an intervening door between the bedrooms and living areas, additional detectors are required in the intervening areas.

As many new homes are larger than older homes, architects incorporate features not ordinarily found in dwelling construction, such as peaked, sloping, and tray ceilings. Section 29.8.3.1 pertains to peaked ceilings. This section requires smoke alarms or detectors to be located within three feet (1 m) of the peak, measured horizontally, but not closer than four inches, measured vertically. **See Figure 10-5.**

Sloped ceilings are defined as having a rise to run greater than one in eight. Smoke and other products of combustion tend to rise to the highest point in a room. Therefore, Section 29.8.3.2 requires smoke alarms or detectors to be placed within three feet (1 m) of the high side of the ceiling but not closer than four inches from the wall. **See Figure 10-6.**

Ceiling mounting is acceptable at any location on flat ceilings. However, efforts should be made to maintain a healthy distance from HVAC air diffusers. A distance of three feet (1 m) is recommended. Section 29.8.3.3 permits smoke alarms and detectors to be wall mounted. This is especially desirable when ceiling surfaces are not accessible or are decorative. When smoke alarms or detectors are wall mounted, they must be mounted so the top of the detector or alarm is 12 inches (300 mm) or less from the ceiling. **See Figure 10-7.**

Section 29.8.3.4 contains specific location requirements for smoke alarms and detectors. Smoke alarms or detectors must not be located where temperature or humidity can exceed the manufacturer's published instructions. The maximum allowable humidity level is usually limited to a maximum of 93% relative humidity. Temperature limitations cannot exceed 40°F (4°C) and 100°F (38°C), unless specifically permitted by the manufacturer. Unfinished attics, crawl spaces, and garages are not usually suitable for smoke

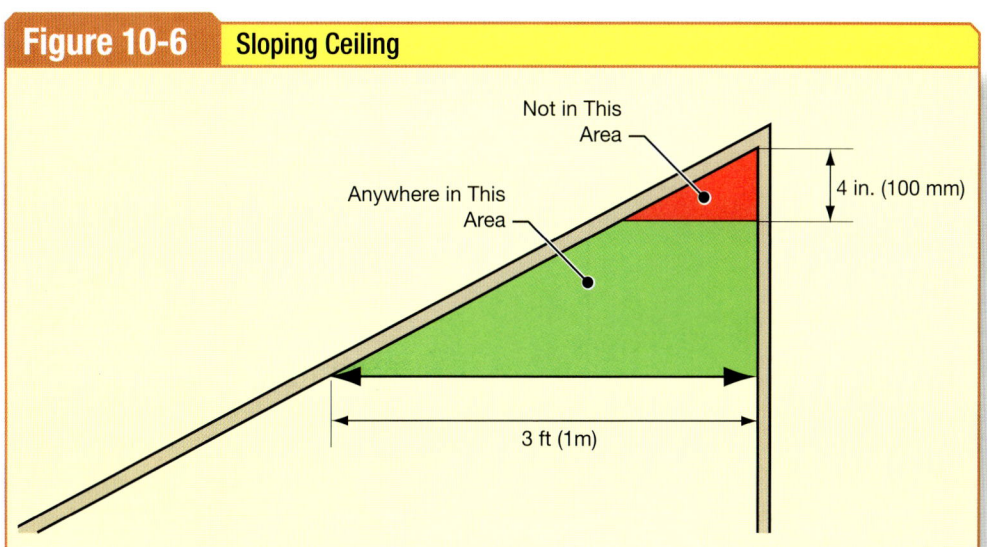

Figure 10-6. Smoke detectors positioned on a sloped ceiling must be positioned more than 4 inches from a wall.

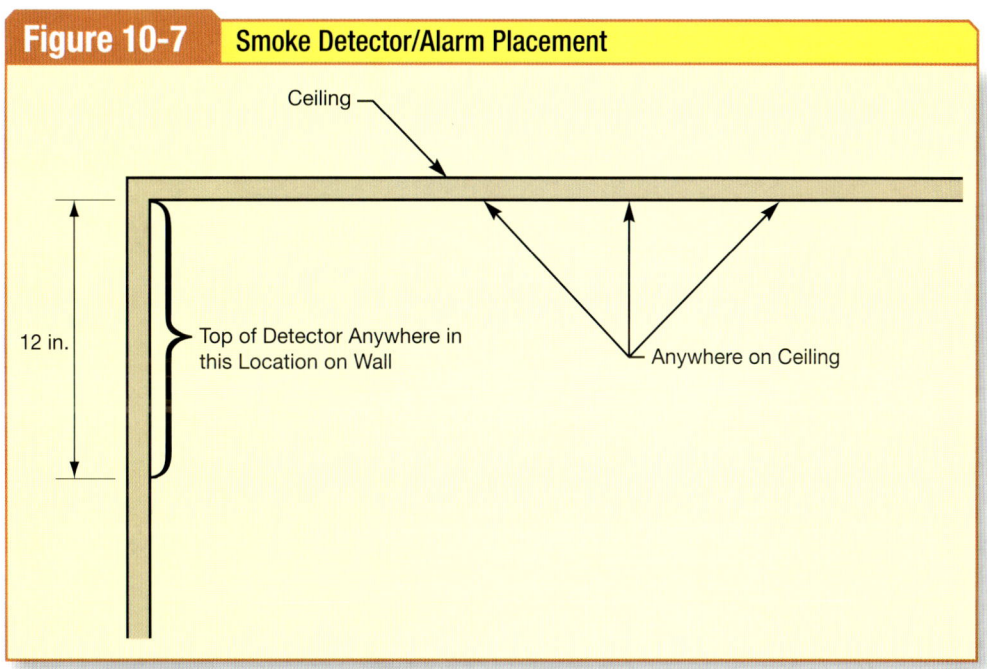

Figure 10-7. Smoke detectors positioned on a wall must be placed within 12 inches of the wall-ceiling connection.

detection. Moisture or temperature extremes can cause nuisance alarms or failure to operate.

The mounting surface can become very warm or cold during seasonal temperature extremes. Always attempt to locate detectors on interior partitions or on well-insulated ceilings.

Smoke alarms and detectors cannot be located within 10 feet (3 m) from a stove or cooktop. Smoke alarms and detectors located between 10 feet (3 m) and 21 feet (6.1 m) of a cooktop or range must be of the photoelectric type or must be provided with a silence feature (hush button). Cooking vapors can cause nuisance

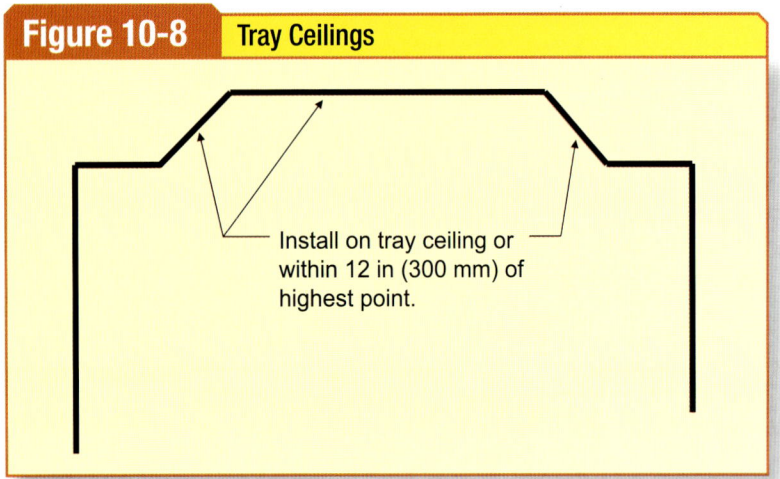

Figure 10-8. Smoke detectors must be mounted within 12 inches of the highest point in a tray ceiling.

delays in detection and helps keep the device cleaner. For tray-shaped ceilings, smoke alarms or detectors must be located on the high portion of the ceiling because smoke will congregate in that location. Section 29.8.3.4(11) specifically requires the smoke detector or alarm to be located within 12 inches from the highest point. **See Figure 10-8.**

Smoke and heat detectors or alarms on joisted ceilings must meet the requirements of Sections 17.7.3.2.4 and 17.6.3, respectively.

Multiple-level dwellings are required to have smoke alarms or detectors on each level, even levels without sleeping rooms. Section 29.8.3.4(9) requires smoke alarms and detectors to be placed near stairs to other occupied levels to intercept smoke before it reaches the next level. **See Figure 10-9.**

Smoke alarms or detectors must also be installed in basements, as on any other level of the home. However, they should not be installed in the stairwell because the air at the top of the stairwell acts as a barrier to smoke flow. Always locate smoke alarms or detectors on the ceiling near the bottom of basement stairs, as required by Section 29.8.3.4(10).

Heat Alarm/Detector Location and Mounting

Because most fires produce measurable quantities of smoke, Chapter 29 of *NFPA*

alarms, but photoelectric-type smoke alarms are more resistant to cooking vapors than ionization smoke alarms.

Effective January 1, 2016, Section 29.8.3.4(5) will require that smoke alarms and smoke detectors used in household fire alarm systems installed between six feet (1.8 m) and 20 feet (6.1 m) along a horizontal flow path from a stationary or fixed cooking appliance must be listed for resistance to common nuisance sources from cooking. This requirement will be effective on January 1, 2016, only where the 2013 edition of *NFPA 72* is adopted.

Smoke alarms or detectors must be installed at least three feet (1 m) from a door to a room containing a tub or shower. Moisture can cause nuisance alarms. A new Section 29.7.3 effective January 1, 2019, will require that smoke alarms and smoke detectors used in household fire alarm systems are listed for resistance to common nuisance sources. This requirement will be effective on January 1, 2019, only where the 2013 edition of *NFPA 72* is adopted.

Smoke alarms and detectors must be installed at least three feet (1 m) from supply air diffusers because fresh air can prevent actuation of the device. This condition can create delays that are unacceptable and will require more frequent cleaning of the detector(s).

Smoke alarms and detectors must be installed at least three feet (1 m) from the tips of paddle fan blades. This prevents

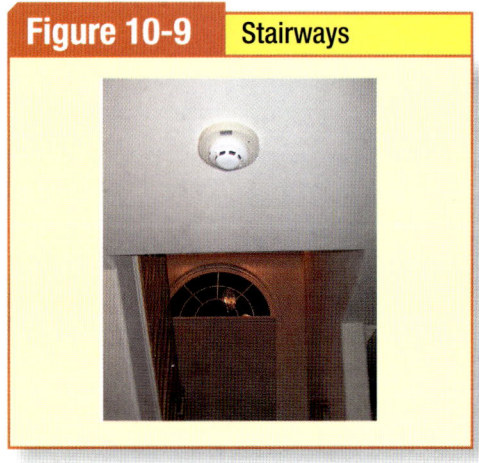

Figure 10-9. Alarms and detectors should be placed near stairs to intercept smoke before it moves to the next level.

72 requires the use of smoke alarms or detectors in most areas. However, smoke alarms and detectors are not suited for harsh environments, such as unheated areas or areas subject to false stimuli. In these areas, heat alarms or heat detectors are a better choice because they do not respond to smoke, dust, and moisture like smoke alarms and detectors do. Heat alarms or detectors are used where smoke alarms or detectors are not permitted because of environmental concerns. Heat detectors are commonly used in attics, kitchens, garages, and even in crawl spaces. However, the selection of the temperature rating of the detector is critical to avoid nuisance alarms. Section 29.7.4.2 requires the selection of the temperature rating for heat alarms and detectors to be between 25°F (14°C) and 50°F (28°C) above the maximum expected ambient temperature.

Section 29.8.4.1 requires heat alarms and detectors to be installed according to their listed spacing on smooth ceilings. Section 29.7.4.1 requires all heat alarms and detectors used in dwellings to have a listed spacing of not less than 50 feet (15.2 m). Heat alarms and detectors must be mounted in the same manner as smoke alarms and detectors. Section 29.8.4.2 requires heat alarms or detectors to be mounted within three feet (1 m) of the peak, measured horizontally, when installed on a peaked ceiling. Additional rows of detectors, if required, are based on a horizontal projection of the room.

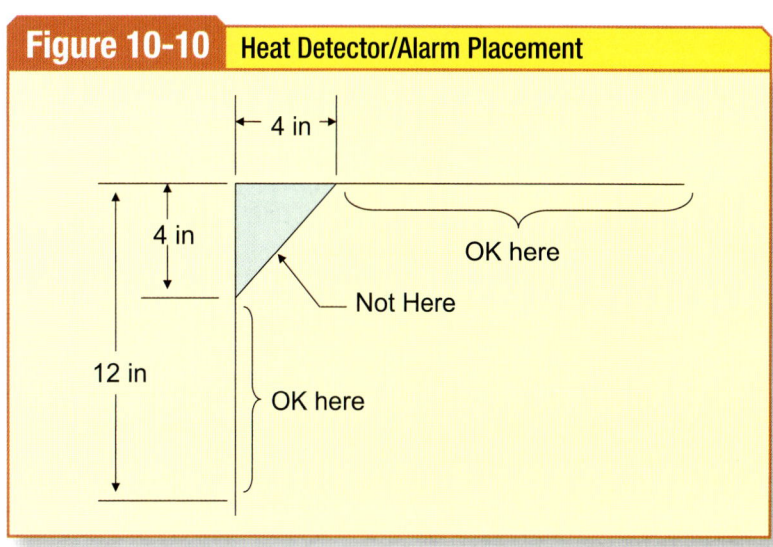

Figure 10-10. Heat alarms and detectors must be mounted on ceilings at least four inches from the wall and on walls between four inches and 12 inches from the ceiling.

Heat alarms and detectors must be mounted on ceilings at least four inches (100 mm) from a wall or on a wall, between four inches (100 mm) and 12 inches (300 mm) from the ceiling. **See Figure 10-10.**

Section 29.8.4.4 requires heat alarms or detectors installed on ceilings that consist of joists or beams to be installed on the bottoms of the joists or beams. Further, Section 29.8.4.5 requires the spacing of the heat alarm or detector to be reduced by one half of the listed spacing in the direction perpendicular to the joists. These requirements prevent delays in detection caused by "spill and fill" of the joist/beam pockets. **See Figure 10-11.**

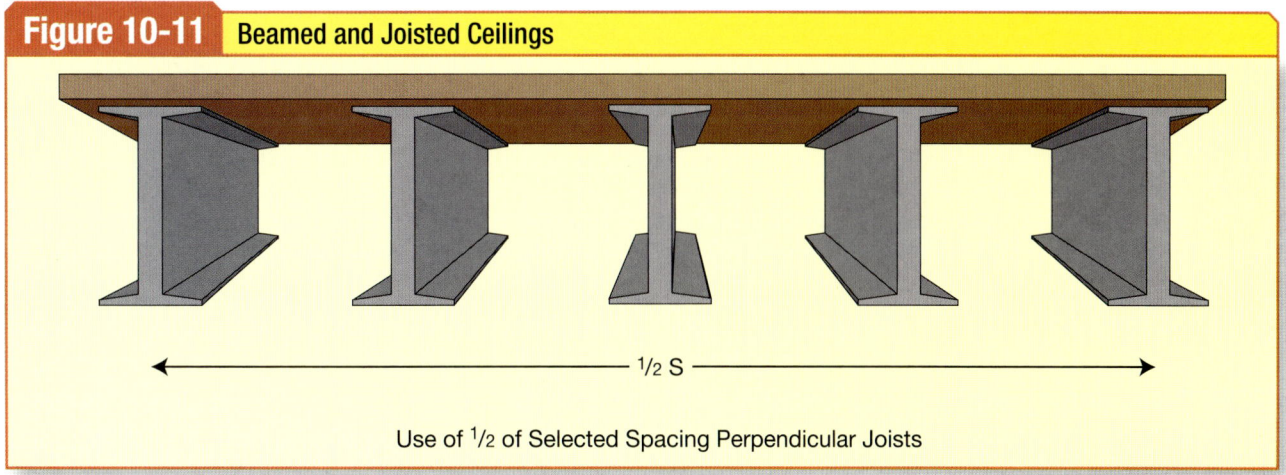

Figure 10-11. Beamed and joisted ceilings have special spacing requirements for smoke and/or heat detectors.

WORKMANSHIP AND WIRING

Multiple-station smoke alarms require a third conductor between the first and last interconnected devices to provide the signal to actuate all devices when one device actuates. Single-station smoke alarms are not capable of this feature and do not require any connecting conductors. Interconnecting smoke alarms between two or more dwelling units is not permitted.

Multiple-station alarms have limitations on the number of devices that can be safely interconnected. Section 29.8.2.2 limits the maximum number of interconnected smoke or heat alarms to 18, no more than 12 of which can be smoke alarms, where they are not monitored for integrity. Where they are monitored for integrity, 64 initiating devices can be interconnected, provided that not more than 42 are smoke alarms. Where a building requires more than this section of the *Code* permits, a system may be required. Additionally, alarms from different manufacturers cannot be interconnected unless they are listed as compatible.

Good workmanship is as important for dwelling fire-warning equipment as it is for protected premises fire alarm equipment. Section 29.8.1.2 requires devices to be located and mounted so that accidental operation is not caused by vibration or jarring of the surface. Heavy doors can, and often do, cause spurious alarms on nearby smoke or heat alarms or detectors when they are slammed shut. Additionally, Section 29.8.1.3 does not permit devices to be mounted by their conductors. This is the same requirement as for protected premises systems and exists because copper is a soft material not suited for device support.

Section 29.8.5 provides the reference to the *NEC* for wiring of all dwelling fire alarm equipment. Single- and multiple-station smoke alarms that are powered by AC primary sources are wired according to the requirements of Chapter 3 of the *NEC*. In many dwellings, the power for smoke alarms is provided on an unswitched portion of a 120-volt alternating current (VAC) branch circuit. Placing the power for these devices on a branch circuit used for lighting and power helps ensure it is monitored for integrity. Some jurisdictions have unique requirements, such as placing the power on the refrigerator branch circuit. Local requirements should always be consulted.

OCCUPANT NOTIFICATION

Notification may be by means of an audible signal or a visible signal. Visible signals are only required in single-family dwellings where the occupants require them. Building codes require visible signaling in a percentage of hotel sleeping rooms, dormitories, and lodging houses.

Audible Notification

Signals from fire-warning equipment and systems must only cause the signal to sound in the dwelling unit of origin, unless otherwise required by another *Code*. Section 29.5.2.2 does not permit the alarm signal from one dwelling unit to cause evacuation signals in the entire building. However, some local codes require an alarm signal within a single dwelling unit to cause a supervisory signal to be annunciated at a constantly attended location (such as a front desk) in some occupancy types (such as hotels).

Section 29.5.2.1.1 requires occupant notification to be audible throughout the occupiable areas of the dwelling unit. In order to achieve this in new construction, multiple-station smoke and heat alarms must be arranged so the actuation of any smoke or heat alarm in the dwelling causes actuation of the other interconnected smoke alarms. Multiple-station smoke alarms require a three-conductor cable to accomplish this feature. Again, fire alarm systems may be used in lieu of smoke alarms provided that the placement and quantity of detectors satisfy the requirements of Section 29.5.2, and the audible signals meet Chapter 18 requirements.

Section 29.3.5 requires dwelling fire alarm evacuation signals to comply with *ANSI S3.41, American National Standard, Audible Emergency Evacuation Signal*. This signal is the same Temporal-3 signal required by Chapter 23 for protected premises systems. The Temporal-3 fire alarm evacuation and burglary signal must not

Fact

Most dwelling fires create untenable conditions within two to four minutes from ignition. Having working smoke alarms or smoke detectors cuts the chances of dying in a fire by half.

be the same signal. Section 29.3.5.1 permits the use of the standard evacuation signal for other uses only if the intended response is evacuation. Therefore, use of the standard evacuation signal for a burglary event is not permitted.

Section 29.3.5.2 permits voice signal notification for dwellings. This is intended to apply to smoke alarms that have recordable message features. This technology is designed to allow owners to record their own voice messages on their smoke alarms. Studies suggest that children wake faster when alerted by a family member than by a beeping tone from a smoke alarm. Where used, the entire recorded message must be contained in a one-and-one-half-second period. These voice signals must still be preceded by the Temporal-3 signal.

Section 29.3.8.1 provides requirements for persons with mild to severe hearing loss. This category includes persons with hearing loss up to 90 decibels and requires audible notification appliances to produce a 520-hertz square wave. The 520-hertz square wave is easier for persons with hearing loss to hear because it is in the lower range of human hearing where hearing loss is not as significant. Section 29.3.8.2 requires visible signaling for persons with profound hearing loss (greater than 90 decibels). As the population ages, there will be many more people with hearing loss.

All signals used in dwelling units must meet the audibility requirements of Sections 18.4.3 and 18.4.5. Section 18.4.3 requires non-sleeping areas to have an audible signal of at least 15 decibels A-weighted (dBA) above the average sound pressure level (SPL) or five dBA above the maximum SPL lasting 60 seconds or more, whichever is greater. This measurement must be taken at five feet above the floor.

Section 18.4.5 requires the SPL in sleeping areas to be at least 15 dBA above the average SPL, five dBA above the maximum SPL lasting 60 seconds or more, or 75 dBA, whichever is greater. This measurement must be taken at the pillow. Additionally, Section 18.4.5.2 requires any barriers, such as doors, between the notification appliance and the pillow to be in place (closed) when the signal SPL is measured. Effective January 1, 2014, Section 18.4.5.3 will require audible appliances provided for the sleeping areas to awaken occupants shall produce a low-frequency (520 hertz) alarm signal. This requirement will be effective on January 1, 2014, only where the 2013 edition of *NFPA 72* is adopted.

Visible Signaling

Visible signals are required in privately owned dwellings only when an occupant is hearing impaired. Section 29.3.7 requires the occupant to notify the appropriate person, such as the building owner, of the impairment. Where the occupancy is required to have visible notification, the equipment must meet the requirements of Section 18.5. However, Section 29.3.9 does not require synchronization of visible appliances in dwellings. It should be noted that the *International Building Code* and other local codes contain requirements for visible signaling in occupancies subject to public use, such as hotels. The *IBC* contains a sliding scale, which requires a certain number of rooms (based on total number of rooms) to be outfitted for visible signaling.

Where required, Section 18.5 covers the spacing for visible appliances. Specifically, these requirements are exactly the same as those used in protected premises systems. Daytime (non-sleeping) areas can take advantage of lower intensity appliances used in protected premises systems. However, sleeping areas must conform to Section 18.5.5.7, which requires higher intensity visible appliances.

Sleeping room visible notification appliances mounted less than 24 inches (610 mm) from the ceiling must have a rating of not less than 177 candela. The higher rating is necessary to penetrate a

Fact

As the population ages, hearing loss will become more prevalent. The National Institute on Deafness and Other Communication Disorders (NIDCD) estimates there are 36 million Americans with hearing loss.

potential smoke layer. If the visible notification appliance is mounted more than 24 inches (610 mm) from the ceiling, a 110-candela appliance may be used. In either case, the visible appliance must be within 16 feet (4.87 m) of the pillow. **See Figure 10-12.**

Visible notification appliances in sleeping areas are not otherwise required to conform to the mounting heights in Section 18.5.5.1 for non-sleeping areas.

POWER SUPPLIES

Section 29.6 of *NFPA 72* provides requirements for power supplies for smoke and heat alarms and dwelling fire alarm systems. Generally speaking, new installations will use smoke and heat alarms that are powered by an AC primary power source, with a battery backup. Some existing occupancies are permitted to have batteries as the means of primary power because of the higher relative cost to install household wiring where smoke alarms are needed.

Smoke and heat alarm power supply requirements are found in Section 29.6.1. Smoke and heat alarms must be powered by one of the following methods:

- Commercial light and power with a secondary battery source capable of operating the device for at least seven days of standby (normal) followed by at least four minutes of alarm output
- A non-commercial (private) AC power source, with a secondary power source capable of providing power to the device for a period of at least seven days of standby (normal) followed by at least four minutes of alarm output, where a commercial AC power source is not available
- A non-rechargeable non-replaceable battery that is capable of providing power to the device for at least 10 years, followed by four minutes of alarm signals, and seven days of trouble signals

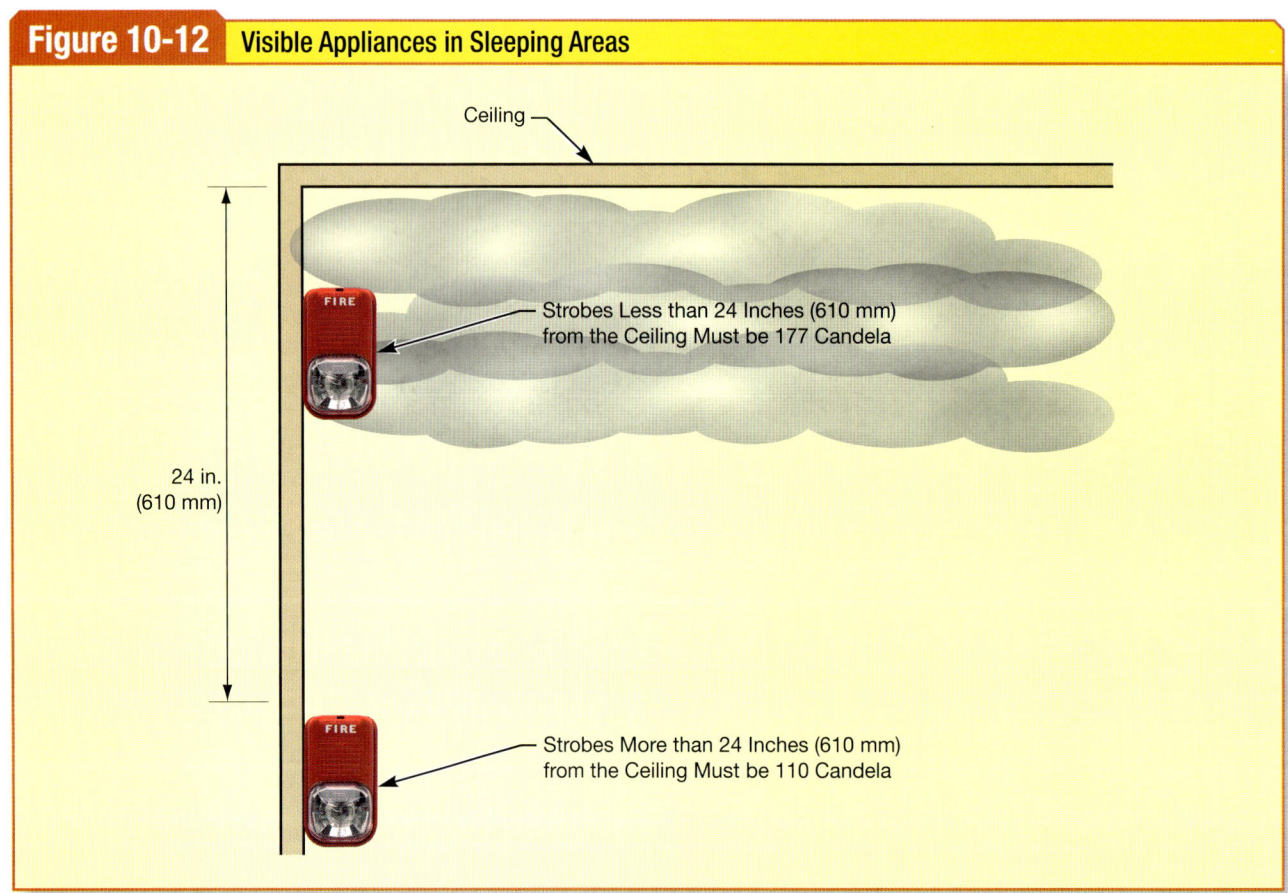

Figure 10-12. Visible alarms must be within 16 feet of the pillow in a sleeping area.

- A primary power supply consisting of either a rechargeable or non-rechargeable battery meeting the requirements of Sections 29.6.6 or 29.6.7

Household fire alarm systems must be supplied with two means of operating power. Section 29.6.2 requires the primary supply to consist of AC power from a commercial light and power source and a rechargeable battery as the secondary source. The secondary source must be capable of providing not less than 24 hours of standby power followed by four minutes of alarm.

AC Primary Power Source

AC power sources generally consist of either a commercial light and power source (utility) or a privately generated supply from a generating plant. The latter is sometimes found in remote areas, islands, and military facilities where commercial utilities are not available and dwelling fire-warning equipment is required.

Section 29.6.3 of *NFPA 72* requires AC power sources and equipment using them to comply with the following requirements:

1. The equipment must be provided with a visible POWER ON indication. This is usually an LED, which remains lit steadily.
2. All electrical equipment that is not designed to be installed by a qualified electrician must be power limited (as defined in Article 760 of the *NEC*) and must have operating voltages of 30 volts or less.
3. All cord-connected AC power supplies, such as plug-in transformers or cord caps, must have a restraining means to prevent accidental disconnection.
4. AC primary power can be supplied from either a dedicated branch circuit or an unswitched portion of a branch circuit used for light and power.
5. Operation of a switch or a ground-fault circuit interrupter (GFCI) cannot cause loss of power to fire-warning equipment, except where the entire dwelling unit is served by a GFCI. Smoke and heat alarms powered by a branch circuit that is protected by an arc-fault circuit interrupter (AFCI) must have a battery backup.
6. Loss and restoration of the primary supply is not permitted to cause an alarm signal for more than two seconds.
7. The primary power supply must be capable of providing sufficient energy to operate the equipment without loading of the secondary power supply.

Secondary Power Source

Secondary power sources are covered by Section 29.6.4 and, where used, the following requirements must be satisfied:

1. The secondary source must be monitored. Disconnection or removal of a charging source must cause a visible or audible signal. Additionally, a low voltage on the secondary supply must cause a trouble signal. Many people interpret this chirping sound as an alarm and disconnect the battery rather than replace it.
2. Acceptable battery sources must be identified in the battery compartment of fire-warning equipment.
3. Rechargeable batteries used for secondary power supplies must meet strict criteria, including the following:
   a. The battery must be automatically recharged by the AC power source (and the fire-warning equipment).
   b. The battery must recharge in four hours or less where the circuit is switched or 48 hours where the circuit is unswitched.
   c. The battery must produce sufficient energy to generate an audible trouble signal for not less than seven days before the battery is incapable of operating the device or system.
   d. The battery must contain sufficient energy to produce not less than four minutes of alarm.
   e. The trouble signal (chirp) must be produced at least once a minute for at least seven consecutive days.

> **Fact**
> The *National Electrical Code* now dictates that all electrical outlets, including smoke alarms, be powered by a branch circuit protected by an arc-fault circuit interrupter (AFCI).

Visible notification appliances are sometimes used in conjunction with multiple-station smoke alarms to alert the hearing impaired. These appliances operate on 120 VAC and have no secondary power supply. Section 29.6.5 does not require these visible notification appliances to be provided with a secondary power source.

Non-Rechargeable Battery Primary Power Source

A replaceable, non-rechargeable power source, such as a nine-volt battery, is permitted in single-station smoke alarms. However, smoke alarms with a non-rechargeable power supply cannot be used in most new construction because they cannot be interconnected. They are suitable for some existing construction. They have several important requirements, but the most important requirement is to replace batteries not less than annually. Many fire departments have instituted a "change your clocks, change your batteries" campaign to raise awareness of this issue.

Non-replaceable, non-rechargeable batteries are found in single-station smoke alarms that are intended for use in existing dwellings. These smoke alarms are not permitted in most new construction because they cannot be interconnected. The battery is designed to last at least 10 years of standby followed by four minutes of alarm. At the end of the 10-year period, the smoke alarm is simply replaced with a new unit.

Rechargeable Battery Primary Power Source

Some newer smoke alarms are supplied with a rechargeable battery as the primary source of power. Section 29.6.7 provides the requirements for rechargeable batteries used as a primary power supply. The smoke alarm must be connected to an AC power source that can recharge the battery within four hours if the power supply is switched and 48 hours if the power supply is unswitched. As with other battery powered smoke alarms, the battery must be capable of delivering enough energy to provide seven days of low-battery power warnings, followed by four minutes of alarm.

HOUSEHOLD FIRE ALARM SYSTEMS

Requirements for household fire alarm control equipment used in dwellings are found in Section 29.7.6. This section requires controls to be self-restoring following restoration of electrical power. Section 29.7.6.2 requires control equipment to lock in alarm, but smoke detectors themselves can be self-resetting. Additionally, any reset switches used on controls must be momentary contact-type switches.

Dwelling systems are permitted to contain many of the same features that are found in protected premises systems. For example, Section 29.7.6.5 permits an alarm notification appliance to be deactivated, provided that the means for turning off the appliance is locked, in a locked enclosure, or otherwise protected against unauthorized use. Additionally, the means of deactivation must generate a visible trouble condition on the control unit when the switch is in the OFF position. It should be noted that the switch or means to deactivate alarm signals cannot be field installed on the control's enclosure without violation of the listing. In other words, a toggle switch used to silence a notification appliance circuit violates the *Code*.

Monitoring for integrity is required for all system-type smoke detectors, initiating devices, and notification appliances. Section 29.7.6.6 requires the occurrence of a single open or ground fault, which interferes with operation to create a trouble signal. Finally, controls used in dwellings must conform to UL985: *Standard for Household Fire Warning System Units*, or UL1730: *Standard for Smoke Detector Monitors and Accessories for Individual Living Units of Multifamily Residences, and Hotel/Motel Rooms*.

Many homeowners prefer to install combination burglary/fire alarm systems. *NFPA 72* permits this arrangement, provided that the requirements of Section 29.7.7 are satisfied. Of primary importance are the requirements in Sections 29.7.7.3 and 29.7.7.6, which require distinctive signals for fire alarm, trouble, and supervisory notifications. Fire alarm signals cannot be used for burglary or intrusion alarms because of the confusion

---

Fact

The smoke alarm is responsible for saving countless lives since it was introduced into the market in the 1970s. Residential fire deaths were nearly cut by 50% since their use began.

that arrangement may create. It should be noted that common notification appliances are permitted, provided that they meet these requirements.

Section 29.7.7.4 requires the fire alarm portion of the system to operate even if there are faults on the non-fire portion of the system. If common wiring is used on a combination system, short circuits, opens, or ground faults on the non-fire alarm components or the wiring between the non-fire alarm components and the fire alarm system cannot cause loss of signals or interfere with the fire alarm system.

Multiple-station alarms are sometimes used in conjunction with other initiating devices like heat detectors, waterflow switches, or manual fire alarm boxes. This other equipment usually operates on a contact closure principle. This arrangement is permitted, provided that an open or short on the wiring that connects this equipment does not prevent operation of the smoke alarm.

Section 29.8 applies to the installation requirements for dwelling fire-warning equipment. The installation of dwelling fire-warning equipment must follow the manufacturer's instructions, equipment listings, and applicable codes, such as *NFPA 70, National Electrical Code*.

Non-Supervised Interconnected Wireless Alarms

Wireless fire alarm systems are available for household use and are an excellent way of retrofitting an older home without extensive wiring. Section 29.7.8.2.1 requires low-power wireless alarms to reliably transmit at least 100 feet (30.5 m) and provides equations for calculating attenuation through typical materials found in household construction. Sections 29.7.8.2.2, and 29.7.8.2.3 require fire alarm signals to have priority and to take no longer than 20 seconds to be indicated at the controls.

Supervising Station Connections

Supervising stations for dwellings can consist of remote supervising stations and central stations. A supervising station connection assumes the use of a system and control unit arrangement. The most common method of connecting a dwelling fire alarm system to a supervising station is through the use of a DACT.

Where a DACT is used, Section 29.7.9.1.1 requires the DACT to be connected to a single telephone line and programmed to call a single DACR number. Unlike protected premises fire alarm systems, a DACT used for a dwelling fire alarm system need only send test signals monthly, rather than daily. If another method of transmission is used, such as an Internet-based means, Section 29.7.1.4 requires failure of the transmission means to be annunciated at the protected premises and the supervising station within seven days. Finally, Section 29.7.9.1.5 does not require dwelling central station fire alarm systems to have placards or certificates displayed or on file.

Unlike any other supervising station system, dwellings are permitted to take advantage of a verification call as permitted by Section 29.7.9.2. A supervising station operator is permitted to place a verification call to the dwelling prior to retransmission to the fire service dispatcher, provided that the delay does not exceed 90 seconds. This does not mean that retransmission is not required. The operator must retransmit the signal but may report that it was a nuisance alarm.

Fact

Wireless equipment is especially well suited to retrofit applications where wiring is difficult to install. Wireless equipment is battery operated and the batteries are supervised.

Many dwellings are connected to a supervising station. The most common connection is to a remote supervising station.

Summary

Single- and multiple-station smoke alarms are the single most significant factor in the reduction of dwelling fire deaths. Since the early 1970s, annual fire deaths in dwellings have been cut by more than half. Unfortunately, more than 80% of all fire deaths in the United States still occur in dwellings, but having just one working smoke alarm or detector in a dwelling cuts the chances of dying in a fire by half. This requires a good installation, of course, followed up with good maintenance.

Many of the requirements for dwelling smoke and heat alarms are similar to those for commercial applications and are either duplicated or referenced in Chapter 29. In terms of saving lives, Chapter 29 of *NFPA 72* is probably the most important chapter in the *Code*. Understanding and properly applying these requirements will undoubtedly save lives.

Review Questions

1. Having a working smoke alarm in your dwelling cuts your chances of dying in a fire by __?__.
    a. 1/4
    b. 1/3
    c. 1/2
    d. 3/4

2. Chapter 11 of *NFPA 72* applies to manufactured housing.
    a. True
    b. False

3. Single-station smoke alarms are suitable for use in the corridor of an apartment building.
    a. True
    b. False

4. Heat alarms or detectors mounted on a ceiling must be located at least __?__ from a wall.
    a. 1"
    b. 2"
    c. 3"
    d. 4"

5. Dwellings with an area more than __?__ require additional smoke alarms or detectors.
    a. 900 ft^2
    b. 1,000 ft^2
    c. 1,500 ft^2
    d. 2,000 ft^2

Review Questions

6. According to Chapter 29, sloping ceilings have a rise/run of __?__.
   a. 1 in 4
   b. 1 in 6
   c. 1 in 8
   d. 1 in 10

7. Smoke alarms or detectors in dwellings must be located at least __?__ from a room containing a tub or shower.
   a. 1'
   b. 2'
   c. 3'
   d. 4'

8. Smoke alarms or detectors in dwellings must be located at least __?__ from a cooktop.
   a. 12'
   b. 15'
   c. 21'
   d. 30'

9. Heat detectors installed on a peaked ceiling of a dwelling must be installed so they are __?__ from the peak.
   a. 1'
   b. 2'
   c. 3'
   d. 4'

10. The minimum sound pressure level (SPL) required for a sleeping room that has an average SPL of 50 dBA is __?__.
    a. 55
    b. 65
    c. 75
    d. 85

11. According to Chapter 29 of *NFPA 72*, smoke alarms must be connected to a dedicated branch circuit.
    a. True
    b. False

12. A DACT on a dwelling fire alarm system must send __?__ test signals to the supervising station.
    a. daily
    b. monthly
    c. quarterly
    d. weekly

Plans and Specifications

Plans, drawings, and specifications are an essential package for the installation of a fire alarm system. Having a good understanding of plans, drawings, and specifications is critical to a good installation. This understanding is also important for the contractor because it enables him or her to correctly estimate job costs. In a cost-conscious world, this may mean getting or losing a job. This chapter provides a tutorial on drawing content and organization, as well as *Code* requirements.

Objectives

- Describe the format and content of typical specification documents
- Explain the difference between a computer-aided drafting (CAD) drawing and a blueprint
- Describe different contracting methods and explain the role of specification documents in the bid process
- Differentiate design drawings from shop drawings from "as-built" drawings
- Understand and recognize standard symbols and notations used in plans and specifications
- Locate new documentation requirements in Chapter 7

Chapter 11

Table of Contents

Specifications 246
    MasterFormat® System 247
    Specification Content 247

Plans and Drawings..................................248
    CAD versus Blueprints 248
    Types of Drawings 249
    Scale...252
    Title Block..254
    Lines and Line Weights 254
    Notes 254
    Plans Review and Approvals 255
    Requests for Information (RFI)...........255

Standardized Fire Protection Plans and Symbols 256
    Firewalls and Other Barriers 256

Contracting Methods.................................257
    Design/Bid/Build 257
    Design/Build 257

Summary 258

Review Questions 258

SPECIFICATIONS

Specifications are generally developed by an engineer, technician, or other person with experience and qualifications in designing fire alarm systems. They are the designer's opportunity to expand on *Code* requirements, require unique methods and materials, define operating characteristics, and set fire protection goals. The designer usually develops specifications after meeting with the authority having jurisdiction (AHJ), the building owner, the insurance interests, and other stakeholders.

Specifications will generally reference plans and drawings to graphically convey concepts to the contractor/installer. Plans and drawings reflect the type, quantity, and location of equipment to be installed. Plans and drawings contain many details not easily provided by the specification text. They can also provide wiring detail, conductor and conduit schedules, battery calculations, interface connections, a matrix of operation, and other details necessary to properly install the system. Standardized symbols help the AHJ and others and should be used at all times.

Project specifications are not developed in a vacuum and are based upon input from all concerned stakeholders. Stakeholders should include, as a minimum, the following:
- Authority having jurisdiction
- Building owner
- Insurance interests, if applicable
- Engineering disciplines
- Other trades (fire suppression, HVAC, elevators, security, etc.)
- Architect

The requirements in fire alarm specifications generally reflect the fire protection goals developed by the designer,

---

Figure 11-1 MasterFormat® Specifications

Division 00 – Procurement and Contracting Requirements
Division 01 – General Requirements
Division 02 – Existing Conditions
Division 03 – Concrete
Division 04 – Masonry
Division 05 – Metals
Division 06 – Wood, Plastics, and Composites
Division 07 – Thermal and Moisture Protection
Division 08 – Openings
Division 09 – Finishes
Division 10 – Specialties
Division 11 – Equipment
Division 12 – Furnishings
Division 13 – Special Construction
Division 14 – Conveying Equipment
Division 15 – Mechanical
Division 16 – Electrical

Facility Services Subgroup:
Division 21 – Fire Suppression
Division 22 – Plumbing
Division 23 – Heating, Ventilating, and Air-Conditioning
Division 25 – Integrated Automation
Division 26 – Electrical
Division 27 – Communications
Division 28 – Electronic Safety and Security

Site and Infrastructure Subgroup:
Division 31 – Earthwork
Division 32 – Exterior Improvements
Division 33 – Utilities
Division 34 – Transportation
Division 35 – Waterway and Marine Construction

Process Equipment Subgroup:
Division 40 – Process Integration
Division 41 – Material Processing and Handling Equipment
Division 42 – Process Heating, Cooling, and Drying Equipment
Division 42 – Process Gas and Liquid Handling, Purification and Storage Equipment
Division 44 – Pollution Control Equipment
Division 45 – Industry-Specific Manufacturing Equipment
Division 48 – Electrical Power Generation

Figure 11-1. The Construction Specifications Institute (CSI) develops the list of specifications, called the MasterFormat® numbering system. These are standardized numbers to delineate different specifications into special categories.

based upon the input collected from the other stakeholders. Project specifications may contain other essential details, such as system operation, interface requirements and details, spare equipment, and special wiring methods. Specifications may, and often do, exceed codes and standards, depending on the fire protection goals for the project.

MasterFormat® System

Each engineering discipline and trade has a separate project specification document. Project specifications are developed according to a strict numbering system developed by the Construction Specifications Institute (CSI). This numbering system was revised in 2004 to use a new MasterFormat® numbering system. The new MasterFormat® numbering system contains a separate number for each trade or discipline for the building. There are gaps in the system for future expansion. **See Figure 11-1.**

Before 2005, fire alarm specifications were usually located within Division 16. This is because fire alarm systems contain a large amount of electrical wiring and because there were no separate division numbers for fire protection. Many fire alarm specifications are still included in Division 16 because many system designers do not yet use the new format. Under the new MasterFormat® system, most fire protection systems are found in Division 28. This includes fire alarm and detection systems.

Specification Content

Not all designers use this new system. Some designers still use the old numbering system, and the contractor must pay particular attention to these details. It should also be noted that Division 1 applies to all specification divisions. Division 1 sometimes contains requirements for the type of electrical connectors, conduit, hardware, or methods.

Not all project specifications will contain a section from each division. Project requirements for a new high-rise building may contain specifications from several divisions, but a fire alarm retrofit may contain a single specification from Division 28. Many fire alarm specifications also contain other requirements for other project needs, such as patch and paint (for retrofit applications). These requirements are usually incorporated into an appendix, which is referenced by the main body of the specifications.

Individual specifications usually contain several main sections, including an introduction, intent (scope) of work, referenced codes and standards, working conditions, product submittals, equipment, wiring, testing and acceptance, spare parts, closeout documentation, patch and paint, training, and warranty. Some specifications may also contain bidder's requirements, alternate bid information, and other contractual text, which must be followed very carefully.

Project specifications are generally written in a legal language similar to codes and standards. Good specifications are very precise and use the word "shall" for all contractor requirements. Specifications are referenced in the contract documents and, therefore, become part of the contract itself. Failure to comply with specifications on the contractor's part may result in legal action and/or non-payment for services provided by the contractor.

Most AHJs will only enforce their locally adopted and amended codes and standards and generally will not enforce project specifications. Many owners are turning to the designer or another independent third party to enforce the requirements contained in project specifications. These third-party verifiers act as a "private AHJ" on behalf of the owner to ensure the system is properly installed. On smaller jobs, the third-party verifier will generally inspect the installation at

Fact

The content of specifications will vary, depending on whether the specifications are used in a design/bid/build or design/build format. Design/build specifications are usually developed by an engineering professional and are suitable for bidding by a contracting firm.

the 40% and 80% completion points and will witness tests with the AHJ at completion. Larger projects may require a third-party verifier on site at all times.

Typical examples of requirements found in project specifications include the following:
- Equipment/product requirements, including restrictions or exclusions
- Sequence of operation
- Type of equipment and circuits used (for example, Class A versus Class B)
- Submittal requirements
- Shop drawing content
- Design criteria
- Wiring requirements beyond those found in the *National Electrical Code*
- Unique operational requirements
- Testing requirements
- Training requirements
- Patch and paint requirements

Some building owners employ the services of a consultant specializing in fire alarm work when developing specifications. Other building owners may enlist the services of an architectural/engineering (AE) firm to develop plans, drawings, and specifications. Still other building owners may opt to use a design/build firm to develop the entire package of specifications and plans.

Larger organizations often employ their own engineers or technicians who develop specifications for their own needs. An example of this arrangement is the United States Government General Services Administration (GSA). The GSA is primarily responsible for construction and maintenance of government office buildings. Other agencies within the government, such as the Architect of the Capitol, U.S. Postal Service, Department of Defense, Department of Veterans Affairs, Department of Homeland Security, and Department of State, all have similar specifications that are tailored to their unique needs. Many multinational corporations follow American standards and use standardized specifications for all of their installations, both domestic and overseas.

Each organization typically has a generic set of fire alarm specifications, which are used for the installation of fire alarm systems in all buildings constructed and maintained by that organization. The specifications for each building may be slightly different because of security, building configuration, or other unique conditions, but most are standardized and build on lessons learned from one project to the next.

PLANS AND DRAWINGS

For the installer, drawings are probably the single most important document he or she uses. The drawings show device and appliance placement, floor plans, a riser diagram, and typical installation details. They may also provide other details such as conduit/cable routing or a matrix of operation.

CAD versus Blueprints

Drawings are almost always created in a computer-aided drafting (CAD) software package. Working drawings are generally copies of a drawing that was created by a large format printer, referred to as a plotter.

Blueprints are made by using a light-sensitive paper coated with a solution of potassium ferricyanide and ferric ammonium citrate. The coated side is a light yellow color, and it is affixed with the original tracing over the light-sensitive paper. The original tracing has opaque lines, which keep the light-sensitive paper from developing when exposed to a light source. The paper and original are exposed to a light ammonia vapor, which causes a chemical reaction to bring out the images. This is called the diazo process. This process is seldom used because it requires chemicals and is more labor intensive than drawing development in CAD.

Blueprints are negatives of the drawings, where lines are white on a blue background. Blueprints are not used much these days because of faster more economical reproduction methods. However, it is still common to refer to drawings as blueprints, even though they are technically "whiteprints."

Fact

Codes are the minimum, and may not directly apply to all situations. Specifications represent the designer's opportunity to exceed codes. They also contain legal language which holds the contractor responsible for not only the installation, but also for commissioning of the system and training.

Drawings come in a variety of standard sizes. These standard sizes include the following:
- **A Size:** 8 ½" by 11" (21.6 cm by 27.9 cm)
- **B Size:** 11" by 17" (27.9 cm by 43.2 cm)
- **C size:** 17" by 22" (43.2 cm by 55.9 cm)
- **D Size:** 22" by 34" (55.9 cm by 86.4 cm)
- **E Size:** 34" by 44" (86.4 cm by 111.8 cm)

Types of Drawings

On most projects, at least three types of drawings are developed for the installation of fire alarm systems: design (layout) drawings, shop drawings, and record (as-built) drawings. Design drawings are sometimes referred to as "conceptual" drawings and may be prepared and submitted at the 50%, 90%, and 100% phases. This allows the fire alarm design to be refined as the architectural plans are finalized. Design drawings contain no details necessary to install the system. They only contain device and appliance locations for *Code* compliance.

A new Chapter 7, Documentation, was added to the 2013 edition of *NFPA 72*. Chapter 7 provides requirements for all system documentation, including drawings. Sections 7.2, 7.6, and 7.7 are mandatory for all systems. Sections 7.3 through 7.5 are only enforced when required by another code, standard, or chapter of *NFPA 72*. Section 7.2.1 contains the minimum required documentation for all systems. Section 7.2.1 requires the following items to be provided for all fire alarm systems and emergency communications systems:
- Written narrative providing intent and system description
- Riser diagram
- Floor plan layout showing location of all devices and control equipment
- Sequence of operation in either an input/output matrix or narrative form
- Equipment technical data sheets
- Manufacturer's published instructions, including operation and maintenance instructions
- Battery calculations (where batteries are provided)
- Voltage drop calculations for notification appliance circuits
- Completed record of inspection and testing
- Completed record of completion
- Copy of site-specific software, where applicable
- Record (as-built) drawings
- Periodic inspection, testing, and maintenance documentation
- Records, record retention, and record maintenance

Design (Layout) Drawings. Section 7.3 provides requirements for design layout drawings, where they are required by another code, standard, or chapter of the *Code*. Design drawings provide details for *Code* compliance but lack the details for installation. Many specifications also contain the minimum content of drawings for fire alarm systems. Although not required by Section 7.3, the following is a recommended list of items that should be contained on all fire alarm system conceptual plans:
- Title block
- Name and address of the protected property
- Name and address of the contractor/installer
- Name and address of the equipment supplier
- Name and address of the designer
- Floor plans
- Location of all devices, appliances, and equipment
- Riser diagram
- Legend of all symbols used
- North arrow
- Scale

The title block contains the name of the CAD operator, revisions, dates of issue, and sheet number. Drawings and plans generally show the location of initiating devices, notification appliances, control and monitor modules, and controls on the floor plans. Drawings must

Fact

Conceptual design drawings are developed in stages as the architectural drawings are refined and details are finalized. It is not uncommon to have 35%, 65%, 85%, and 100% conceptual drawings. Shop drawings may have one or two reviews, usually by an engineer, and a final review by the AHJ.

also have a riser diagram to show vertical wiring and the arrangement of device order on each circuit. Conceptual design drawings provide details for *Code* compliance but do not generally provide installation details.

Notes are an important part of any drawing. Notes provide critical details for the installer, such as applicable *Code* references, specific methods, use of materials, interface information (elevator/HVAC), and operating instructions. Drawings may also contain battery calculations, voltage drop calculations, and a conduit/wire schedule. A matrix of operation may also be required to be located on fire alarm drawings to facilitate programming and testing.

There is a tremendous amount of coordination between trades. Drawing notes may provide some or all of the necessary information, but every effort to communicate with other trades should be made.

These trades include:
- Fire service personnel
- Sprinkler fitters/plumbers
- Mechanical installers (HVAC)
- Security
- Architects
- Engineers

Design drawings are the products of an engineering firm or individual and do not typically contain manufacturer-specific information. Design drawings usually indicate equipment locations for *Code* compliance, such as smoke detector locations. These drawings are not suitable for the installation of a system because they lack details for the field installer.

The designer is responsible for the proper interface of the fire alarm with other systems such as HVAC, smoke control, and suppression systems. Therefore, design drawings will provide locations and quantities of monitor and control modules. Design drawings usually undergo several iterations as floor plans and other building system plans before they are finalized. It is common to have 50%, 90%, and 100% drawings, as well as final construction (shop) drawings on many projects. Design drawings are usually developed in conjunction with specifications, which provide greater detail for the supplier to use in development of shop drawings.

Shop Drawings. Where concept drawings provide the information necessary for code compliance, shop drawings provide much more information. Concept drawings generally include equipment locations on floor plans and may include a riser diagram. Shop drawings are essential to the proper installation of the system. They contain much more information than conceptual drawings. This specifically includes wiring connections, mounting requirements, and a matrix of operation.

Shop drawings provide additional specific wiring details for the contractor. Generally, the fire alarm contractor or equipment supplier will provide details specific to the equipment used in the system. The contractor may also add detail for conduit/cable routing as part of the

Drawings provide information that words alone cannot provide. Drawings contain the details necessary to ensure a proper installation which is Code *compliant.*

development of shop drawings. This additional information will provide the installer with sufficient information to route the wiring, connect the conductors, and install the system.

Other typical details, such as firestopping, device/appliance mounting, and graphic annunciator details can be very helpful in assisting the installer. These details are generally found on shop drawings and are necessary for the correct installation of the system.

Shop drawings must show other details critical for a successful installation. Section 7.4 provides required content of shop drawings, where they are required by another code, standard, or chapter of the *Code*. Most building codes will require shop drawings for fire alarm systems. Where shop drawings are required, Section 7.4.4 provides a list of items that must be provided on shop drawings as follows:

- Name of protected premises, owner, and occupant (where applicable)
- Name of installer or contractor
- Location of protected premises
- Device legend and symbols in accordance with *NFPA 170* or other symbols acceptable to the AHJ
- Date of issue and any revision dates

Section 7.4.5 provides a list of items that must be provided for all floor plans, where required by another code, standard, or chapter of the *Code*. This list is as follows:

- Floor or level identification
- Point of compass (indication of North)
- Graphic scale
- All walls and doors
- All partitions extending to within 15% of the ceiling height (where applicable and when known)
- Room and area descriptions
- System devices/component locations
- Locations of fire alarm primary power disconnecting means
- Locations of monitor/control interfaces to other systems
- System riser locations
- Type and number of system components/devices on each circuit, on each floor or level
- Type and quantity of conductors and conduit (if used) for each circuit
- Identification of any ceiling over 10 feet (3 m) in height where automatic fire detection is being proposed
- Details of ceiling geometries, including beams and solid joists, where automatic fire detection is being proposed
- Where known, acoustic properties of spaces

Section 7.4.6 requires riser diagrams to contain the following:

- General arrangement of the system in building cross section
- Number of risers
- Type and number of circuits in each riser
- Type and number of system components/devices on each circuit, on each floor or level
- Number of conductors for each circuit

Section 7.4.7 contains requirements for control unit diagrams, as follows:

- Identification of the control equipment depicted
- Location(s) of control equipment
- All field wiring terminals and terminal identifications
- All circuits connected to field wiring terminals and circuit identifications
- All indicators and manual controls
- Field connections to supervising station signaling equipment, releasing equipment, or emergency safety control
- interfaces, where provided

Section 7.4.8 requires typical wiring diagrams for all initiating devices, notification appliances, remote indicators, annunciators, remote test stations, and end-of-line and power supervisory devices.

Section 7.4.9 requires a narrative description or input/output matrix of operation shall be provided to describe the sequence of operation. Section 7.4.10 also requires system calculations shall be included as follows:

- Battery calculations
- Notification appliance circuit voltage drop calculations

- Other required calculations, such as line resistance calculations, where required

Appendix C contains samples of shop drawings that show sufficient detail for the system to be installed by experienced individuals. These drawings show device location but also contain additional information, such as device addresses and wiring details. Other information found on these drawings includes wiring schedules, visible notification appliance candela rating, approximate wire routing, and notes.

The cover sheet contains a legend for all devices and appliances used on the system. The matrix of operation provides the information necessary to program the system. The cover sheet also contains a wiring legend, general notes, and other information useful to the installer.

Each manufacturer will use a different type of device and loop numbering (addressing) scheme. The drawings in Appendix C clearly show device addresses in rectangular boxes. For example, the Telephone Room smoke detector on Sheet FA-3 has an address of L1D22, or Loop 1 (this is a signaling line circuit), Device 22. Devices and modules used on this set of drawings use different addresses. Devices have a "D" where Modules (including manual fire alarm boxes) use an "M" in the address. Again, other manufacturers use different terminology when describing signaling line circuits.

All shop drawings must include device and appliance wiring connection details, such as those found on Sheets FA-9 through FA-11 of the appendix. This information shows the field installer the correct polarity, mounting height, and other details necessary to complete the installation. These details are commonly provided by the manufacturer and are added to the drawing by the equipment supplier or party that develops the shop drawings.

In addition to the items described above, drawings must contain a riser diagram. This portion of the drawing shows the wiring of the system between floors and the relationship between the circuits and the floors they serve. It also shows the actual sequence of the wiring of each device on the circuit. Sheet FA-8 provides a riser diagram for this system. Each floor is protected by one or more Class B notification appliance circuits (NACs), and two signaling line circuits (SLCs) cover the entire building.

Record (As-Built) Drawings. Record (as-built) drawings are the final set of drawings developed by the installing contractor. Since no system is installed exactly as shown on shop drawings, the installing contractor must make minor changes to reflect actual conditions. These changes are referred to as "red lines." The installers must keep a set of the shop drawings on-site while installing the system, and actual conditions must be shown on the record drawings. Accurate record drawings are essential for proper maintenance of the system over its lifespan. Unfortunately, many contractors choose not to provide these drawings at the end of the process because of time and expense. Section 7.5.5.4 of *NFPA 72* and most specifications require delivery of accurate record drawings to the owner upon completion of the system installation.

Scale

Drawings are drawn to scale so the placement of initiating devices and notification appliances can be determined on floor plans. Without this scale, it would be impossible to determine if the design meets *Code* requirements. A common scale is 1/4" = 1'. On a drawing using this scale, a line four inches long would

Fact

Record (as-built) drawings are essential for proper maintenance of the system after it is installed. Record drawings are usually developed by marking shop drawings in the field, as the system is installed by the contractor.

Chapter 11 Plans and Specifications

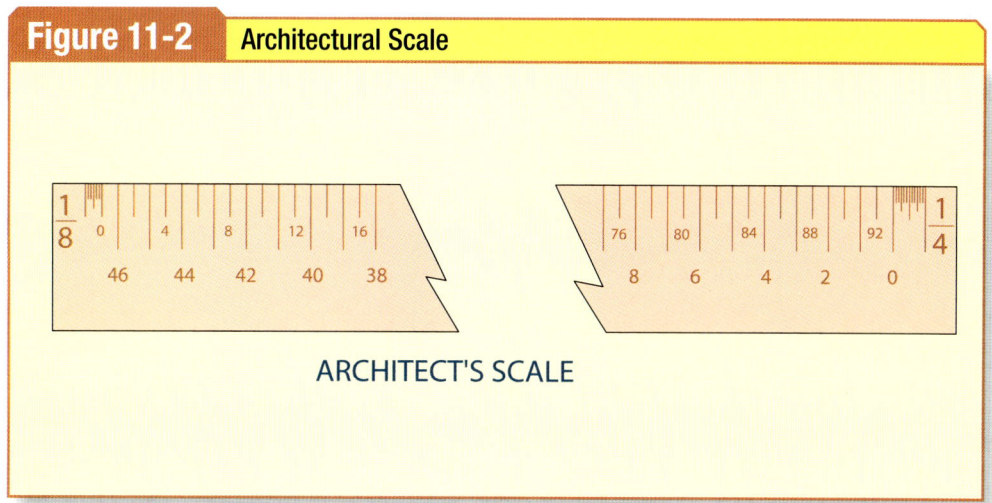

Figure 11-2. An architect's scale is commonly used to determine exact placement locations for mounting fire alarm devices.

Figure 11-3 Typical Scales

Labeled Scale	Scale	Actual Size
3	3"=1'	1/4
1 1/2	1 1/2"=1'	1/8
1	1"=1'	1/12
3/4	3/4"=1'	1/16
1/2	1/2"=1'	1/24
3/8	3/8"=1'	1/32
1/4	1/4"=1'	1/48
3/16	3/16"=1'	1/64
1/8	1/8"=1'	1/96
3/32	3/32"=1'	1/128

Figure 11-3. These scales are used on drawings. The most common scale is the 1/8" scale.

actually represent a distance of 16 feet. **See Figure 11-2.**

The most commonly used drawing scale is 1/8" = 1'. A drawing made with this scale is 1/96 actual size. Various scales are used on an architect's scale. **See Figure 11-3.**

Architect's scales are triangular shaped rulers and have up to 12 different overlapping reduced size scales. Each beveled edge may have two different scales, one running in each direction. Flat scales may have four or eight reduced size scales on a single rule. Rolling scales

Fact
One of the most common problems when using computer-aided design programs is scaling. Many CAD operators forget to properly scale their drawings before plotting them.

Figure 11-4 — Typical Architect's Scale

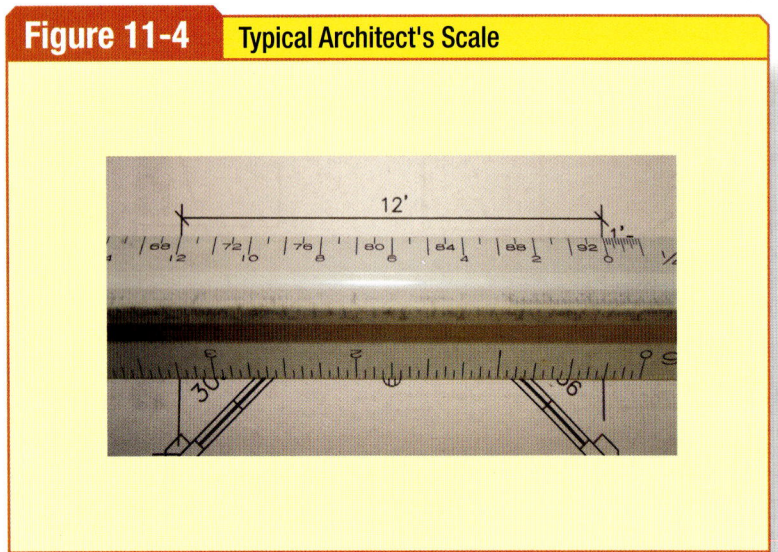

Figure 11-4. Architectural scales have a variety of scales, from 1/32" = 1' to 1" = 1'. Electronic rolling scales are available and are excellent for estimators when calculating wire or conduit length on drawings.

are less accurate but are good for estimating wire lengths by project estimators. **See Figure 11-4.**

Engineering scales use a slight variation from the architect's scale. Instead of using inch dimensions, engineering scales use a reduction of distance by multiplier ratios. Common engineering scales are:
- 1:50
- 1:100
- 1:200

In the 1:100 scale, one foot on the scale equals 100 feet. Engineering scales are available in triangular, flat, and rolling versions, just like architect's scales.

Dimensions are often found on architectural drawings and fire protection plans. Properly interpreting the scale will be especially important when conducting detector or notification appliance layouts. It is always a good idea to check the scale's accuracy, too, before beginning installation.

Title Block

All drawings must have a title block. The title block contains essential information for the installer, maintainer, approver, and owner. Title blocks usually contain the following:
- Name and address of designer
- Name and address of installing contractor
- Name and address of the protected premises
- Name of CAD operator
- Sheet number
- North arrow
- List of revisions and dates of revisions
- Location for professional engineer (PE) seal, if applicable
- Scale used

Lines and Line Weights

Drawings contain several different types of lines. These include border lines, object lines, dimension lines, hidden lines, and match lines. Border lines are heavy lines that encompass the working area of each sheet. Object lines are used to show details of objects, as the name implies. Object lines have a lighter line weight (they are thinner) than border lines. Dimension lines, as the name implies, show distances or dimensions of objects. These lines have a lighter line weight than object lines. Hidden lines show objects that are hidden behind other objects or lines. Hidden lines are usually dashed lines. Match lines are heavy weight lines used when objects are too large to fit on a single sheet. Match lines usually have zigzags to indicate a break and have letters on each end to make the match, such as A-A or C-C.

Lines may be solid, dashed, dotted, or have another type of identifier. These unique line types may be used to indicate different types of wiring method, different fire rating of walls, or hidden objects. Unique lines, such as those used for wiring types or rated walls are usually shown in the legend. Good legends are critical to a proper understanding of drawings.

Notes

Notes are used to clarify certain items or to provide a reminder for the installer. Notes may provide details not easily captured on a drawing or may provide a reference to a particular code or standard. Notes can be used to indicate a typical method or to give the contractor latitude for installing devices or appliances. Notes can also be used to remind the installer of the process for handling unique situations. Notes can be used to clarify the installation requirements and may be located on any sheet of the drawings.

Plans Review and Approvals

Some AHJs prefer to review conceptual design drawings in order to prevent misunderstandings and *Code* violations. These reviews may occur at the 35%, 50%, and 90% completion stages of the development process. Prudent designers will involve the AHJ from the very beginning to avoid issues.

Final approvals of drawings by the AHJ usually take place at the shop drawing stage. Most AHJs will require submittals of shop drawings, specifications, equipment product sheets, calculations, matrix of operation, and other pertinent information before approvals or a permit can be granted. Upon approval, the drawings are usually marked as approved or suitable for construction. Only then can the drawings be used to install the system. Some jurisdictions will impose fines or require removal of equipment installed without a permit or approved drawings. One should always check with the AHJ before beginning an installation.

Before submitting plans and product submittals for review, it is a good idea to conduct a plans check. It is suggested that a party different from the designer/drafter conduct this review, as a means of bringing another set of eyes to the project. The checklist should require (as a minimum) the following items:

1. Name and address of the designer
2. Name and address of installing contractor
3. Name and address of the protected premises
4. Licensing information, if required
5. Insurance information
6. Revisions, if any
7. Proper information in title block (for example, scale, title)
8. Floor plans, showing control panel location
9. Floor plans, showing location of all devices and appliances
10. Point-to-point wiring diagrams
11. Wire or cable data
12. Symbol list
13. Load calculations
14. Transmitters, if used
15. Indication of North
16. Elevator capture
17. Firefighters' telephones, if used
18. Matrix of operation
19. Annunciator details
20. Typical mounting and wiring details for all devices and appliances
21. Operating instructions for each device
22. Sequence of operation
23. Sequence of test and inspection

Requests for Information (RFI)

Drawings contain a great deal of information critical to proper installation. They must be followed. However, many designers allow latitude in the installation, such as for the location of field devices near obstructions. Some designers, particularly on larger projects, require a request for information (RFI) for any change that will alter the information provided in the original design drawings.

Upon project completion, record (as-built) drawings are required by Section 7.5.5.1 of *NFPA 72*. Record drawings must accurately reflect the actual locations of installed equipment, devices, appliances, and wiring routing. Since unforeseen minor changes take place between the design phase and construction phase, the shop drawings and design drawings will almost never reflect the actual installation. For example, smoke detector locations may need to be changed because of an HVAC supply register or luminaire. Sections 7.5.5.4 and 7.7.2 require all system documentation to be stored in a cabinet labeled "system record documents." If this cabinet is not located at the system controls, the location of the cabinet must be identified at the control unit.

Fact

Not all field conditions can be anticipated by the designer. Requests For Information are the process by which the contractor or installer can document changes to the design for unforeseen conditions.

RFIs should be sent to the designer requesting clarification or permission to deviate from the plans, unless permission is granted in the design documents. As minor changes are made, the installer updates shop drawings, as the installation takes place. This ensures the record drawings reflect actual conditions. Copies of all documentation must be kept in the project file for future reference.

Prior to installing any system, the installer or foreman should spend time reviewing the layout, types of wiring, devices used, and locations of all equipment. Becoming familiar with the project before installation will save time and expense.

STANDARDIZED FIRE PROTECTION PLANS AND SYMBOLS

NFPA 72, National Fire Alarm and Signaling Code, requires the use of standardized fire alarm symbols on all fire alarm drawings. Standard symbols used on fire protection plans have two primary sources. *NFPA 170, Standard for Fire Safety and Emergency Symbols*, and *NECA 100, Symbols for Electrical Construction Drawings*, are the primary sources for standardized symbols used on fire alarm plans. Unfortunately, many designers and CAD operators choose to use their own versions of fire alarm symbols instead of standardized drawing symbols and, therefore, symbols can vary from one designer to another.

Most AHJs do not enforce the use of standard symbols, perpetuating this problem. In any event, plans must always have a legend containing all symbols used on the drawings. These symbols are explained in detail and represent most symbols encountered in the field.

Firewalls and Other Barriers

Fire alarm and electrical symbols represent active fire protection equipment. However, passive fire protection features are also shown on drawings. Firewalls, fire barriers, smoke barriers, and smoke-resistant barriers are shown on fire protection or life-safety drawings. These barriers are usually found on life-safety drawings, rather than on electrical or fire alarm drawings, because they pertain to the construction type used. Knowing where these barriers are located is essential because penetrations through them must be fire stopped, ducts may require dampers, and doors usually require self-closing devices. Rated walls and smoke barriers are generally shown as heavy lines for better visibility.

Fact

Every effort should be made to use standard symbols. *NFPA 72, National Fire Alarm and Signaling Code,* requires all drawings to use standardized symbols, per *NFPA 170, Standard for Fire Safety Symbols.*

Smaller projects, which require less engineering, often use the design/build format.

CONTRACTING METHODS

There are two basic methods of contracting the installation of a fire alarm system:
1. Design/Bid/Build
2. Design/Build

Design/Bid/Build

In the design/bid/build format, the owner usually contracts with an architecture and engineering (AE) firm to develop basic plans and specifications. The AE firm may contain all expertise in-house or may subcontract the fire protection design to a specialist. The AE firm and/or its subcontractors develop plans called "conceptual design drawings," which are then sent to the contractor for pricing and subsequent development into "shop drawings." Conceptual design drawings contain device location, device type, and a matrix of operation. However, they do not contain wiring connections and other information necessary for the installation. This is usually because the designer does not yet know what products are to be used.

Design/Build

In the design/build format, the contractor will have sufficient expertise and resources to develop conceptual designs in-house, and an AE firm is not generally used. Design/build formats work well for smaller, specialized projects, but they have been used on large projects as well. In this case, the designer may develop shop drawings directly from the start because he or she will know what products are to be used by the installer.

Summary

Drawings and specifications are essential to a good installation and help ensure that the system can be maintained. Specifications provide requirements beyond codes and standards, furnish owner-specific and insurance requirements, and supply details for interfacing other systems to the fire alarm system. Specifications also provide detailed operating requirements. Drawings offer the installer exact information as to where devices must be located, wiring routed, and typical details for system installation. Standardized details and symbols help the designer and installer save time and resources and should always be used.

Review Questions

1. The document containing standard fire safety symbols is __?__.
    a. NFPA 117
    b. NFPA 127
    c. NFPA 170
    d. NFPA 177

2. The document containing standard electrical symbols is __?__.
    a. NECA 10
    b. NECA 50
    c. NECA 75
    d. NECA 100

3. Specifications often contain __?__.
    a. additional system requirements
    b. patch and paint requirements
    c. training requirements
    d. all of the above

4. Which of the following is not typically involved with the development of specifications?
    a. AHJ
    b. Building owner
    c. Installer
    d. Insurance company

Review Questions

5. Under the new MasterFormat®, fire alarm specifications will be found in Division __?__.
   a. 1
   b. 13
   c. 16
   d. 28

6. Which division of the MasterFormat® covers fire suppression systems?
   a. Division 11
   b. Division 16
   c. Division 21
   d. Division 28

7. A contractor possessing sufficient technical expertise in-house may take on projects using this method.
   a. Bid/Build
   b. Bid/Design
   c. Design/Bid/Build
   d. Design/Build

8. The most commonly used scale is __?__ .
   a. $1/32" = 1'$
   b. $1/16" = 1'$
   c. $1/8" = 1'$
   d. $1/4" = 1'$

9. Issues between actual conditions or discrepancies on shop drawings must be handled through which of the following?
   a. RFI
   b. RIF
   c. FIR
   d. FRI

10. The part of the drawing that shows the relationship of devices to each floor is called the __?__.
    a. floor plan
    b. matrix
    c. reflected ceiling plan
    d. riser

11. Which of the following would be found in a specification?
    a. Intent (scope) of work
    b. Referenced codes and standards
    c. Patch and paint
    d. All of the above

Inspection, Testing, and Maintenance

Testing and maintenance of fire alarm systems is one of the most overlooked aspects of fire protection. Many building owners would rather spend their maintenance budget on things that improve appearance or comfort. However, inspection, testing, and maintenance of fire alarm systems are required by Chapter 14 of *NFPA 72, National Fire Alarm and Signaling Code*. Where *NFPA 72* is adopted by a jurisdiction or referenced by other adopted codes, the requirements carry the same weight as any other law. This chapter reviews troubleshooting methods and inspection, testing, and maintenance requirements for fire alarm systems.

Objectives

- » Describe the requirements for inspection, testing, and maintenance as given in the *Code*
- » Describe requirements for documentation
- » Troubleshoot Class B and Class A circuits
- » Troubleshoot signaling line circuits (SLC)

Chapter 12

Table of Contents

General Requirements 262
    Impairments 262
    Responsibilities 263
    Qualifications 263
    Notification 264
    Tools 265
    Documentation 265
    Releasing Systems and Interfaces 266
Inspections and Tests 268
Initial/Acceptance Tests 268
Periodic Inspections and Tests 268
Reacceptance Tests 269

Testing Methods and Frequency 270
    Control Equipment 270
    Smoke Detectors 270
    Heat Detectors 273
    Special Situations 273
Maintenance 274
Troubleshooting 274
    Class B Initiating Device Circuits 274
    Class A Initiating Device Circuits 280
    Notification Appliance Circuits 281
    Signaling Line (Addressable) Circuits 283
Summary 285
Review Questions 285

GENERAL REQUIREMENTS

No equipment or device can be expected to last forever. Insurance loss data indicates that fire alarm systems have an average life span of about 15 years. The expected life span is directly proportional to how well a system is maintained. Better inspections, maintenance, and testing will lengthen system life, but the reverse is also true. For example, it is not reasonable to expect 100,000 miles from a car without performing basic engine maintenance. The same is true for fire alarm systems.

Studies of electronic equipment manufactured today indicate that most equipment will be subject to a failure rate of approximately 1% to 3% per year. These failure rates are relatively predictable and can be approximated by a bathtub curve. Failure rates at the beginning of the product life are called "burn-in" failures, and the failures at the end of the product life are "end-of-life" failures. **See Figure 12-1.**

Contrary to popular belief, testing itself does not improve overall reliability. Testing is intended to ferret out failures so they may be corrected within a maximum permitted time limit. The frequency of periodic tests is determined by the acceptable maximum failure period of a component. Faulty systems are more subject to nuisance alarms, which tend to desensitize occupants and lower faith that the alarm signals are real. Testing is intended to find and remedy these types of problems so the system meets its goals.

Chapter 14 of *NFPA 72, National Fire Alarm and Signaling Code*, requires fire alarm system testing to ensure the system functions properly. Testing is required by law and is not optional where *NFPA 72* is adopted. New systems must be tested to ensure they meet *Code* requirements and function according to specifications and stated goals. Existing systems must be periodically tested for the same reason.

There are three types of fire alarm system tests:
1. Initial acceptance tests
2. Periodic tests
3. Re-acceptance tests

Section 14.2.9 permits performance-based inspection and testing programs. It is anticipated that a third party would be involved with development of a performance-based inspection, testing, and maintenance program. Under this provision, an alternate means of compliance is required to be approved by the authority having jurisdiction (AHJ). In this case, the test frequencies in Table 14.4.3.2 are not necessarily required to be used.

Each of the three types of test has its own unique requirements and procedures, but some things are common to all of them.

Impairments

Sometimes there is a delay between finding a problem and fixing it. Section 14.2.2.2 of *NFPA 72* requires system defects or malfunctions to be corrected, but not all defects or malfunctions can be

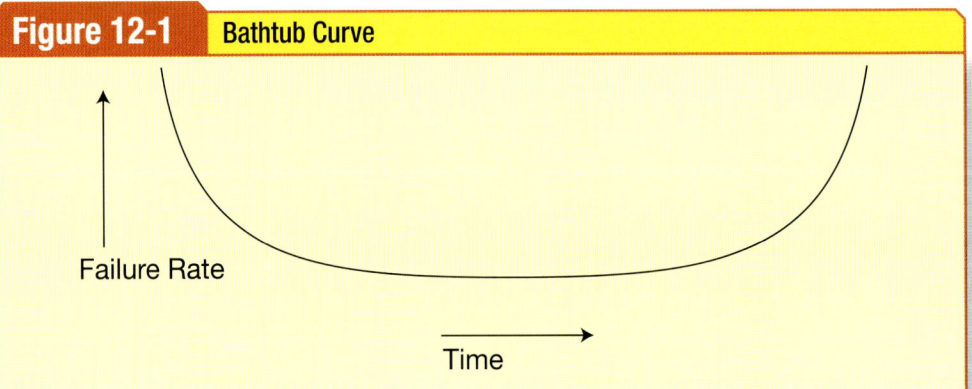

Figure 12-1. *This curve shows relative failure rates of electronic components such as smoke detectors or control circuits. Initial failures are usually experienced during "burn in" periods, but stabilize to failure rates of about 3% per year.*

corrected immediately upon discovery. For example, failure of a motherboard may require new parts that are only available from a distributor. In this case, Section 14.2.2.2.3 requires the owner or the owner's designated representative to be notified in writing within 24 hours. In order to mitigate potential liability, handwritten notification should be delivered before leaving and then followed up with a formal letter (certified mail or registered letter) within 24 hours. A call to the AHJ is also a prudent precaution whenever a system is out of service.

Many AHJs will require some sort of additional protection during a significant impairment. If large portions of the system are out of service, some AHJs will require fire watches in the affected areas. Fire codes, such as *NFPA 1, Uniform Fire Code*, and the *International Fire Code* contain requirements for the handling of impairments. These codes will generally require notification of the AHJ for a system impairment lasting more than four hours.

Responsibilities

The owner of the fire alarm system is responsible for properly testing and maintaining the fire alarm system. Section 14.2.3.1 of *NFPA 72, National Fire Alarm and Signaling Code*, requires the owner or a designated representative to be responsible for proper inspection, testing, and maintenance of the system. This section does not require the actual system owner to conduct required inspections and tests. In fact, most owners are not qualified to conduct such tests. Section 14.2.3.2 permits the owner to appoint a designated representative who is qualified to conduct the tests, and Section 14.2.3.3 permits the inspection, testing, and maintenance to be by a third party, if that party is under a contract. This delegation must, however, be under a written contract and must clearly delineate the responsibilities of the third party.

Some systems, such as those found at a college campus or large hospital, are sufficiently large to warrant full-time staff to service the fire alarm system. Where the owner delegates an employee to service and maintain fire alarm equipment, the duties and responsibilities must be described in the employee's position description. It should be noted that Section 14.2.3.4 also requires a copy of the delegation (in writing) to be provided to the AHJ upon request.

Central station alarm systems are required to be tested and maintained under a written contract, as required by Section 14.2.3.5. Central station alarm systems are the only supervising station fire alarm systems required by *NFPA 72* to include servicing under contract. However, *NFPA 72* requires all systems to be inspected, tested, and maintained.

Qualifications

All qualifications for personnel are located in Chapter 10 of the *Code*. Section 10.5.3 of *NFPA 72* requires all service personnel for fire alarm systems to be qualified and experienced. The *Code* provides examples but does not limit the choices. These examples of qualifications include the following:

1. Factory trained and certified for the specific brand and model encountered in the field
2. Certified by a nationally recognized fire alarm certification program acceptable to the AHJ, such as:
   a. National Institute in Certification in Engineering Technologies (NICET) Fire Alarm Certified
   b. International Municipal Signal Association (IMSA) Fire Alarm Certified
3. Personnel who are registered, certified, or licensed by a state or local authority
4. Trained and qualified personnel employed by a listed organization listed by a national testing laboratory for the servicing of fire alarm systems.

Those individuals who are factory trained and certified on certain products are qualified only on those products for which they have been trained. It is unreasonable to assume they would also be qualified to service systems with another manufacturer's products.

Fact

All persons conducting inspections and tests of fire alarm systems must be qualified and experienced. Both IMSA and NICET certifications are considered adequate to meet this requirement.

Fact

Central station systems must include all inspections, testing, and maintenance. The contract must be between the subscriber and the listed central station or the listed local alarm service company.

NICET certification is a rapidly developing certification program based in the United States. This program involves both work experience and tests to ensure that an individual is qualified and experienced. IMSA also has an interior fire alarm certification program. The organization is primarily concerned with traffic signaling and street signs. However, IMSA also promotes an interior fire alarm certification program, which is widely accepted in the U.S.

Staff employed by a listed organization engaged in the servicing of fire alarms are also considered qualified by *NFPA 72* because the organization will not likely jeopardize its listing through the use of unqualified staff.

Local licenses are not necessarily good indicators of whether an individual is qualified and experienced in the servicing of fire alarms. Many local licenses focus on business and contracting aspects rather than technical issues.

Programming requires special talents and abilities. There is great potential for programs to be incorrectly written, which can place lives and property in danger. Therefore, Section 10.5.3.4 requires programmers to be certified by the system manufacturer.

Finally, Section 10.5.3.5 requires evidence of personnel qualifications to be provided to the AHJ upon request.

Notification

Section 14.2.4.1 of *NFPA 72* requires all occupants and facilities receiving alarms to be notified before any tests begin in order to prevent unnecessary responses to alarm signals. Supervising stations, fire department dispatchers, and security guards must also be called for the same reasons. Supervising stations require a password to prevent tampering by unauthorized persons. It is also wise to call the local fire department non-emergency dispatch number to alert them of tests in case a passerby hears the signal and reports it.

There are a variety of methods used to alert occupants. These include, but are not limited to the following:
1. Signs at all entrances to the building
2. Email
3. Memorandums
4. Bulletin boards
5. Live voice announcements

All posted notification should contain the time that the system will be tested and should instruct occupants what to do in case of a real emergency. **See Figure 12-2.**

The supervising station operator and fire department must also be called when testing is complete to ensure any real signals are retransmitted. Additionally, the persons conducting the test should have an emergency plan in case a real fire emergency is detected.

Testing should always be conducted so as to minimize impact on the occupants and owner. One should limit tests to small areas when possible and use walk-test features when they are available. Audible and visible notification appliances should

Figure 12-2 | Occupant Notification

Fire Alarm Testing

122 Water Street, NW

Washington, DC

When: Monday April 19

Time: 6:00 AM to 8:00 AM

Where: Third Floor

In case of emergency, please call 911

Tests are being conducted by:

ABC Alarms

11199 Lorton Rd.

Suite C

Arlington, VA 22209

(703) 555-1234

Figure 12-2. *Note the specificity of the times the system will be tested and contact information.*

be actuated only long enough to test them. Allowing them to operate every time an initiating device is actuated will cause significant occupant disruption. Off-hours testing is an excellent way to avoid occupant disruption. Nights and weekends are good times to perform long testing programs which require notification appliances to be operated for long periods.

Section 14.2.4.3 requires the system owner or designated representative and service personnel to coordinate with other trades to ensure critical building services, such as HVAC or elevators, are not severely impacted. This step requires good communications skills to ensure the proper personnel are present when needed. Many organizations test fire alarm systems at night or early in the morning to prevent major disruptions to occupants.

Section 14.2.5 requires the owner or the owner's designated representative to provide service personnel with all documentation required to test the system. This documentation includes but is not limited to: record (as-built) drawings, specifications, alterations, floor plans, wiring diagrams, and operations manuals. This documentation is essential to properly test the entire system.

Tools

Testing a fire alarm system will require special tools and equipment. Most projects will require the following:
- Aerosol smoke or smoke-producing instrument
- Volt/ohmmeter (VOM)
- Sound pressure level (SPL) meter
- Screwdrivers
- Flashlight
- Ladders
- Two-way radios
- As-built plans
- Manual pull station and control unit keys
- Operations and maintenance manual
- Sequence of operations
- Specifications
- Clipboard and pen/pencil
- Special wrenches or keys for resetting manual fire alarm boxes
- A copy of NFPA 72

Older non-addressable systems using smoke detection will require a method for testing smoke detector sensitivity. EVACS will also require use of an intelligibility meter to ensure the voice message is clearly understandable.

When conducting tests, at least two service technicians are generally required. Smaller systems can be tested using walk-test features and a system printer. However, most tests require one technician to remain at the control unit while the other actuates initiating devices and verifies outputs. Systems with complicated safety control functions may require numerous technicians to verify fire safety functions operate correctly.

When testing, the technician at the controls should report what device or zone is actuated, in order to ensure accuracy. The technician who is actuating field devices should not report the device he or she just actuated. He or she should verify what is reported from the other technician at the control unit. This prevents inaccuracies in programming which may be missed by the technician at the controls.

Documentation

System documentation is the only way to verify that inspections, testing, and maintenance have been conducted. Section 14.6.1.1 of *NFPA 72* requires fire alarm service organizations to provide permanent copies of maintenance manuals, record drawings, and sequence of operation to the owner or the owner's designated representative at the conclusion of testing.

Before beginning any inspections and tests, it is imperative that all necessary equipment, personnel, and documentation be assembled and ready. At a minimum, the following are required:
- System record (as-built) drawings
- Operations and maintenance (OM) manual
- Copy of software (if needed)
- *NFPA 72, National Fire Alarm and Signaling Code*
- Fire alarm service personnel (at least two)
- Personnel from other trades (HVAC, elevators, fire suppression system)

Fact

The number of technicians required for a test will vary, depending on the size of the job. Large buildings may have multiple crews simultaneously testing different areas, but a technician must be located at the controls.

- Tools (ladders, aerosol smoke, heat source, volt/ohmmeter, hand tools, etc.)
- Radios
- Building keys
- Checklist and pen/pencil

Section 14.6.1.2 requires a copy of the site-specific software to be provided on a non-volatile medium. The preferred method of software transfer is a compact disc (CD) or thumb drive. A floppy disk can easily be erased by electrical noise caused by transformers inside the control unit. Electronic media will facilitate a much faster restoration after a catastrophic event such as a lightning strike.

Section 14.6.1.3 requires the system owner to maintain all records for the life of the system. This section also grants the AHJ permission to examine these records at any time. Failure by the owner to maintain the system may result in fines.

Section 14.6.2.4 requires the service organization to complete the form represented in Figure 7.8.2(g) of *NFPA 72* when conducting inspections, tests, or maintenance of fire alarm systems. A completed copy of this form must be delivered to the owner or the owner's designated representative at the conclusion of testing. Figure 7.8.2(g) may be copied and used; however, electronic databases can also be used. Some service organizations prefer to use custom databases, but the database files must contain all information required by Figure 7.8.2(g). In any event, the owner must be presented with a non-alterable (permanent) record. Electronic (read-only) files are acceptable, but a paper copy is considered as a permanent record. **See Figure 12-3.**

Section 14.6.2.1 of *NFPA 72* requires the organization or individual conducting maintenance, inspections and tests to retain records for a period of one year after the next test. The owner must retain all test records for the life of the system. This includes a copy of the certificate of occupancy from the AHJ.

Some authorities having jurisdiction will require signed copies of the Record of Completion (*NFPA 72*, Figure 7.8.2(a)) and the Record of Inspection and Test (*NFPA 72*, Figure 7.8.2(g)) at the time they witness any acceptance tests.

Accidental discharge of suppression systems can cause extensive damage and may sometimes cause fatalities.

Releasing Systems and Interfaces

Testing a fire alarm system that interfaces with a suppression system, such as a gaseous suppression system, kitchen hood, CO_2, or pre-action system, can be very challenging. These special hazard systems must not be actuated because the suppression agent is expensive, causes damage to the building and its contents, or requires extensive cleanup. There have also been several cases where improper testing of CO_2 systems has resulted in accidental fatalities.

Section 14.2.6.1 requires testing personnel testing interfaced releasing systems to be further qualified and experienced in the hazards associated with inadvertent system discharge. Section 14.2.6.2 requires occupant notification when a fire alarm system configured for releasing service is being tested or maintained. There have been cases where accidental discharge of a carbon dioxide system resulted in occupant deaths. Finally, Section 14.6.4 requires simulated tests of releasing of special hazard systems to record how the tests were simulated, along with a signature and date. The frequency of tests for special hazards systems is found in Item 18 of Table 14.4.3.2 in the *Code*.

Figure 12-3 Inspection, Testing, and Maintenance Form

FIRE ALARM AND EMERGENCY COMMUNICATION SYSTEM INSPECTION AND TESTING FORM

To be completed by the system inspector or tester at the time of the inspection or test.
It shall be permitted to modify this form as needed to provide a more complete and/or clear record.
Insert N/A in all unused lines.
Attach additional sheets, data, or calculations as necessary to provide a complete record.

For additional information, visit qr.njatcdb.org Item #1022

Date of this inspection or test: _____ Time of inspection or test: _____

1. PROPERTY INFORMATION

Name of property: _____
Address: _____
Description of property: _____
Occupancy type: _____
Name of property representative: _____
Address: _____
Phone: _____ Fax: _____ E-mail: _____
Authority having jurisdiction over this property: _____
Phone: _____ Fax: _____ E-mail: _____

2. INSTALLATION, SERVICE, AND TESTING CONTRACTOR INFORMATION

Service and/or testing organization for this equipment: _____
Address: _____
Phone: _____ Fax: _____ E-mail: _____
Service technician or tester: _____
Qualifications of technician or tester: _____
A contract for test and inspection in accordance with NFPA standards is in effect as of: _____
The contract expires: _____ Contract number: _____ Frequency of tests and inspections: _____
Monitoring organization for this equipment: _____
Address: _____
Phone: _____ Fax: _____ E-mail: _____
Entity to which alarms are retransmitted: _____ Phone: _____

3. TYPE OF SYSTEM OR SERVICE

- Fire alarm system (nonvoice)
- Fire alarm with in-building fire emergency voice alarm communication system (EVACS)
- Mass notification system (MNS)
- Combination system, with the following components:
    - Fire alarm • EVACS • MNS • Two-way, in-building, emergency communication system
- Other (specify): _____

© 2009 National Fire Protection Association

NFPA 72 (p. 1 of 11)

Figure 12-3. *This record must be completed each time the system is inspected, tested, or maintained. It can be reproduced in electronic form, provided the information matches the requirements of the form in NFPA 72.*

Section 14.2.7 addresses testing of interface equipment and control systems. This section requires testing personnel to be qualified and experienced in the arrangement and operation of interface equipment and emergency control functions. The test frequency and methods for interface equipment and emergency control functions are found in Item 20 of Table 14.4.3.2.

INSPECTIONS AND TESTS

Visual inspections are designed and required to detect changes to the building or system, which may affect system performance. Damage to devices or equipment must be corrected before tests begin because the repairs may affect other parts of the system or require further testing. Devices or appliances added or deleted after the previous test will also trigger additional tests and should be noted on record drawings.

Visual inspections do not require any manipulation of system equipment or components. Inspections are easily conducted by walking through the building and looking at devices, appliances, and equipment. A careful record should be kept to document all deficiencies prior to testing. Examples of deficiencies include poorly mounted detectors and appliances, loose terminals, smoke detectors too close to HVAC diffusers, and trouble conditions. Inspection checklists can be developed and used from project to project. **See Figure 12-4.**

Small-scale plans are very useful during inspections to ensure that all equipment is inspected.

Section 14.3.1 of *NFPA 72* requires inspections to be conducted, but Table 14.3.1 determines the frequency of the inspections. An "X" in the column indicates a required inspection. Naturally, a 100% inspection is conducted at the initial acceptance of the system. All components are inspected at the initial acceptance of the system. Some devices will require inspections on a weekly basis, but some require annual inspections. Sections 14.3.2 and 14.3.3 permit inspection frequencies to be extended where safety concerns exist, such as in high radiation areas. However, inspection frequencies cannot be extended beyond 18 months in any case.

Table 14.3.1 of the *Code* is organized by equipment type and is simple to use. As an example, Item 2(a) in Table 14.3.1 requires an annual inspection of control equipment that is monitored by a supervising or trained staff for all signals. However, Item 2(b) of Table 14.3.1 requires weekly inspections of control equipment that is not provided with a supervising station connection or directly and continuously monitored by trained staff.

INITIAL/ACCEPTANCE TESTS

Initial/acceptance tests are conducted upon completion of the initial installation, prior to building occupancy. Table 14.4.3.2 requires a 100% test of the system at the time of completion. Section 14.4.1.2 also requires notification of the AHJ prior to the initial acceptance tests. This requirement exists because the AHJ will witness the acceptance test of the system before signing the certificate of occupancy. The AHJ will not accept any system that does not correctly operate. Minor matters such as device labeling might be waived, provided that there is no overriding safety concern.

It is always wise to conduct a complete acceptance test before the AHJ arrives, just in case the equipment of the system does not operate correctly. Failures and subsequent tests require a second visit by the AHJ to ensure correct operation. Tests are usually scheduled weeks in advance. Therefore, testing the system in its entirety before the AHJ arrives will save time and money.

PERIODIC INSPECTIONS AND TESTS

Periodic inspections and tests are conducted to ensure system operability and reliability. Periodic tests must be conducted even if there are no changes to the system. Periodic tests are primarily intended to find and correct any component failures caused by aging. Periodic inspections are designed to detect any

Fact

Inspections never include a physical manipulation of the equipment. Inspections simply identify equipment which is defective or building conditions which prevent proper system operation.

Figure 12-4   Inspection Checklist

Controls:
1. No trouble signals ☐
2. No supervisory signals ☐
3. No alarm signals ☐
4. Power on ☐
5. Free of damage ☐

Batteries:
1. Installed properly ☐
2. Proper rating ☐
3. Proper type (sealed Lead Acid) ☐

Smoke detectors:
1. Installed away from vents ☐
2. Properly supported ☐
3. Clean and free of debris ☐
4. Free of damage ☐
5. Installed outside dead air space ☐

Heat detectors:
1. Installed away from vents ☐
2. Properly supported ☐
3. Clean and free of debris ☐
4. Free of damage ☐
5. Installed outside dead air space ☐

Manual Fire Alarm Boxes:
1. Installed at proper height ☐
2. Glass rod or plate installed ☐
3. Free of damage ☐
4. Installed near exit doors ☐

Audible notification appliances:
1. Installed at proper height ☐
2. Free of damage ☐
3. Properly supported ☐

Visible notification appliances:
1. Installed at proper height ☐
2. Free of damage ☐
3. Properly supported ☐
4. Proper rating ☐

Wiring:
1. Conduit and box fill does not exceed 40% ☐
2. Proper methods ☐
3. Terminations properly made ☐
4. Class A separation (where needed) ☐
5. Survivability (EVAC systems) ☐

Waterflow devices:
1. Properly installed (flow is in direction of arrow) ☐
2. Properly supported ☐
3. Access to internal components ☐

Supervisory devices:
1. Properly installed (does not interfere with valve) ☐
2. Properly supported ☐

Figure 12-4. The checklist is a good reminder of basic inspection criteria. More detailed checklists can be developed and used with experience.

problems with system components, such as device/appliance support. Periodic inspections also help find problems caused by changes to a building, such as a partition built too close to a detector.

Periodic tests must be conducted according to the frequency of tests found in Table 14.4.3.2 of the *Code*. Table 14.4.3.2 is arranged in a similar fashion to Table 14.3.1. An "X" indicates a required test. The methods of testing are also found in Table 14.4.3.2. Each component of the system has a required method of test. Test methods were incorporated into Table 14.4.3.2 in the 2013 edition of *NFPA 72* for ease of use.

REACCEPTANCE TESTS

Reacceptance tests are provided to ensure changes to the system do not adversely affect system performance. Reacceptance tests are required after any changes to the system, including addition or deletion of devices and software changes. Changes to the building, such as the installation of a new wall, may require additional devices or appliances. Any new devices must be tested to ensure the system operates correctly.

Software changes may seem harmless. However, small changes in software can, and often do, severely impact system performance. There have been instances where a seemingly small software change resulted in huge changes in system performance. These changes sometimes result in system failure and large losses for the owner. Every effort should be made to ensure that all software changes are thoroughly tested.

For these reasons, Section 14.4.2 provides requirements for reacceptance testing. Section 14.4.2.1 requires a functional test of newly added devices, appliances, or control relays. When devices or appliances are deleted, other devices or appliances on the same circuit are also required to be tested, as required by Section 14.4.2.2.

Control equipment reacceptance tests are covered by Section 14.4.2.3. This section contains specific requirements for changes to site-specific software. Specifically, all functions known to be affected by the software change must be 100% tested. An additional 10% of initiating devices not affected by the change must also be tested, up to a maximum of 50 devices. These devices should be selected randomly on different input circuits. For example, the addition of a software-controlled relay would require a functional test of the relay, plus a functional test of 10% of the initiating devices on the system.

Changes to executive software, such as BIOS, are generally made by factory-authorized technicians and usually affect the entire system operation. Changes to BIOS are covered by Section 14.4.2.5. Changes to BIOS require a minimum 10% test of the system, including a test of at least one device or appliance on every connected circuit.

TESTING METHODS AND FREQUENCY

Section 14.4.3.2 requires periodic testing in accordance with Table 14.4.3.2. This table provides the frequency and methods of periodic testing, itemized by system component. Table 14.4.3.2 is used in conjunction with manufacturers' instructions and recommended practices. Section 10.4.4.1 permits testing frequencies to be extended only where safety concerns exist. Examples of these areas include high radiation areas or hazardous locations. In no case can testing frequencies be extended beyond 18 months.

The technical committees developed definitions for each frequency of inspection and test, which are included in the definitions section in the glossary at the end of this book and in Chapter 3 of *NFPA 72*. These definitions make the intent of the testing periods clear and set maximum limits on the length of time between each inspection or test.

Control Equipment

Control equipment must be properly tested in accordance with Item 2 in Table 14.4.3.2. **See Figure 12-5.**

This section requires a number of tests to ensure the basic system functions are operating correctly. Item 2(a) of Table 14.4.3.2 requires these tests for control equipment to be completed at annual intervals.

Smoke Detectors

Smoke detectors must undergo two types of tests: functional testing and sensitivity testing. Functional tests involve smoke entry into the sensing chamber. Aerosol smoke products that are listed are permitted by most manufacturers, but magnet tests do not meet the requirements of the *Code*. Smoke detector sensitivity testing must be performed in accordance with Item 17(h) of Table 14.4.3, which requires detector sensitivity to be tested within one year after installation and every alternate year thereafter. If records indicate there are no unwanted alarms, sensitivity tests may be extended to a maximum period of five years. There are a variety of methods

Figure 12-5. Control Equipment Test Methods

2. Control equipment	
(a) Functions	At a minimum, control equipment shall be tested to verify correct receipt of alarm, supervisory, and trouble signals (inputs); operation of evacuation signals and auxiliary functions (outputs); circuit supervision, including detection of open circuits and ground faults; and power supply supervision for detection of loss of ac power and disconnection of secondary batteries.
(b) Fuses	The rating and supervision shall be verified.
(c) Interfaced equipment	Integrity of single or multiple circuits providing interface between two or more control units shall be verified. Interfaced equipment connections shall be tested by operating or simulating operation of the equipment being supervised. Signals required to be transmitted shall be verified at the control unit.
(d) Lamps and LEDs	Lamps and LEDs shall be illuminated.
(e) Primary (main) power supply	All secondary (standby) power shall be disconnected and tested under maximum load, including all alarm appliances requiring simultaneous operation. All secondary (standby) power shall be reconnected at end of test. For redundant power supplies, each shall be tested separately.

Figure 12-5. Item 2 from Table 14.4.3.2 of the Code gives testing methods for control equipment.

Reprinted with permission from NFPA 72®-2013, *National Fire Alarm and Signaling Code*, Copyright © 2012, National Fire Protection Association, Quincy, MA. The information in this figure is intended to be used in conjunction with the requirements of this code. It is not the complete and official position of the NFPA on the referenced subject, which is represented only by the standard in its entirety.

used to conduct sensitivity tests. They are as follows:
1. Calibrated test method
2. Manufacturer's calibrated test instrument
3. Listed control equipment arranged for the purpose
4. Smoke detector/control unit arrangement where the detector causes a signal at the control unit when its sensitivity is outside its listed sensitivity range
5. Other calibrated sensitivity test methods approved by the AHJ

Calibrated test methods include portable test units designed for the purpose, such as the Solo True Test or Gemini Smoke Detector Analyzer, which deliver a measured quantity of aerosol product (a proprietary oil mist) to the detector. **See Figure 12-6.**

This category also includes smoke boxes, such as those used at testing laboratories. Smoke boxes use a small piece of burning cotton wick to produce smoke in known quantities, as measured by a light obscuration instrument inside the box. Listed control equipment arranged for the purpose may include a panel in a laboratory or service shop that is used to test equipment.

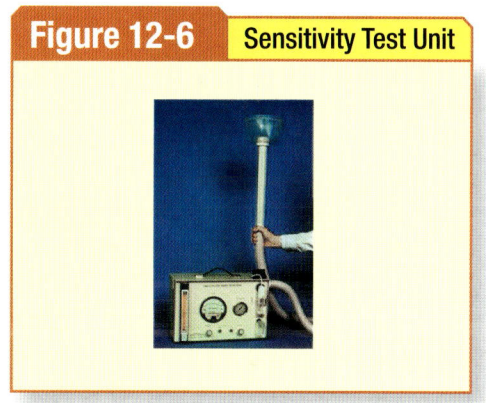

Figure 12-6. The unit shown in the photograph is the Gemini Smoke Detector Analyzer. This unit can be used to test the sensitivity of any smoke detector and conducts a functional test at the same time.

Manufacturer's calibrated test instruments include specially designed meters or piggyback meters designed for use with a digital volt/ohmmeter. These instruments have a plug that simply attaches to the detector under test.

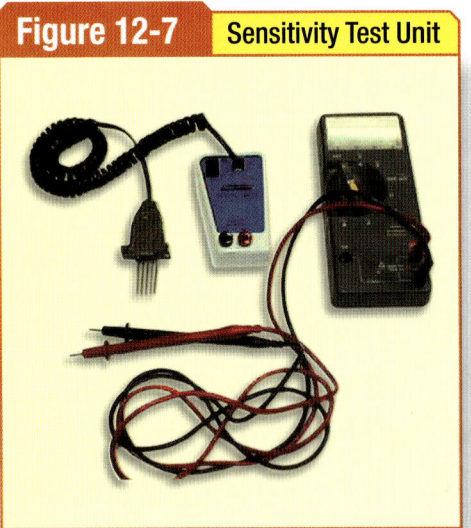

Figure 12-7. This unit is designed to test one make and model smoke detector. It requires a digital volt/ohmmeter, which is used to display the sensitivity of the detector under test.

 Fact

Sensitivity tests are never conducted using an unmeasured quantity of smoke. Most new addressable smoke detectors are tested by their control equipment every few minutes. The sensitivity test reports should be printed annually and retained in the test file.

Sensitivity levels are read on the meter. **See Figure 12-7.**

New analog addressable equipment typically interrogates each device several times per minute. This interrogation includes a determination of the detector sensitivity. Sensitivity tests are not required on analog addressable detectors that report sensitivity levels to the control unit. Sensitivity tests are usually required on conventional (non-addressable) detectors because they cannot report sensitivity levels to the controls. However, some newer conventional smoke detectors can report trouble when they are outside their listed sensitivity range.

Sensitivity tests are never conducted using an unmeasured quantity of smoke or aerosol test product because the concentration of the smoke or aerosol is not known. Therefore, the sensitivity cannot be determined. Smoke detectors and residential smoke alarms that fail to pass required sensitivity tests must be cleaned and recalibrated or replaced as required by Section 14.4.4.3.6.

Aerosol smoke, incense sticks, punk sticks, and other similar products delivered in unmeasured quantities are considered to be functional tests, which are required to ensure smoke entry into the sensing chamber. Table 14.4.3.2 Item 17(g)(1) requires annual functional tests of smoke detectors. Magnet tests of smoke detectors are acceptable for pretesting of smoke detectors but do not satisfy the requirements of *NFPA 72* for functional tests.

Aerosol smoke must be approved by the manufacturer and should always be used sparingly. Short puffs will usually produce a response within a short time. Excessive use of any smoke product or aerosol will take longer to clear, thereby increasing the time required for testing. Compressed air helps to clear smoke from the sensing chamber after testing and is approved by most manufacturers. **See Figure 12-8.**

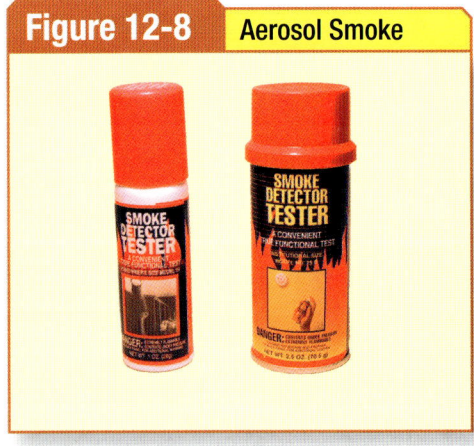

Figure 12-8. These products must be listed for such use. *NFPA 72* requires smoke for the functional test. Magnet tests do not accomplish this function.

Heat Detectors

Non-restorable, fixed-temperature heat detectors are not tested in the field. Table 14.4.3.2 Item 17(d)(3) requires a statistical sample to be tested by a qualified laboratory after 15 years. If any detectors fail the required tests, more detectors must be tested. This section also permits the generally less expensive method of wholesale replacement of these devices after 15 years, rather than testing them.

Section 14.4.4.5 and Item 17(d)(1) of Table 14.4.2.2 require restorable heat detectors to be tested so that at least two devices are tested on each circuit annually, and all detectors are tested at least once every five years. This requires accurate record keeping so that the same detectors are not tested every year. Heat sources may include electric heat guns or hair dryers. Flame-producing devices are not recommended because they may burn the detector housing or start a fire. A bucket of hot water may be used to test restorable heat detectors installed in hazardous (classified) locations.

Special Situations

Item 22(c) of Table 14.4.3.2 requires visible notification appliances to be tested by annually verifying the proper rating and location and then verifying that the appliance operates when the system is in alarm.

Audible appliances are tested using a sound pressure level (SPL) meter to measure the output of the signal throughout the protected area, according to Item 22(a) of Table 14.4.3.2. This section requires SPL measurements to be conducted annually. It is not necessary to measure more than a few points in most average-sized rooms. Obviously large areas will require more test points to ensure requirements are satisfied. For EVACS, only the alert tone SPL is tested; the SPL of the voice message is not measured. Voice intelligibility is required to be tested, however, using a specially designed testing device. This ensures the voice signals are able to be understood.

Other devices requiring special testing equipment may include duct smoke detectors and air-sampling smoke detectors. These types of detectors must be tested to ensure there is sufficient air flow into the detector. In some cases, a manometer or magnehelic gage is used to test the pressure differential. The duct smoke detector sensor is tested the same way as open-area smoke detectors.

Wet pipe sprinkler systems are tested by flowing water through the waterflow switches. Inspector test stations are commonly located in stairways on each level and have a valve that allows water to flow through an orifice sized to the smallest sprinkler on the branch line. An alarm must be received by the FACU within 90 seconds of water flow. Dry pipe systems use an alarm bypass valve in order to prevent actual tripping of the dry valve. In some jurisdictions, only sprinkler technicians are permitted to operate the sprinkler system. One should be aware of local laws before testing the sprinkler valves and supervisory devices. Test frequency and methods for waterflow switches are found in Item 17(k) of Table 14.4.3.2, which requires semiannual tests.

Safety control functions are usually tested using the same methods and frequencies as the initiating devices that operate the function. For example, elevator recall features are tested by introducing smoke into each smoke detector, which causes recall. The control function tests may be incorporated with the functional tests of the smoke detector, rather than testing the detector twice.

Safety control functions should be tested "end to end." Testing in this fashion ensures each input will result in the correct safety function output. For example, an elevator lobby smoke detector should be functionally tested, and the elevator response (firefighter's recall) to the actuation of the smoke detector should be verified. Item 18 of Table 14.4.5 requires annual tests of safety control functions.

Systems must be restored promptly after testing. A system left in silent walk-test mode overnight is a liability that

Fact

Manufacturers often specify required maintenance beyond the *Code* requirements. This information is located on the installation sheets that accompany the equipment. These sheets must be retained to create the Operations and Maintenance (OM) manual.

must be avoided. When testing is completed, the supervising station and fire department dispatcher must be informed that testing is complete. If an EVAC system is used, occupants may be alerted by use of the system.

MAINTENANCE

Maintenance can vary from changing paper in a printer to cleaning smoke detectors. Manufacturers generally prescribe specific maintenance frequencies and practices because the *Code* cannot address every make and model of device, appliance, and equipment. Maintenance is specific to equipment type and model, and Section 14.5.1 of *NFPA 72* refers the user to manufacturer's recommended maintenance schedules and practices. Environmental conditions, such as a dusty room, may warrant more frequent maintenance of devices or equipment.

TROUBLESHOOTING

Troubleshooting fire alarm systems is not difficult, provided common sense and proper tools are used. Good troubleshooting skills are essential for both new and existing installations. Wiring faults are the most common type of fault. Software problems are unique to each manufacturer and require knowledge of the software involved.

Class B Initiating Device Circuits

Wiring faults are the most common fault. Class A and Class B circuits are the two types of circuits used for input of signals to the fire alarm control unit. Class B circuits operate in a daisy-chain fashion, and they operate in a parallel arrangement. On a Class B circuit, conductors extend to the end-of-line device and do not extend back to the controls. **See Figure 12-9.**

Initiating device circuits and notification appliance circuits are terminated with an EOL device, usually a resistor. Some manufacturers, however, use a capacitor as an EOL. Monitoring for integrity is provided by a small electrical current passing through the circuit, from the panel to the EOL and back to the

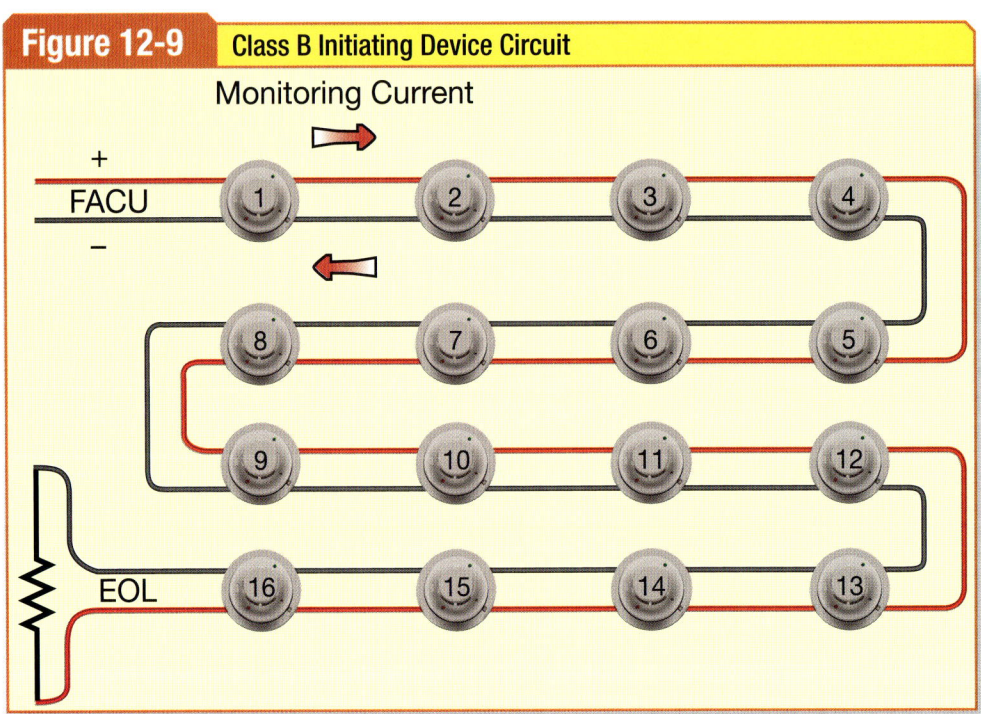

Figure 12-9. The monitoring current circulates through the EOL and back. Any interruption, such as a loose terminal, will cause a trouble signal.

panel. Conventional initiating devices are open contact devices and current does not pass through them unless they are in the alarm state.

Initiating devices are connected in parallel across the circuit. A contact closure creates a short, which causes a larger current flow through the circuit. This is perceived as an alarm by the circuit controls. Some initiating devices, such as smoke detectors, require power to operate. A short by an actuated detector near the panel may prevent downstream detectors from operating. Other devices, such as heat detectors and manual fire alarm boxes, do not require power to operate and are immune to this problem.

There are three different conditions that occur on initiating device circuits:
1. Normal – No troubles (opens, shorts, or ground faults) and no alarms
2. Alarm – Short-circuit condition
3. Trouble – Open circuit or ground-fault conditions

Some initiating device circuits and notification appliance circuits can, however, detect short circuits through the use of a resistor in series with the alarm contacts.

Proper monitoring for integrity will be achieved when all devices are properly connected, including the EOL device. Devices that receive power from the circuit are polarity sensitive and must be properly connected. Once all devices are properly connected, the circuit should be checked for opens, shorts, and grounds using a meter. Table 14.4.2.2, Item 12, of the *Code* requires these tests on new circuits only.

When conducting the initial conductor tests required by *NFPA 72*, there are a number of measurements that must be taken and recorded. These measurements include stray voltage from each conductor to ground (AC and DC), resistance to ground, and loop resistance. *NFPA 72* requires less than one volt when measuring stray voltage. The system manufacturer will specify minimum ground resistances, but resistance to ground should really be no less than one million ohms (1 megohm). Loop resistance should be within manufacturer's tolerances.

Records must be kept as part of the final documentation. It is suggested these tests be conducted before devices and appliances are installed.

The following troubleshooting tools should be added to one's tool bag:
- High impedance multimeter (digital volt/ohmmeter)
- Various resistors with six-inch leads and alligator clips on leads
- Six-inch jumpers with alligator clips

All troubleshooting will be conducted with the circuit disconnected from the FACU. Three readings will be taken during troubleshooting: positive to negative, positive to ground, and negative to ground. A troubleshooting value table can be used to provide expected and recorded values. **See Figure 12-10.** In this case, the fictitious EOL is 10,000 (10k) ohms. In a real situation, the manufacturer's actual EOL value would be substituted for the 10k ohms.

Troubleshooting open circuits on an initiating device circuit (IDC) is actually

Figure 12-10	Troubleshooting Value Table for a Normal Class B IDC		
Zone #1	Circuit Type: IDC	EOL Value: 10,000 Ohms	Date:
Meter Reading Point		Desired Results	Actual Results
Positive to Negative		EOL Value	10,000 ohms
Positive to Ground		Infinity	Infinity
Negative to Ground		Infinity	Infinity

Figure 12-10. Values were measured with the circuit disconnected from the controls.

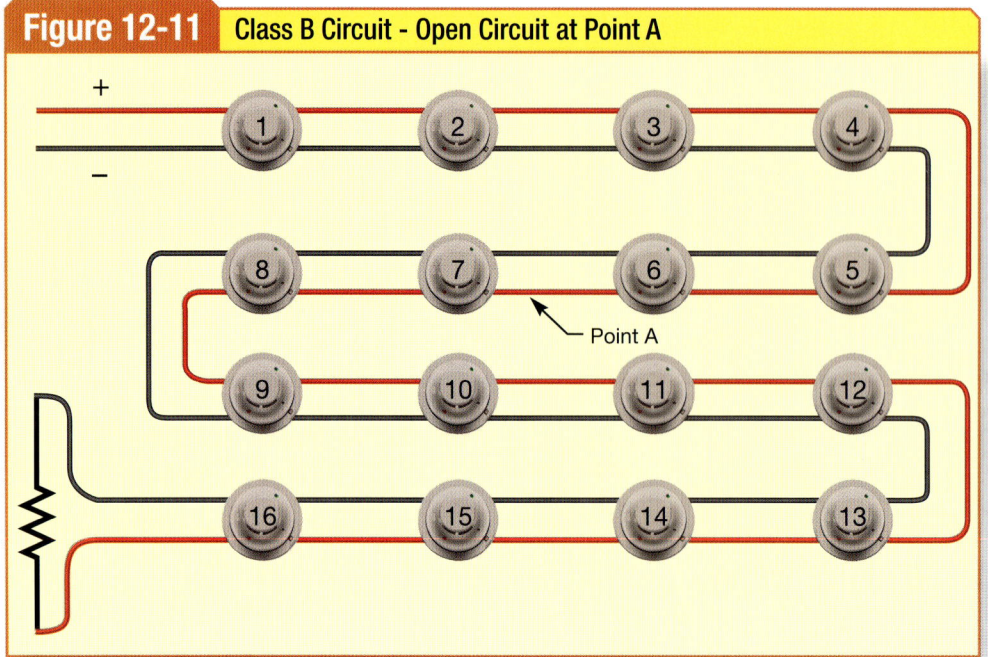

Figure 12-11. The monitoring current cannot circulate through the EOL and back, causing a change in voltage at the controls, which in turn creates a trouble signal.

Figure 12-12. Values were measured with the circuit disconnected from the controls.

Figure 12-12	Troubleshooting Value Table for an Open Class B IDC		
Zone #1	Circuit Type: IDC	EOL Value: 10,000 Ohms	Date:
Meter Reading Point	Desired Results		Actual Results
Positive to Negative	EOL Value		Infinity
Positive to Ground	Infinity		Infinity
Negative to Ground	Infinity		Infinity

quite simple. Consider a Class B circuit with an open circuit. **See Figure 12-11**.

All readings are taken with the meter set to the resistance scale and the circuit disconnected from the FACU. The readings on the circuit at the FACU should be recorded in the expected values table. **See Figure 12-12**.

The meter reading at the FACU should have been 10,000 ohms, which is the same value as the EOL device. An infinite reading on the meter indicates an open circuit between the FACU and the EOL device.

The basic approach to troubleshooting is to divide and conquer the circuit. Wiring is generally accessible at device or appliance locations. These make good locations to easily open the circuit or to isolate devices and appliances. Cut the circuit in half and take additional readings on each side of the midpoint: one set looking toward the FACU and the other at the EOL. In this case, the circuit is opened at the midpoint (Device #8), dividing the circuit in half in order to more easily isolate the open circuit. By connecting a known resistance (5,000 ohms) to the circuit at the FACU, the open can be isolated more quickly. **See Figure 12-13**. If the open is between the FACU and the new test point, infinite resistance will be measured. If not, a 5,000-ohm resistance will be measured.

In this case, the resistance looking at the EOL is 10,000 ohms, which is the correct value. Looking at the FACU, resistance measures infinite, indicating an open circuit. At this point, the device should be reconnected at Device #8 and opened at the new mid-point (Device #4). This process is repeated until the open is located.

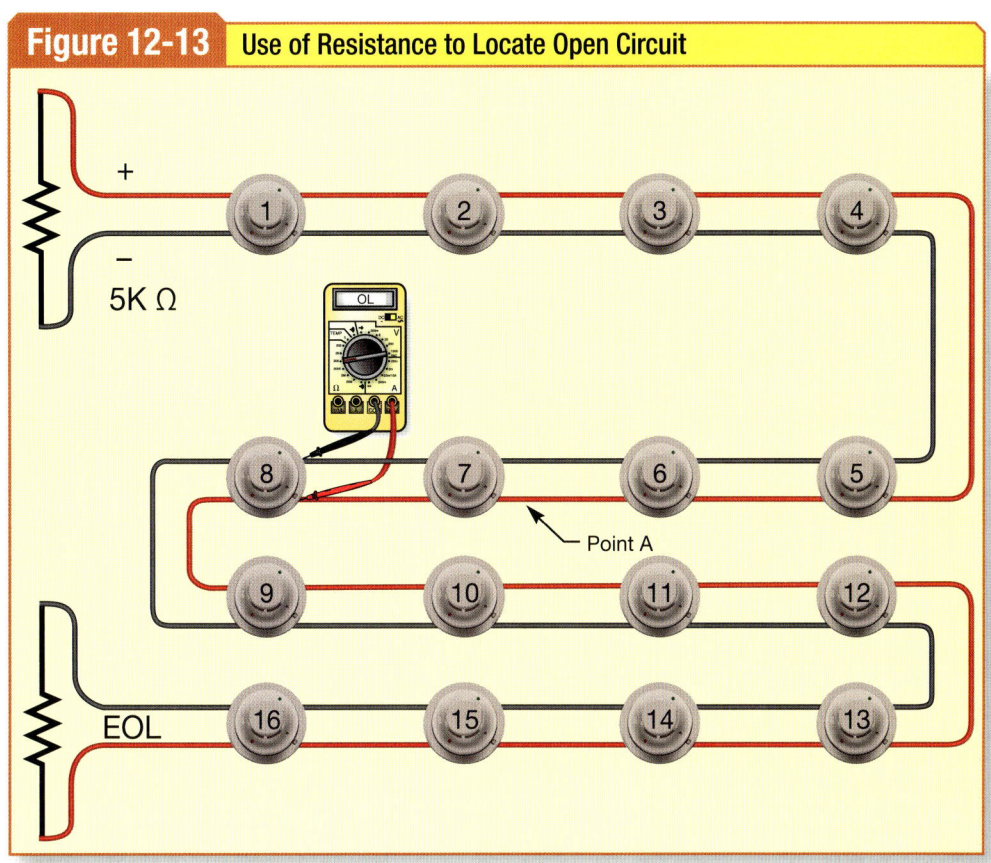

Figure 12-13. In this circuit, the known resistance is used to assist with troubleshooting. The open is located at Point A, and measurements on each side of the open yield different values, which helps locate the open circuit.

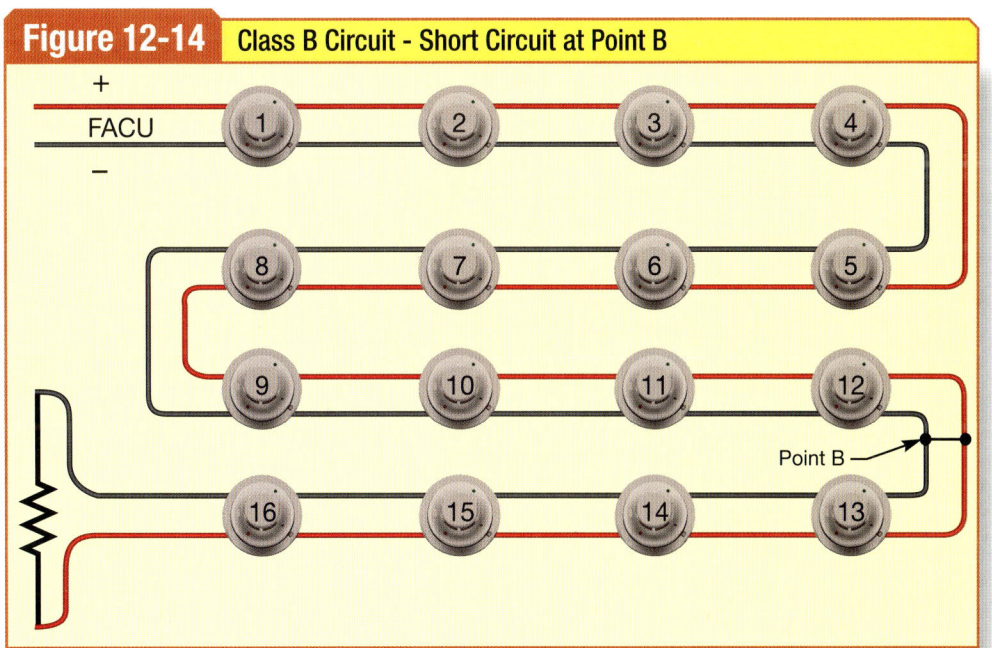

Figure 12-14. The monitoring current is shunted through the short and the controls sense this as either a trouble or an alarm, depending on the make and model of controls.

Troubleshooting short circuits on an IDC is similar to finding open circuits. The circuit should be divided into manageable pieces to isolate the fault. This will help to accelerate the troubleshooting of faults. Consider a short circuit on a Class B circuit. **See Figure 12-14.**

Figure 12-15 — Troubleshooting Value Table for a Shorted Class B IDC

Zone #1	Circuit Type: IDC	EOL Value: 10,000 Ohms	Date:
Meter Reading Point		Desired Results	Actual Results
Positive to Negative		EOL Value	Zero
Positive to Ground		Infinity	Infinity
Negative to Ground		Infinity	Infinity

Figure 12-15. *Values were measured with the circuit disconnected from the controls.*

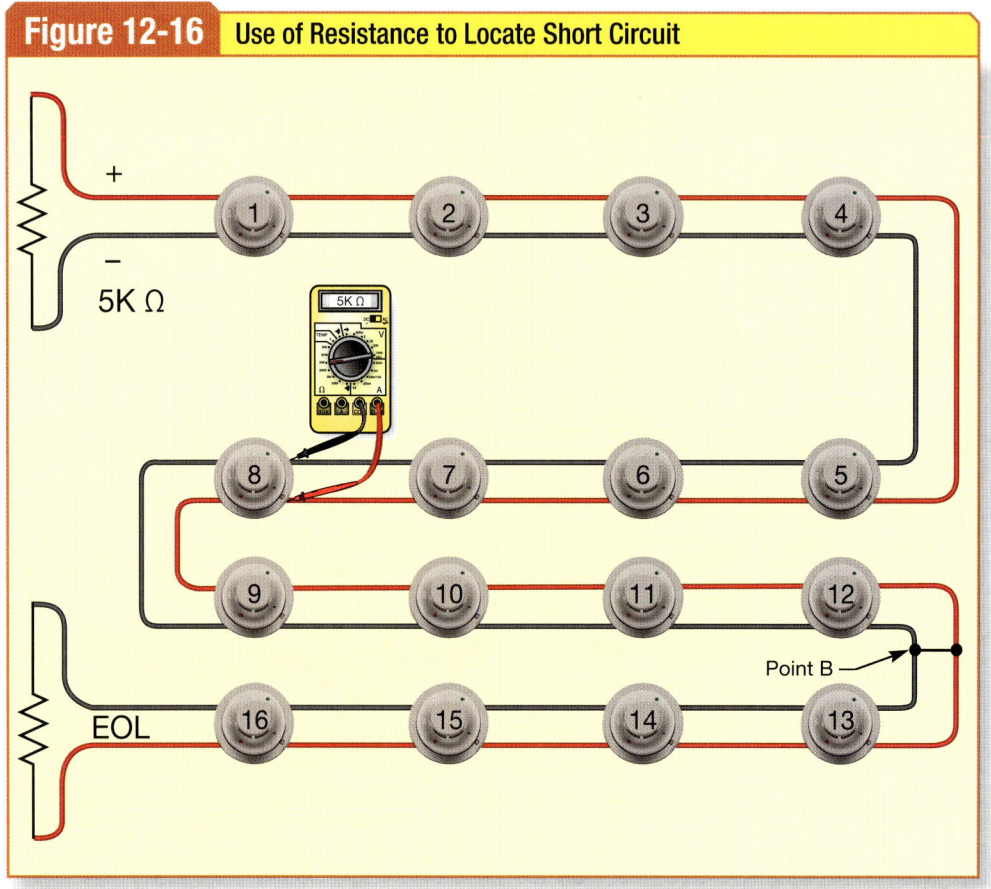

Figure 12-16. *In this circuit, the open is located at Point B and measurements on each side of the open yield different values, which helps locate the short circuit.*

The initial meter readings taken at the FACU should be measured and recorded on the expected values table. **See Figure 12-15.**

As in the previous example, the circuit leads are disconnected and the 5,000 ohm resistor is installed at the FACU. The circuit can then be divided into two circuits at Device #8. First the meter is connected and measurements are taken looking toward the FACU. **See Figure 12-16.**

Next the meter is connected and measurements are taken looking toward the EOL. A short circuit will be measured, indicating the short is between Device #8 and the EOL.

If the open at Device #8 is reconnected and the circuit at Device #12 is opened, another meter reading looking at the

FACU will indicate 5,000 ohms. The meter is then placed on the leads looking toward the EOL, which will indicate a short. The circuit is reconnected and opened at Device #14, and the process is repeated until the short is located.

Ground faults are frequently encountered in new work or underground runs. Many ground faults occur when wire is improperly pulled or if conduit is not properly de-burred. They can also occur when the incorrect insulation type is used in wet locations (such as THHN versus THWN). When planning underground runs, it is advised that outside plant cable be used. Outside plant cable is designed for wet locations and ground faults are especially rare on this type of cable. Whenever possible, optical fiber cable should be used so ground faults are avoided altogether.

Ground faults are among the most difficult to locate. This is because the fault may be an intermittent fault caused by moisture or vibrations. Underground circuits can be especially challenging because the weather can cause intermittent ground faults. Additionally, underground runs tend to be run over long distances, which makes troubleshooting even more difficult. Ground faults may also be a soft fault with resistances that are less than a direct short to ground. The zone will indicate trouble because not all of the monitoring current is returning to the FACU. Some circuits may actually indicate ground-fault conditions in addition to trouble conditions. However, some do not, which makes locating this type of fault more time consuming.

Consider a Class B that is grounded at Point C, between Devices #10 and #11. A meter is used at the FACU to measure the resistance between each circuit conductor and ground. **See Figure 12-17.**

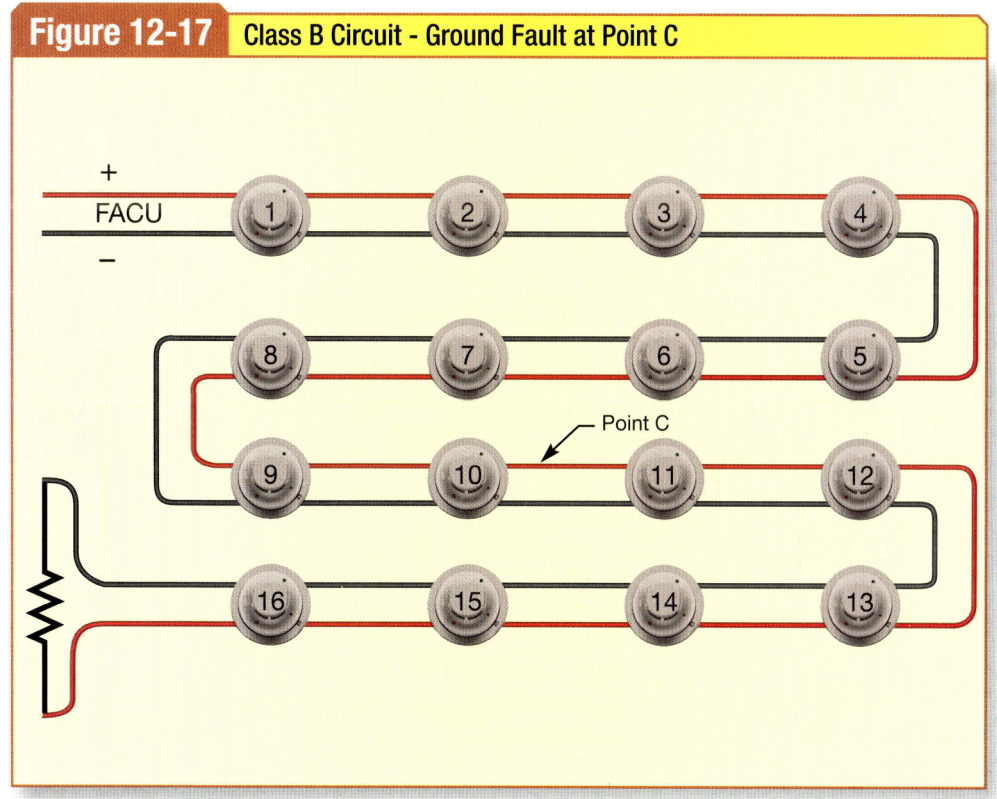

Figure 12-17. *The monitoring current is shunted through the ground fault and the controls sense this as a trouble.*

Figure 12-18. Troubleshooting Value Table for a Grounded Class B IDC

Zone #1	Circuit Type: IDC	EOL Value: 10,000 Ohms	Date:
Meter Reading Point		Desired Results	Actual Results
Positive to Negative		EOL Value	10,000 ohms
Positive to Ground		Infinity	10,000 ohms
Negative to Ground		Infinity	zero

Figure 12-18. Values were measured with the circuit disconnected from the controls.

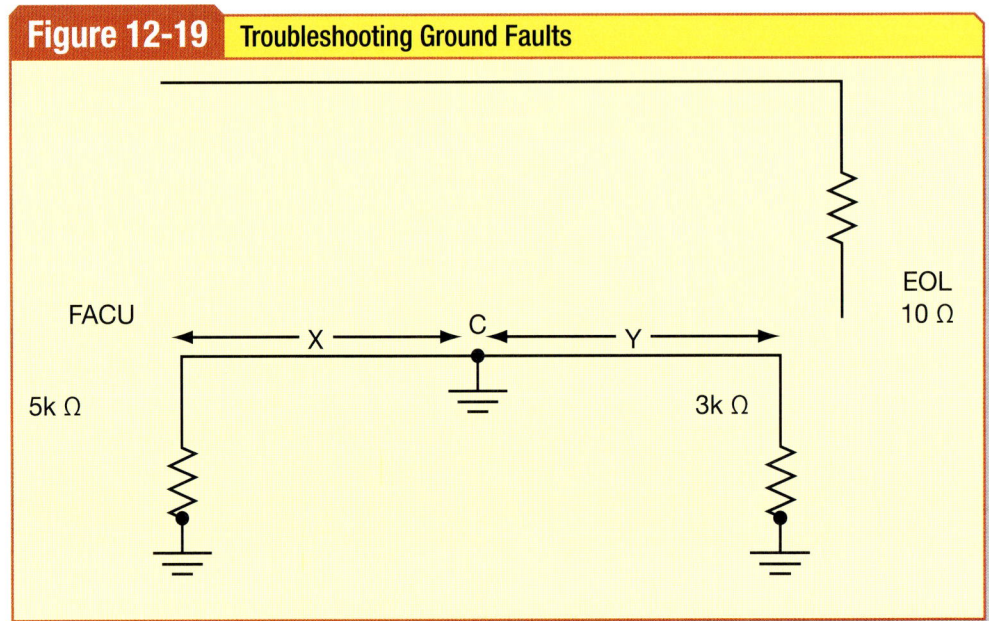

Figure 12-19. In this circuit, the ground fault is located at Point C and measurements on each side of the open yield different values, which helps locate the ground fault.

The resistance values are recorded. **See Figure 12-18.** The readings indicate the ground fault is on the positive leg of the circuit.

Isolate the positive leg from the EOL and connect a 3,000-ohm resistance from the field end of the positive leg to ground and a 5,000-ohm circuit on the FACU end. **See Figure 12-19.**

The positive leg of the circuit can be divided at Device #8, as in other troubleshooting techniques. Meter readings will be taken between the positive leg and ground. The meter reading on the FACU end will indicate 5,000 ohms, while the reading looking toward the EOL will be close to zero ohms. The circuit can be reconnected and divided again between Device #12 and #13. The "divide and conquer" process is repeated until the ground fault is located and corrected.

Class A Initiating Device Circuits

Class A circuits have a pair of return conductors from the last device to the FACU. Therefore, there are no end-of-line devices on Class A circuits. Both pairs of conductors can receive signals from field devices. Return conductors make the circuit inherently more reliable than Class B circuits. Class A circuits are sometimes easier to troubleshoot because both ends of the circuit are located at the FACU.

In a typical Class A arrangement, Terminals 1 and 2 are the return pair and

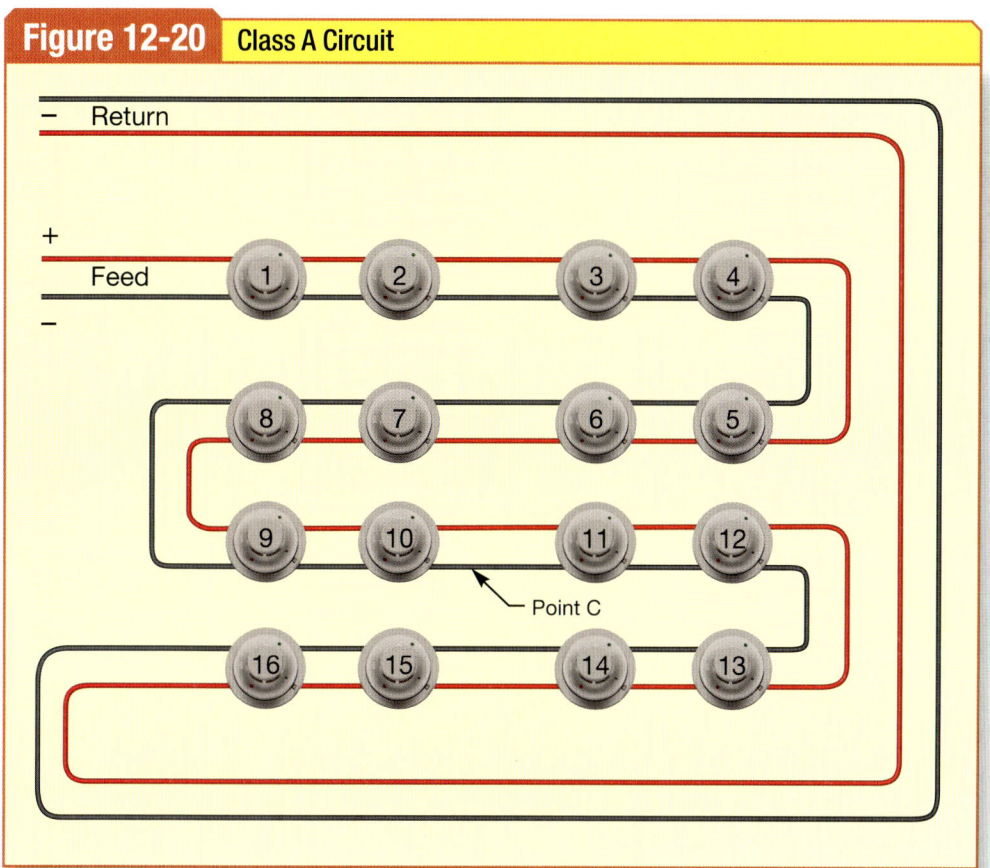

Figure 12-20. *Class A circuits have return conductors from the last device to the controls. End-of-line devices are generally not used on Class A circuits.*

Figure 12-21	Troubleshooting Value Table for an Open Class A IDC		
Zone #1	Circuit Type: IDC	EOL Value: 10,000 Ohms	Date:
Meter Reading Point	Desired Results	Actual Results	
Positive to Negative	10,000 ohms	10,000 ohms	
Positive to Ground	Infinity	Infinity	
Negative to Ground	Infinity	Infinity	

Figure 12-21. *Values were measured with the circuit disconnected from the controls.*

Terminals 3 and 4 are the feed pair. **See Figure 12-20.**

Start by installing a 10,000-ohm resistor on the feed pair and taking readings. **See Figure 12-21.**

Troubleshooting a Class A circuit is much the same as for a Class B circuit. Divide the circuit and take readings in each direction.

Notification Appliance Circuits

Troubleshooting notification appliance circuits is very similar to troubleshooting initiating device circuits. Although notification appliance circuits and initiating device circuits have similarities, the appliances do not have open contacts. The monitoring current for notification appliance circuits does not pass through the appliances because of a blocking diode on each appliance. The normal circuit polarity is such that the diode is reverse biased, which acts as an open circuit. Notification appliance circuits simply reverse the polarity, which causes the blocking diodes to short (forward bias), and allow current to pass through the appliances. Notification appliance circuits have blocking diodes, which require special care when using a volt/ohmmeter. **See Figure 12-22.**

There are three different conditions that occur on notification appliance circuits:
1. Normal – No troubles (opens, shorts, or ground faults)
2. Alarm – Reverse polarity condition
3. Trouble – Open circuit or ground-fault conditions

The first reading is taken with the meter polarity such that the meter leads will not forward bias the internal diodes. **See Figure 12-23.**

The meter will not forward bias the diodes, and the reading should be roughly the value of the EOL device. The second reading is taken with the meter leads swapped. In this case, the diodes may become forward biased by the meter voltage and may indicate a short condition. The meter diode check feature should be used, rather than the resistance scale. An indication of 0.3 to 0.6 volts is expected, and the diode check feature can help differentiate between a short and a forward-biased diode. If the diode check is positive on both polarities, one or more appliances are connected backwards. It should be noted that speakers use a blocking capacitor instead of a diode. Troubleshooting speaker circuits is the same as for other notification appliance circuits, except that the diode check feature is not used.

If appliances are connected backwards, they can be located by energizing the

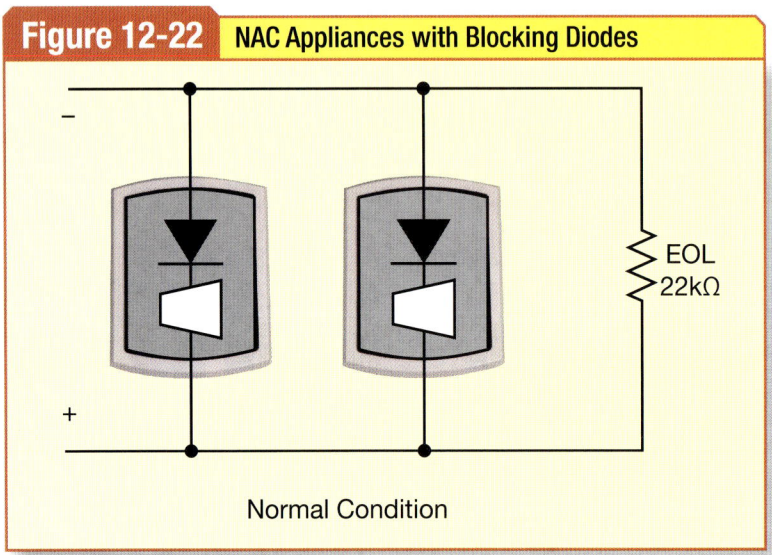

Figure 12-22. The blocking diodes are internal to the appliances and prevent monitoring current from energizing the appliances.

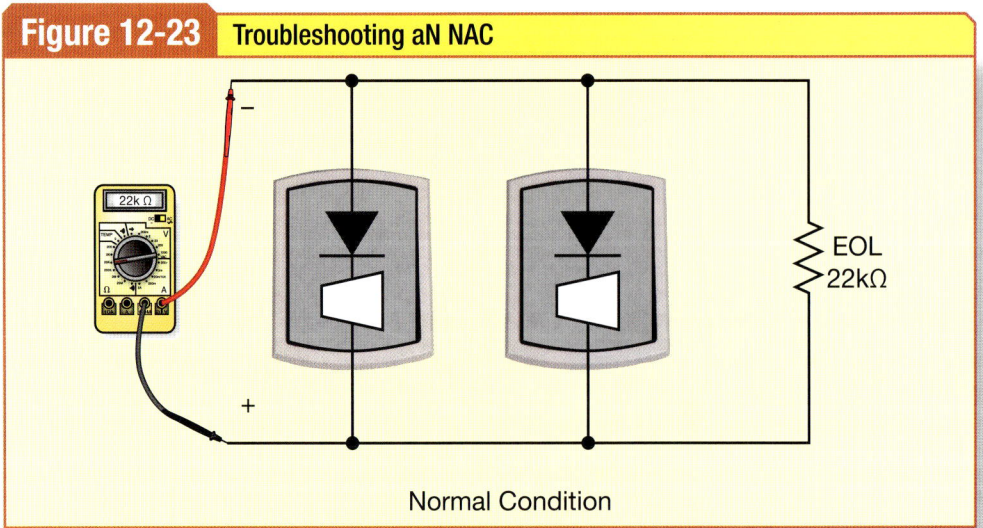

Figure 12-23. The polarity will not bias the internal diodes and affect the reading.

circuit and finding the appliance that is not operating. However, energizing a shorted notification appliance circuit will cause damage to the power supply. Once all appliances are properly connected, troubleshooting may proceed much the same as for initiating device circuits. Again, the divide and conquer approach works well for most types of faults.

Signaling Line (Addressable) Circuits

Signaling line circuits may be configured in either a Class A, Class B, or Class X arrangement. Addressable devices communicate by digital means, and do not create a short on the circuit when a device is placed in the alarm condition.

The meter reads through the electronics of the devices, and actual readings may vary, depending on loading of the circuit. The following conditions must be satisfied to ensure a trouble-free loop:
1. The loop is terminated to the FACU.
2. All device addresses are programmed.
3. The circuit is free of opens, shorts, and grounds.

Open-circuited signaling line circuits are easy to troubleshoot. Signaling line circuits operate between 12 and 36 volts direct current (VDC), depending on the manufacturer. This voltage provides power, but the data carrier rides on top of the DC power. Troubleshooting opens on addressable circuits can be accomplished without the use of a meter. Those devices that have no operating power will be reported as missing or not reporting by the FACU. A typical message on the controls is:

ADDRESS 002: DEVICE MISSING/NO REPORT

Devices downstream of the open will not report to the FACU. A review of the drawings will provide a good clue to where the open circuit may be located. However, devices that are not programmed or programmed with a duplicate address will also not report to the FACU.

Troubleshooting shorts on a signaling line circuit is more difficult than troubleshooting opens. Most addressable controls report short circuits but cannot indicate the location. Some addressable controls simply crash or cease to operate under short-circuit conditions. Therefore, the same procedure is used as for locating shorts on an initiating device circuit. A shorted or grounded SLC will cause most addressable systems to crash, which means that most devices will not respond. Most panels will indicate a communications failure and short circuit, followed by the loop number. Since Class B circuits can be T-tapped, troubleshooting may be more difficult. Class B signaling line circuits should not be T-tapped for this reason.

Consider a Class B signaling line circuit, with T-tapped circuit branches and a short circuit at Point D. **See Figure 12-24.**

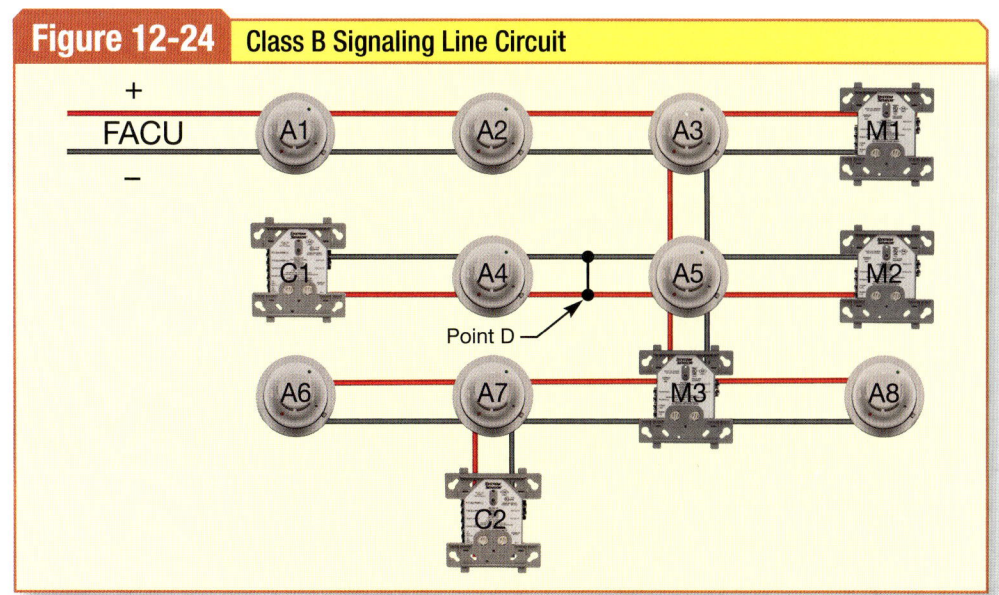

Figure 12-24. This circuit is T-tapped, which is permitted for Class B SLC, but may make troubleshooting a bit more difficult.

A good place to start is to isolate branches of the circuit, starting at each T-tap. As branches are isolated, reset the FACU to see if communications are restored. Disconnected devices will cause a trouble signal because they will not report; however, the absence of a crash will be an indicator of the location of the fault. Reconnect the conductors and continue.

Troubleshooting methods for grounds and shorts on an SLC is the same as for IDCs and notification appliance circuits. Class A signaling line circuits cannot be T-tapped, so troubleshooting faults on Class A circuits is somewhat easier. However, the procedures are the same as for a Class B circuit. Troubleshooting may be a little easier because:
- Both the feed and return are at the same location.
- The circuit can be divided.
- There are no T-taps.

Some manufacturers will require twisted, shielded pairs for SLCs to prevent unwanted electromagnetic interference with communications. The shield must be connected for the entire length of the circuit and must be grounded only at the panel. Grounding of the shield at points in the field will cause the shield to act as an antenna, resulting in erratic operation. Shields are often pinched in back boxes and must be insulated at all connection points.

Summary

Inspection, testing, and maintenance of fire alarm systems are essential steps to the proper operation and longevity of the system. They are also required by law in most jurisdictions. Chapter 14 of *NFPA 72, National Fire Alarm and Signaling Code,* provides the requirements and methods for inspection, testing, and maintenance of fire alarm and signaling systems. Any changes to the system must be tested in order to ensure the system operates correctly at the conclusion of the installation. Periodic inspection and testing of a fire alarm system is essential to ensure the system is operating over its entire life cycle.

All fire alarm system equipment is required to be inspected not less than once per year. Some equipment may be inspected as often as once per week. These frequencies of tests are based upon the maximum period of an outage. Following the requirements found in Chapter 14 of *NFPA 72* will help ensure systems correctly operate for many years.

Review Questions

1. A technician who is factory trained on Brand X is qualified to service and maintain Brand Y.
   a. True
   b. False

2. Who is responsible to ensure periodic inspections, tests, and maintenance of fire alarm systems are conducted?
   a. The AHJ
   b. The contractor
   c. The owner
   d. All of the above

3. Which of the following could be considered examples of ways that individuals could become qualified and experienced?
   a. Factory trained and certified
   b. IMSA
   c. NICET
   d. All of the above

4. What percentage of a fire alarm system is tested at the time of acceptance?
   a. 100
   b. 75
   c. 50
   d. 25

5. Inspections require a physical manipulation of the equipment.
   a. True
   b. False

Review Questions

6. **Before tests can begin, which of the following steps must be taken?**
   a. Call the fire department dispatcher
   b. Call the supervising station
   c. Notify the occupants
   d. All of the above

7. **Changes to software functions must be 100% tested and an additional __?__ of initiating devices must be tested, up to a maximum of __?__ devices.**
   a. 10% / 50
   b. 10% / 100
   c. 20% / 50
   d. 20% / 100

8. **Non-restorable heat detectors must be tested or replaced after __?__ years of service.**
   a. 10
   b. 15
   c. 20
   d. 30

9. **Testing the interface between a carbon dioxide suppression system requires the suppression to be actuated.**
   a. True
   b. False

10. **Failure to de-burr conduit may result in __?__ .**
    a. ground faults
    b. open circuits
    c. shorts
    d. all of the above

11. **There are __?__ types of fire alarm system tests.**
    a. 2
    b. 3
    c. 4
    d. 5

12. **When a fire alarm system becomes disabled because of an impairment that cannot be repaired immediately, *NFPA 72* requires that the owner be notified in writing within __?__ hours.**
    a. 4
    b. 8
    c. 16
    d. 24

13. **Which of the following is not typically documented on an inspection checklist (audible notification appliances)?**
    a. Free of damage
    b. Installed at proper height
    c. Properly supported
    d. SPL measurement

Appendix

Table of Contents

Appendix A..289
  Certification
Appendix B..299
  One Prosperity Place Specifications
Appendix C.....................................Enclosed
  One Prosperity Place Drawings

Appendix A

Certification

INTRODUCTION

Certification in professional or technical competency is one of the most rapidly developing fields in the United States. Many jurisdictions now require contractors and installers to provide proof of a minimum level of competency. In many jurisdictions, proof of competency takes the form of a local license. This type of certification is not voluntary, and all persons practicing the trade are required to be licensed.

Two examples of nationally recognized certification are those provided by the National Institute for Certification in Engineering Technologies (NICET) and the International Municipal Signaling Association (IMSA).

IMSA and NICET certification exams can be accessed at qr.njatcdb.org
Item #1083

PURPOSE OF CERTIFICATION

The primary purpose of becoming certified in a program such as the IMSA or NICET certification program is to protect one's liability and that of one's employer. Personal and professional liability in all areas of public safety has changed over the past decades. Federal regulations are setting more restrictive standards in all areas of public safety, and the courts are finding employees accountable to meet these new standards. Fire alarm systems are life safety systems, and a minimal amount of qualification is necessary to avoid litigation.

The technology used on fire alarm systems has changed dramatically over the past few decades, and the rate of change continues to accelerate. Fire alarm technology mirrors technological advances in communications and microprocessors and is growing more complex by the day. Finding qualified and experienced employees to install and maintain these systems is becoming increasingly difficult.

The IMSA and NICET programs are designed to keep individuals current with state-of-the-art technology. They are also designed to increase professionalism in the industry through distribution of knowledge relating to inspections and *Code* enforcement.

Practical applications are stressed in these programs, which also include formal review processes. Certification processes usually require the applicant to provide proof of work experience in his or her field in addition to any testing requirements that may be imposed. Once a person is certified or licensed, there may also be additional burdens of continuing education to maintain the certification. Continuing education may include formal classroom training or distance learning.

CODE REQUIREMENTS AND QUALIFICATIONS

Section 10.5.2.1 of *NFPA 72, National Fire Alarm and Signaling Code*, requires fire alarm systems and emergency communications systems installation personnel to be qualified and experienced in the installation, inspection, and testing of the systems. Section 10.5.3 also requires service personnel to be qualified and experienced in the inspection, testing, and maintenance of the systems under the scope of *NFPA 72*.

Section A.10.5.3.3(2) provides examples of what qualified and experienced is intended to mean. Examples found in Section A.10.5.3.3(2) include:
- IMSA certification
- NICET certification
- Certified factory training in fire alarm system design for the brand encountered in the field
- Certification or licensing by state or local authority

None of these sections in the *Code* require IMSA, NICET, or any other specific certification or license. However, they all require the designer, installer, and maintainer to be qualified and experienced. There are many ways to meet the intent of the *Code*, but two important ways are IMSA and/or NICET certification.

IMSA INTERIOR FIRE ALARM CERTIFICATION

The International Municipal Signal Association is a privately operated, not-for-profit, non-governmental organization. IMSA was founded in 1896 and was originally called the International Association of Fire and Police Telegraph Superintendents. The primary purpose of the organization was to keep its membership informed of the proper procedures for construction and maintenance of municipal signaling systems and new technology. In 1900, the name of the organization changed to the International Association of Municipal Electricians, which remained its name until 1935. At the 1935 conference in Richmond, Virginia, the organization changed its name to the National Municipal Signal Association but then changed it to the International Municipal Signal Association in 1937. The IMSA headquarters is currently located in Newark, NY.

Today, IMSA membership is divided into 22 geographic regions covering North America and other regions. There is also a "sustaining section" comprising individuals, firms, and organizations that share a common goal with IMSA. Each IMSA section has officers and a director that serve on the IMSA International Board of Directors. All officers are elected by the membership through section level elections. IMSA currently has over 10,000 members and has issued over 80,000 certifications.

IMSA serves on technical committees of several organizations, including the Automatic Fire Alarm Association, National Fire Protection Association, Congressional Fire Service Fire Institute, and the National Electrical Manufacturers Association. These committee memberships help keep IMSA informed on all aspects of fire alarm advancements. This information is passed to IMSA members through a bi-monthly magazine called the *IMSA Journal*.

The primary purposes of IMSA fire alarm certification programs are to increase public safety and reduce unnecessary legal action. However, IMSA also publishes specifications for wire and cable. Today, IMSA is concerned with interior and exterior fire alarm signaling and offers certification programs related to both types of systems. There are currently nine IMSA certification programs that include the following:
1. Fire Alarm Interior Systems
2. Fire Alarm Municipal Systems
3. Fiber Optics for ITS
4. Flagging and Traffic Control
5. Public Safety Telecommunicator
6. Roadway Lighting
7. Signs and Pavement Marking
8. Traffic Signal Inspection
9. Traffic Signal Systems

Work Zone Safety Exterior fire alarm systems include public emergency alarm reporting systems, as covered by Chapter 27 of *NFPA 72*. All IMSA programs are developed by industry experts to provide technology fields with a qualified workforce. Experts are selected from all aspects of the industry. IMSA certification programs are an increasingly popular way for employers to measure job skills and knowledge. IMSA membership is not a requirement for certification. IMSA certification is usually not a substitute for any required local licenses. One must be sure to understand any local requirements before working in any jurisdiction.

IMSA Certification in Interior Fire Alarm Systems – Level I

There are two levels of IMSA certification for interior fire alarm technicians. IMSA Level I is intended for entry-level technicians with limited work experience of less than two years. Level I certification is a prerequisite to Level II certification. Level II is designed for more seasoned applicants with fire alarm experience over two years. Level II is a continuation of the Level I program and involves more advanced topics, such as detection device design. IMSA certification programs are based on established fire protection curricula from accredited institutions, as well as established regulations, safety procedures, and codes/standards.

The Level I certification program covers the following areas:
- **Fire Detection Systems Design I.** Fire alarm terminology is introduced. This program covers wireless detectors, UL fire tests, detector spacing, and sensitivity.
- **Understanding Blueprints and Specifications.** An overview of working drawings and blueprints is presented. This program covers scaling and dimensional practices, reading blueprints for trade information, symbols, floor plans, riser diagrams, and point-to-point wiring diagrams. Specifications and a plans check review are also discussed.
- **Fire Alarm Wiring Methods.** Wiring methodology is explained. This program covers Ohm's Law, resistor identification, series circuits, parallel circuits, fire alarm wiring, plenum cable, and twisted pair cable.
- **Testing Techniques.** Codes require proper testing. This area covers smoke detector testing methods, sound metering, duct detectors, fire alarm system testing, inspections, notification, inspection procedures, records, retention of records, permits and checklists, and proper maintenance of fire alarm systems.

IMSA Certification in Interior Fire Alarm Systems – Level II

Level II is designed for more seasoned applicants with fire alarm experience over two years. Level II is a continuation of the Level I program, and involves more advanced topics. It is necessary to complete IMSA Level I certification before testing for Level II Certification.

The Level II certification program covers the following areas:
- **Fire Detection System Design II.** A fire alarm and detector overview are presented. Detector types, listings, and proper placement are reviewed. A system overview encompassing microprocessors with emphasis on multiplexes, addressable systems, and analog detection systems is discussed in detail. System wiring classifications and styles and 2-wire and 4-wire detectors are explored.
- **Understanding Computers and Computer Aided Design.** Basic computer technology is introduced for those not totally comfortable with computer hardware and software. Basic computer components are reviewed and explained, and an introduction to computerized fire alarm systems is covered.
- **Sprinkler/Standpipe Design and Connections.** Sprinkler concepts including wet pipe, dry pipe, deluge, and pre-action systems are reviewed. Straight pipe riser and

alarm check valve systems are examined. Specific emphasis is placed upon alarm, supervisory and trouble switches, monitoring methods, and electrical wiring.
- **The Americans with Disabilities Act and Understanding Notification Appliances.** The need, intent, coverage areas, and application of the ADA in places of public accommodations and commercial facilities are explained. Requirements for medical care, transient lodging, equivalent facilitation, and audible/visible applications are explored in detail.

IMSA Exam

The participant must complete an application for examination, which is retained at the local Section and International levels. Applications are processed through each Section Certification Committee, which also arranges for local testing sites and times, appoints examiners, and sends results to the participant. A minimum grade of 70% is required on each examination before an applicant is qualified for certification.

A copy of the IMSA membership application and exam application may be obtained from IMSA at www.imsasafety.org.

The NJATC Fire Alarm Systems curriculum and other NJATC curricula are intended to prepare students for the IMSA Certification examinations. The NJATC is now an IMSA Section, and exams can be scheduled through NJATC Headquarters.

As with most certification programs, continuing education is required to maintain the certification level obtained. Certified technicians must obtain a minimum number of continuing professional development (CPD) points every three years. This type of continuing education is easily obtained by attending classroom or online training, such as that provided by the NJATC, NFPA, Automatic Fire Alarm Association, or International Association of Electrical Inspectors. All CPDs must be relevant to fire alarm systems in order to qualify for IMSA approval.

The IMSA-certified interior fire alarm technician certification exam involves taking a multiple-choice exam. The exams must be proctored by an IMSA-certified installer. These exams may be arranged at one's local JATC or at another location determined by IMSA.

All exams are open book in nature. In fact, candidates are encouraged to bring and use reference materials. The minimum number of references will depend on the specific level. The complete list of references is as follows:
- Ugly's guides
- *NFPA 13, Standard for the Installation of Sprinkler Systems*
- *NFPA 70, National Electrical Code*
- *NFPA 72, National Fire Alarm and Signaling Code*
- *NFPA 75, Standard for the Fire Protection of Information Technology Equipment*
- *NFPA 90A, Standard for the Installation of Air-Conditioning and Ventilating Systems*
- *NFPA 101, Life Safety Code*
- *NFPA 110, Standard for Emergency and Standby Power Systems*
- *NFPA 170, Standard for Fire Safety and Emergency Symbols*

Level II certification primarily requires the use of *NFPA 70*, *NFPA 72*, *NFPA 90A*, and *NFPA 170*. Some of the important requirements of *NFPA 90A* and *NFPA 170* can also be found in the *NFPA Pocket Guide to Fire Alarm and Signaling System Installation (third edition)*. Other references such as those in the following list are permitted and encouraged, although not required.
- *Fire Alarm Signaling Systems*
- *NFPA Pocket Guide to Sprinkler System Installation*
- *The SFPE Handbook of Fire Protection Engineering*
- *NFPA Fire Protection Handbook*
- Typical cut sheets or catalogs for equipment used in the field

NICET FIRE ALARM SYSTEMS CERTIFICATION

The National Institute for Certification in Engineering Technologies is a privately operated, not-for-profit, non-governmental organization. NICET is a division of the National Society of Professional Engineers (NSPE), founded in 1961, and is located in Alexandria, Virginia. All NICET certification programs are voluntary.

There are approximately 30 NICET programs, covering such things as construction materials testing, underground utilities construction, industrial instrumentation, and fire sprinkler system certifications.

There are several NICET fire protection certification programs:
1. Inspection and Testing of Water-Based Systems
2. Special Hazards Suppression Systems
3. Water-Based Layout
4. Fire Alarm Technician

The application process is similar for all of these programs.

The NICET certified fire alarm technician certification program requires the applicant to meet minimum criteria, such as work experience, recommendations, and successful completion of a computer-based exam.

The NICET certified fire alarm technician program requires the applicant to meet minimum work experience criteria for each level of certification, to have a required number of recommendations, and to successfully complete a computer-based exam. Certified engineering technicians (CETs) are members of the engineering team who typically work in the field under the guidance of engineers or certified technologists. They must be familiar with fire alarm equipment, methods, and codes. Certified engineering technicians typically engage in the following areas:
- System layout (plan preparation)
- System equipment selection
- System installation
- System installation testing
- System troubleshooting
- System service
- System sales

In the certified engineering technician program, there are four levels, each increasingly difficult to obtain.
1. Level I is intended for entry-level technicians with a limited work experience of about three months.
2. Level II is designed for more seasoned applicants with two to five years of fire alarm experience.
3. Level III is for technicians with between five and 10 years of experience.
4. Level IV is intended for technicians with more than 10 years of experience.

Since February 2011, all NICET fire alarm certification testing is computer based. Each level is strictly pass/fail.

Application Requirements

Applicants must apply for examination dates and testing center locations online at www.nicet.org. All tests are offered through proctored regional testing centers around the U.S. and Canada.

Applying for NICET certification is a two-part process. Each applicant must complete a test application package and an experience application package. The applicant will be able to schedule an exam upon receipt of the application and payment. Upon successful completion of the exam, NICET staff review the applicant's work experience application to ensure he or she meets the requirements for that particular level. Each level requires the applicant to complete both parts of the application because each level has its own experience and testing requirements.

Work history is a very important part of the NICET certification process and cannot be overemphasized. Passing a multiple choice exam does not guarantee that the candidate has any relevant field experience or knowledge, nor will it guarantee certification. Therefore, applicants must provide a written work history as part of the application process. The work history must include duties and responsibilities of the applicant for each relevant experience.

Work history is measured by the months or years a candidate has been working specifically with fire alarms. For example, an electrician who spends 10% of his or her time working with fire alarm systems will take 10 years to attain a single year of relevant experience. Furthermore, the experience must be specifically in fire alarms, not a related field such as fire suppression systems. A candidate must pass the exam and achieve the required field experience before NICET will grant certification. This emphasis on experience is the reason engineers are frequently unable to achieve certification because they lack the field experience. Finally, all experience must be recent.

It is also necessary to provide written verification from a supervisor. Applicants without supervisors (for example, a small business owner) are permitted to use a non-supervisory verification reference. References must be very familiar with the applicant's work, and non-supervisory references must themselves have technical expertise in the area of fire alarms.

Level IV certification involves 10 or more years of experience. Applicants seeking Level IV certification must provide a separate work history experience sheet for a major project. A major project usually includes a large building and would typically require extensive involvement on the candidate's part. Large project involvement should include design, layout, and project management activities for NICET Level IV. The narrative for a major project must include the scope of work involved, type of occupancy, time involvement, and supervisory duties of the applicant.

Certification Renewal

Once certified, the technician or technologist will receive a wallet card and certificate that proves he or she has successfully achieved a particular level. Certificates are valid for three years. If a certified technician fails to renew (or chooses not to renew) his or her certification, the certificate and all records will be deleted by NICET. Once certification is forfeited, individuals must reapply and retest in order to recover a certification.

As with most certification programs, continuing education is required to maintain the certification level obtained. Certified technicians must obtain a minimum number of Continuing Professional Development (CPD) points every three years. This type of continuing education is easily obtained by attending classroom or online training, such as that provided by the National Joint Apprenticeship and Training Committee, National Fire Protection Association, Automatic Fire Alarm Association, or International Association of Electrical Inspectors. All CPDs must be relevant to fire alarm systems in order to qualify for NICET approval.

PE versus Certification

It should be noted that NICET certification does not entitle the certificate holder to practice engineering. Most jurisdictions in the United States regulate the practice of engineering and require a Professional Engineer (PE) license to practice. Failure to abide by these laws is a violation of state or local law. Heavy fines can be, and often are, imposed by local boards of registration for persons who practice engineering without a PE license. However, some jurisdictions have exemptions that permit certain qualified and experienced individuals to design systems of a limited size or occupancy type without a PE license.

NICET Exam

The exam tests one's knowledge of typical duties and job tasks. The NICET web pages and program manual for each program offer descriptions of the areas that will be covered on the exam. The questions, all multiple choice, are designed to be answered with little difficulty by those who regularly perform the tasks within those covered areas.

NICET test questions are submitted and reviewed by volunteer committees of technical experts. Pilot tests are used to check performance before opening an exam to the public.

All exams are open book in nature. However, NICET restricts the use of some materials. It is always wise to inquire about these restrictions well in advance of the test. Candidates are encouraged to bring and use approved reference materials. References allowed include:
- *NFPA 13, Standard for the Installation of Sprinkler Systems*
- *NFPA 70, National Electrical Code*
- *NFPA 72, National Fire Alarm and Signaling Code*
- *NFPA 75, Standard for the Protection of Information Technology Equipment*
- *NFPA 80, Standard for Fire Doors and Other Opening Protectives*
- *NFPA 90A, Standard for the Installation of Air-Conditioning and Ventilating Systems*
- *NFPA 92A, Standard for Smoke Control Systems*
- *NFPA 101, Life Safety Code*
- *NFPA 110, Standard for Emergency and Standby Power Systems*
- *NFPA 170, Standard for Fire Safety and Emergency Symbols*
- *NFPA 230, Standard for the Fire Protection of Storage*
- *NFPA 2001, Standard on Clean Agent Fire Extinguishing Systems*

Level II certification primarily requires the use of *NFPA 70, NFPA 72, NFPA 90A*, and *NFPA 170*. Other references such as those in the following list are also permitted and encouraged:
- Ugly's guides
- *Fire Alarm Signaling Systems*
- *NFPA Pocket Guide to Sprinkler Systems*
- *The SFPE Handbook of Fire Protection Engineering*
- *NFPA Fire Protection Handbook*
- Typical cut sheets or catalogs for equipment used in the field

Items not allowed are:
- Preparatory and/or training manuals specifically addressing NICET certification
- Practice or sample tests of any kind
- Handbooks and/or training manuals with practice/sample tests
- Questions and answers of any kind
- Loose-leaf printed text, charts/tables, formulas/equations, etc., with no specific publication reference, author(s), and/or publisher
- Loose-leaf blank paper
- Handwritten notes of any kind

Each test session begins with a tutorial allowing one to get used to the process. During the test, applicants will see one question at a time but can move forward or backwards to view or review other questions. There may be a graphic or text box to be viewed by clicking a button. Correct answers will usually be selected by clicking the boxes next to the answers. In a few questions, one may be presented a picture and asked to click on the part of the picture that correctly answers the question. Questions may have more than one correct answer, but in those cases, one will be told how many answers to choose. The applicant will receive his or her score at the conclusion of the test session.

AFTER THE EXAM
Exams and applications may take several weeks to several months to process. IMSA and/or NICET staff must process all applications, which may number in the hundreds. Examination results are generally reported within two to three weeks. For the NICET exams, candidates can call 1-888-476-4238 to check on applications or exam results if delays in notification appear excessive.

SUMMARY
Certification provides the technician with a way of measuring his or her skills. Certification provides verification to employers that the employee is qualified to install and maintain fire alarm systems. Certification opens doors and provides opportunities for those achieving certification. It also helps prevent problems in the field.

Exam Strategies
Nobody can memorize the entire content of a single code, standard, or reference. Therefore, candidates are highly encouraged to use a tab system on each reference. Placing tabs on the pages will allow the user to quickly find information without thumbing through large volumes. This strategy will save time and stress during the examination. Candidates are also encouraged to be familiar with *NFPA 72*.

Preparation and review are critical to success. Candidates cannot expect to walk into an exam and pass it with a minimal level of preparation.

The IMSA and NICET exams are intensive tests that require sharpness and stamina. It is strongly recommended that all candidates obtain a good night's rest prior to taking the examination. If the test site is far away from home, it is recommended that candidates travel to the test site the day before the exam so they do not spend several hours driving before the test. Global positioning systems (GPS) are recommended when driving to an unfamiliar area.

Candidates should also try to locate the test site before the day of the exam. Parking lots or garages and places to eat can also be located during this trip. This practice will help eliminate wasted time and also help reduce stress levels on the day of the exam.

Things to bring to the exam include the following:
- Photo identification
- IMSA/NICET exam approval correspondence
- Address and local map of test location
- Wristwatch
- Eyeglasses (if needed)
- Telephone number of test site if available

- Ballpoint pens
- Pencils
- Architect's scale or ruler
- Highlighters
- Post-it notes
- Calculator
- Spelling dictionary
- Eraser
- All reference materials
- Cell phone
- Money for parking and lunch
- Bottled water
- Aspirin or other pain reliever

Test management skills will help applicants pass the exam. IMSA exams are multiple-choice and a minimum number of points are required for successful completion of the exam. Candidates are not penalized for wrong answers but must attain the required points to pass the exam.

The first step in taking the exam should be to quickly read through the exam before working on any questions. This strategy will not only allow one to know what questions lie ahead but will also put more difficult questions in the back of one's mind. This strategy helps sort easier questions from more difficult ones.

Easier questions should always be answered first when taking multiple-choice exams. This strategy will allow candidates to score quick points and will allow more time to focus on more difficult questions when there is time.

Some questions will require more time to answer. The proper strategy for these difficult questions is to eliminate obvious wrong answers and narrow choices. This increases the odds of selecting the correct answer. Test developers often like to provide answers that appear plausible upon first inspection. These incorrect answers are called distracters and account for many low scores. Distracters are developed by incorporating incorrect assumptions or other similar mistakes into the answer set.

Some questions will be more challenging than others, and some candidates will have difficulty even knowing where to begin. No penalty is imposed for wrong answers, but of course, points are not awarded for incorrect answers. Therefore, candidates are encouraged to guess when there is insufficient time to complete the exam.

In summary, the reader is highly encouraged to obtain certification in one of the aforementioned certification programs. Doing so will boost confidence and prove to others that one is serious about life safety.

Appendix B

One Prosperity Place Specifications

1.0 GENERAL

1.1 GENERAL DESCRIPTION

A. Provide all materials and labor for the installation of an automatic fire detection and alarm system, hereafter referred to as the Work (or System), in One Prosperity Place.

B. All work shall be performed in accordance with these specifications and approved drawings. No modifications to these specifications will be accepted without the express written approval of the Owner's Representative. It is the Contractor's responsibility to document Designer's approval of any such modifications prior to the execution of work.

1.1.1 INTENT OF SPECIFICATIONS

A. Work performed pursuant to these specifications shall be complete in every respect, resulting in a System installed entirely in accordance with the applicable codes, standards, manufacturer's recommendations, and Underwriters Laboratories, Inc. (UL) listings.

B. Upon completion of this Work, the Contractor shall provide the Owner with:

1. Complete information and Contractor's fire alarm record drawings describing and depicting the entire System as installed, including all information necessary for maintaining, troubleshooting, and/or expanding the system at a future date.

2. Complete documentation of system testing.

3. Certification that the entire system has been inspected and tested, is installed entirely in accordance with the applicable codes, standards, manufacturer's recommendations and UL listings, and is in proper working order. The Contractor shall follow the requirements of *NFPA 72, National Fire Alarm and Signaling Code,* and provide a completed Record of Completion form and a completed Inspection and Testing form.

4. A copy of the final software program both in written and disk format, including any and all necessary passwords.

1.1.2 WORK INCLUDED

A. Provide and install a new system consisting of:

1. New FireLite® analog/addressable fire alarm system control unit(s).

2. Manual fire alarm stations at each exit on each floor, at every elevator lobby landing, and such that the maximum travel distance to any manual fire alarm box is not greater than 200 feet, measured horizontally on the same floor.

3. Devices, equipment, and wiring necessary to monitor special fire suppression systems, including kitchen hood extinguishing systems and ventilation system suppression systems.

4. Wiring and modules necessary to monitor the actuation of the fire pump, sprinkler, and standpipe system alarm and supervisory devices.

5. Devices, wire, and conduit necessary to perform fan shutdown as required by *NFPA 90A, Standard for the Installation of Air-Conditioning and Ventilating Systems.*

6. Devices, relays, wiring, and conduit necessary to initiate elevator recall terminating at the elevator controller in the elevator machine room.

7. Devices, relays, wiring, and conduit necessary to provide shunt trip of power to elevators per *ANSI A17.1*.

8. Visible notification appliances throughout the lobby, meeting spaces, ballrooms, restrooms, corridors, kitchens, restaurants, common office areas, laundry, and other common work spaces. Visible notification appliances shall not be installed in individual offices but shall be installed in common office areas.

9. Remote annunciator located at the main entrance.

B. Test and adjust all new equipment and systems.

C. Prepare and submit contractor record drawings and other submittals required herein.

D. Guarantee all new equipment and systems during installation and for one year after final acceptance of the System by the Owner's Representative.

E. Coordinate all work with other Contractors working in the building (for example, adjusting waterflow alarm switch retards by the Sprinkler Contractor).

F. Provide electrical surge protection for all control equipment, including primary power supplies.

G. Provide lightning protection, including grounding rods for all control equipment. Lightning protection shall include the installation of UL listed surge protection for all power supplies and grounding of primary power to fire alarm control panels.

H. All devices installed within areas exposed to atmosphere or wet locations, such as equipment located outdoors or near indoor pools and spas, shall be provided with NEMA 4 or equivalent enclosures.

1.1.3 OWNER'S REPRESENTATIVE

A. All queries shall be directed to the Owner's Representative, hereafter referred to as the Owner:

> Mr. John Jones
> Chief Engineer
> Virginia Management Co., Inc.
> 22 Center St.
>
> Stafford, VA 22554
> Telephone (540) 555-1776
> Fax (540) 555-1812

B. The Owner will issue all approvals and instructions required for this work. No other person may issue an approval or instructions to the Contractor without the written authorization of the Owner.

1.1.4 WORKING CONDITIONS

A. It shall be the Contractor's responsibility to inspect the job site and become familiar with the conditions under which the work will be performed.

1.2 QUALITY ASSURANCE

1.2.1 CODES, STANDARDS, ORDINANCES, AND PERMITS

A. All work shall conform to the requirements of the applicable portions of the National Fire Protection Association (NFPA) codes, standards, and other guidelines or standards listed herein:

1. *NFPA 70 - 2014, National Electrical Code (NEC)*

2. *NFPA 72 - 2013, National Fire Alarm and Signaling Code*

3. *NFPA 90A - 2012, Standard for the Installation of Air-Conditioning and Ventilating Systems*

4. *ANSI A17.1, Safety Code for Elevators and Escalators, 2010*

5. *UL 864, Standard for Control Units and Accessories for Fire Alarm Systems, 2011*

6. *UL 268, Smoke Detectors for Fire Alarm Systems,* 2009

7. *UL 464, Audible Signal Appliances,* 1996

8. *UL 1971, Signaling Devices for the Hearing Impaired,* 2008

B. The Contractor shall be responsible for compliance with referenced standards, product installation sheets, and these specifications.

C. All work and materials shall conform to all federal, state, and local codes and regulations governing the installation, including the current editions of the *International Building Code,* as modified or interpreted by Stafford County Officials to permit use of current NFPA standards

D. If there is a conflict between the referenced NFPA standards, federal, state, or local codes and this specification, it shall be the Contractor's responsibility to immediately bring the conflict to the attention of the Owner for resolution. The Contractor shall not attempt to resolve conflicts directly with the local authorities unless specifically authorized by the Owner.

E. The Contractor shall be responsible for filing of all documents, paying all fees, securing all permits, inspections, and approvals necessary for conducting this work. Upon receipt of approved drawings, executed permits, or inspection reports from the authority having jurisdiction (AHJ), the Contractor shall immediately forward two sets to the Owner. These documents shall either be drawings stamped approved or a copy of a letter from the AHJ stating approval of this work.

F. All devices, systems, equipment, and materials furnished and installed shall be of types or models approved by the County of Stafford Officials for use in systems and occupancies of this type.

G. All devices, systems, equipment, and materials furnished and installed as part of the covered scope shall be new and listed by Underwriters Laboratories, Inc. (UL) for the intended use. All equipment shall be installed in accordance with the manufacturer's recommendations and the UL listing limitations. Compatibility listing requirements for separate voice, fire alarm systems, and smoke detectors shall be met. The Contractor shall provide evidence, with the submittals, of listings of all proposed equipment and combinations of equipment.

1.2.2 QUALIFICATIONS OF CONTRACTOR

A. The Contractor shall:

1. Hold all licenses and obtain all permits necessary to perform work in the Commonwealth of Virginia.

2. Provide a job site supervisor who is to be present full-time on-site each day that work is actively in progress. This individual shall be the same

person throughout the course of the project, unless written authorization is obtained from the Owner.

1.3 SUBMITTALS

1.3.1 CONTRACTOR RECORD DRAWINGS

A. The Contractor shall provide and maintain on the site an up-to-date record set of approved shop drawing prints, which shall be marked to show each and every change made to the approved shop drawings. This shall not be construed as authorization to deviate from or make changes to the shop drawings approved by the Owner without written instruction from the Owner in each case. This set of drawings shall be used only as a record set.

B. Upon completion of the work, the record set of prints shall be used to prepare complete, accurate final record drawings reflecting any and all changes and deviations made to the fire alarm system and other work covered by this specification.

C. Upon completion of the work, two sets of blueline record drawings shall be submitted to the Owner for review.

D. Upon review of the blueline record drawings, before final approval, one package of corrected, or amended as necessary, reproducible mylars and AutoCAD release 2000 CAD diskettes, and four additional sets of blueline record drawings shall be submitted to the Owner.

E. The Contractor record drawings are required to show and to identify quantities of junction boxes, spare conductors, splices, device back boxes, and terminal strips. This submittal shall include a schedule of all connections/terminations, indexed by junction box, device back box, and terminal strip and shall reference wire tag numbers as installed.

1.3.2 APPROVALS BY CODE AUTHORITY

A. The Contractor shall provide the Owner with one copy of all documents that are reviewed and approved by the local code authorities. These documents shall include, but not be limited to, the following:

1. Site inspection forms

2. Permits and permit drawings

3. Final inspection forms

B. All documents must include all required approval stamps, signatures, or other information necessary to properly certify that the installation has been reviewed and accepted by the County of Stafford Building and Fire Departments and all other required regulatory agencies or departments.

1.3.3 TEST RECORD

System certification and documentation of system testing required by Specification Section 3.4.1 shall be submitted to the Owner for review and approval at least 14 days prior to the final acceptance test. At a minimum, the Fire Alarm System Record of Completion form depicted in Figure 7.8.2(a) of *NFPA 72*, 2013 edition, shall be completed and submitted to the Owner for review.

1.4 GUARANTEE

The Contractor shall guarantee all materials and workmanship during the installation period and for a period of one year, beginning with the date of final acceptance by the Owner. The Contractor shall be responsible during the design, installation, testing, and guarantee periods for any damage caused by the Contractor or Subcontractors, or by defects in the Contractor's or Subcontractors' work, materials, or equipment.

1.5 SPARE PARTS AND SPECIAL TOOLS

1.5.1 SPARE PARTS

The Contractor shall supply as part of this contract, the following spare parts:

A. Manual fire alarm stations — 2%, but not less than five of the installed quantity of each type.

B. Audible and visual devices — 2%, but not less than five of the installed quantity of each type.

C. Light bulbs and LEDs — five of each for each type of lamp or field replaceable LED used in the system.

D. Fuses — five each for each type, rating, and size of fuse used in the system.

E. Keys — A minimum of three sets of keys shall be provided and appropriately identified.

F. Notification appliances — 2%, but not less than five of the installed quantity of each type.

1.5.2 SPECIAL TOOLS

The Contractor shall supply the Owner with three complete sets of any special tools or keys necessary for normal operation and maintenance of the system.

1.6 FINAL APPROVAL AND ACCEPTANCE

Final approval and acceptance of the work will be given by the Owner when:

A. The complete system has been inspected, tested, and approved in writing by the Owner and the AHJ.

B. All required submittals, including system operation and maintenance manuals, accurate contractor record drawings, test reports, spare parts, special tools and training have been provided to, reviewed by, and accepted in writing by the Owner.

2.0 DESCRIPTION OF THE SYSTEM

2.1 GENERAL

A. The system shall be of the distributed, addressable type. All system components shall be of the FireLite® product line.

B. All equipment and system components furnished and installed shall be new and listed by UL for their intended use. The equipment and system components shall be installed in accordance with the applicable codes and standards, the manufacturers' recommendations, and within the limitations of the UL listings. All equipment and system components shall be the standard product of a single manufacturer. Evidence of UL listing is required.

C. System components shall be modular in design to provide future expansion capability of the system. Expansion capability shall pertain to capacity and quantities of devices, circuits, primary and secondary power supplies, conductor ampacities (size), and lengths.

D. The system shall have spare installed capacity enabling it to support a 20% increase in initiating, control, and notification appliance circuits. Spare circuit capacity shall be evenly distributed throughout the system.

2.2 SYSTEM CONFIGURATION

2.2.1 INITIATING DEVICE CIRCUITS AND SIGNALING LINE CIRCUITS

All initiating device circuits and signaling line circuits shall be Class "B" as described in 12.3.2 of *NFPA 72*, and as shown on the shop drawings.

2.2.2 NOTIFICATION APPLIANCE CIRCUITS

All notification appliance circuits shall be Class B, 2-wire circuits as described in 12.3.2 of *NFPA 72*.

2.3 POWER SUPPLIES

A. Except where otherwise required by local code, all AC power connections shall meet the requirements of *NFPA 72*, Chapter 10, and *NFPA 70*, Article 760. Shop and record drawings shall indicate panel number and location as well as circuit breaker number where terminated for each such connection.

B. BATTERIES/SECONDARY POWER SOURCES

1. System control equipment shall receive secondary operating power from batteries integral to the equipment. Such batteries shall supply operating power for a period of not less than 24 hours.

2. Any portion of the system operation on secondary power shall annunciate as a trouble signal, identifying the inoperable power supply(ies). Low-capacity conditions of secondary power supplies shall immediately annunciate as a trouble signal and shall identify the power source. All standby batteries shall be continuously monitored by the system. Low-battery conditions shall immediately annunciate as a trouble signal, identifying the deficient batteries.

3. Secondary operating power provided shall be capable of operating all notification appliances simultaneously for a minimum of not less than five minutes after a period of 24 hours on secondary power. Batteries shall be sized to 120% of calculated capacity, per *NFPA 72*, Section 10.6.7.2.1(1). The date of installation shall be marked on the batteries.

4. Design load connected to any power supply, amplifier, or secondary power source shall not exceed 80% of its rated capacity.

5. Upon failure of normal (AC) power, the affected portion(s) of the system shall automatically switch over to secondary power without losing any alarm, trouble, or operator acknowledgment signals.

2.4 ANNUNCIATION

2.4.1 GENERAL

A. The system shall be designed and equipped to receive, monitor, annunciate, and retransmit signals from devices and circuits installed throughout the building.

B. The system shall be designed and equipped to provide inputs and outputs as described in the input/output matrix contained on the system shop drawings.

C. The system shall recognize and annunciate the following signals:

1. Fire alarms

2. Supervisory signals

3. Trouble signals

4. Operator acknowledgment of annunciated signals.

D. All alarm signals, supervisory signals, and trouble signals shall be annunciated by the control panel(s) and by each remote annunciation device. Operator acknowledgment of smoke detection signals and system reset shall be annunciated by the control panel(s) and by each remote user interface.

2.4.2 FIRE ALARM SIGNALS

A. Fire alarm signals shall be distinctive in sound from all other signals and shall consist of an alert tone followed by prerecorded voice instructions. The system shall also be capable of reproducing live voice instructions.

B. Actuation of the following devices shall be recognized and annunciated by the system as fire alarms:

1. Manual fire alarm stations
2. System-type smoke detectors, including in-duct smoke detectors
3. Heat detectors
4. Sprinkler system waterflow switches
5. Devices monitoring actuation of special fire suppression systems

2.4.3 SUPERVISORY SIGNALS

The following conditions shall be recognized and annunciated by the system as supervisory signals:

A. Valve supervisory switch actuation

B. Trouble and supervisory signals on suppression system control units

C. Fire pump supervisory signals

1. Pump running
2. Common trouble
3. Controller in off position or common trouble

2.4.4 TROUBLE SIGNALS

A. The system shall recognize and annunciate initiating device and signaling line circuit trouble conditions as required by Sections 10.10, 10.17, and 12.6 of *NFPA 72*. All other fire alarm circuits shall be supervised for opens or grounds.

B. The system shall also recognize and annunciate the following trouble conditions:

1. Power supply trouble conditions
2. Remote annunciation device trouble conditions

2.4.5 OPERATOR ACKNOWLEDGEMENT SIGNALS

Silencing of integral audible devices required by Specification Section 2.4.1.C shall be recognized and annunciated by the system as operator acknowledgment of the signal(s) displayed.

2.5 FIRE ALARM CONTROL PANEL(S)

2.5.1 DESCRIPTION OF EQUIPMENT

The fire alarm control panels shall be designed and equipped to provide the following:

A. A visual display or displays indicating current status of the entire system.

B. Standby power supplies capable of supporting all dependent devices and equipment as required by Specification Section 2.3.

C. Auxiliary relays to effect the following functions:

1. Fan or damper control
2. Elevator recall
3. Elevator shunt trip
4. Door unlocking
5. Smoke door releasing service

D. Devices or controls to effect reset of the system.

E. The control panel(s) shall not be capable of being reset until all alarm conditions have been cleared.

2.5.2 NOTIFICATION APPLIANCE CIRCUIT POWER EXTENDER PANEL

A. Description of Equipment. The notification appliance circuit power extender panels shall be designed and equipped to provide the following:

1. Standby power supplies capable of supporting all dependent devices and equipment as required by Specification Section 2.3.

2. The panel(s) shall supply a minimum of four Class B notification appliance circuits.

3. The panel's notification appliance circuits shall be actuated, controlled, and supervised via the main fire alarm panel.

4. The main fire alarm panel shall provide a trouble signal upon fault condition detected on the panel(s) power supply, notification appliance circuits, and control circuit.

5. Automatic smoke detection shall be provided at the location of each panel in areas that are not continuously occupied.

2.6 ANNUNCIATION DEVICES

2.6.1 DESCRIPTION OF EQUIPMENT

2.6.1.1 INDICATING LAMPS DISPLAYS

A. Indicating lamp displays shall indicate all fire and supervisory signals received until such time as the alarm condition is cleared and the display manually reset.

B. Indicating lamp displays shall indicate all trouble signals received until such time as the trouble condition is cleared.

C. Indicating lamp displays shall be clearly labeled using nomenclature approved by the Owner.

Operating (electrical) power shall be derived from an integral power supply meeting the requirements of Specification Section 2.3. Power for the indicating lamp display panel shall be supplied by the building's emergency circuit via a permanent (non-plug) connection.

2.6.1.2 REMOTE ANNUNCIATORS

A. Each remote annunciator panel shall indicate all addresses of actuated area smoke detector, duct type smoke detector, water flow, special suppression system, and manual fire alarm stations fire alarm signals by floor of alarm and location received until such time as the alarm condition is cleared and the display reset.

B. Each remote annunciator shall indicate all supervisory signals received until such time as the condition is cleared and the display reset.

C. Each remote annunciator shall indicate all smoke detector, water flow, and manual fire alarm stations trouble signals received until such time as the trouble condition is cleared and the display reset.

D. Each remote annunciator shall be point back-lighted and use socket-mounted LEDs and/or liquid crystal displays. LEDs shall be available in four colors: red, orange, yellow and green. Incandescent lights are not acceptable.

E. Each remote annunciator shall be UL listed.

F. All electronics shall be factory mounted to a metal backplate that will be mounted to the back box(es) at the time of final annunciator installation. All field connections to the backplate shall be made through the system-standard terminal strips. All front panel LEDs shall be connected to the backplate by quick-disconnected ribbon cables. All front panel switches shall be supervised.

G. Operating (electrical) power shall be derived from an integral power supply meeting the requirements of Specification Section 2.3. Power for the graphic annunciator shall be supplied by the building's emergency circuit via a permanent (non-plug) connection.

2.7 MANUAL FIRE ALARM STATIONS

2.7.1 DESCRIPTION OF EQUIPMENT

A. Manual fire alarm stations shall be of the addressable, single-action, non-coded type. They shall consist of a housing, fitted with a pull-down lever, which when operated, locks in position after releasing a spring-loaded contact switch to effect activation of an alarm circuit. Resetting the station after operation shall require the use of a special tool or key. The manual station shall be suitable for surface, flush or semi-flush mounting. Manual stations shall be flush or semi-flush mounted in all public areas. Manual stations shall be semi-flush mounted in all back-of-house areas unless mounted on concrete or masonry surfaces where surface mounting is permissible. Where mounted in back-of-house areas, manual fire stations shall be protected from physical damage by steel brackets or other approved means that shall not be attached to the station itself.

B. Manual stations installed outdoors, in wet locations, and in damp locations shall be UL listed for such use and shall be so marked. They shall be installed using NEMA Type 4X rain-tight enclosures and raceways.

2.8 SMOKE DETECTORS

2.8.1 DESCRIPTION OF EQUIPMENT

2.8.2 AREA SMOKE DETECTORS

A. All area smoke detectors shall be system-operated, photoelectric type plug-in detectors that mount to a twist lock base. To prevent unauthorized removal of smoke detector heads, the detector shall contain an alarm initiating LED, which will illuminate to signal activation of the detector. Detectors shall be listed by UL as "Smoke-Automatic Fire Detectors" and tested according to *UL Standard 268*. Detectors listed as "Single and Multiple Station Smoke Detectors" and tested according to *UL Standard 217* shall not be used.

B. Smoke detectors shall be of the addressable analog type with each device individually annunciated on the system.

2.8.3 IN-DUCT SMOKE DETECTORS

A. In-duct photoelectric type smoke detectors shall be installed in all new and existing air-handling systems over 2,000 cubic feet per minute (cfm), installed in conformance with the requirements of *NFPA 72*, *NFPA 90A*, local requirements, and in accordance with the detector manufacturer's installation instructions. In-duct smoke detectors shall be suitable for the full range of air velocity conditions in the air-handling systems in which they are installed.

B. In-duct smoke detectors shall be of the analog, addressable type, consisting of a plug-in type detector head in a duct-mounted housing equipped with air-sampling tubes providing air flow through the detector housing. In-duct smoke detectors shall be listed or approved for that application. In-duct smoke detector operating voltage shall be 24 volts direct current (VDC).

C. Each in-duct smoke detector shall be monitored individually via an integral addressable element.

D. In-duct smoke detectors shall be arranged to initiate shutdown of their associated fan and air-handling unit or initiate smoke-control functions on alarm from their associated addressable control relays.

E. The Contractor shall provide the necessary interlock wiring to fan and air-handling unit starters and motor control centers for shutdown of fans, as well as smoke control when smoke is detected.

2.9 HEAT DETECTORS

2.9.1 DESCRIPTION OF EQUIPMENT

A. Heat detectors shall be low profile, combination rate-of-rise and fixed temperature type.

B. Heat detectors shall actuate when the temperature either increases at a rate exceeding 15°F per minute or reaches 135°F except where otherwise required by *NFPA 72* (180°F where ambient temperature exceeds 100°F).

C. Heat detectors shall be of the addressable type with each device individually annunciated on the system.

D. Heat detectors for elevator power interrupt shall be provided and installed in elevator machine rooms in accordance with *ANSI A17.1*.

2.10 SPRINKLER ALARM SUPERVISORY EQUIPMENT

A. All sprinkler alarm and supervisory devices exist or shall be installed by others. These devices will be located as indicated on the drawings and include:

   1. Waterflow switches

   2. Valve supervisory switches

   3. Fire pump controller supervisory contacts

B. Sprinkler supervisory devices shall be connected to the fire alarm system and monitored under this contract.

C. Waterflow switches shall be monitored as fire alarms. All other sprinkler supervisory devices shall be monitored as supervisory signals.

D. Fire pump controller supervisory contacts shall be individually monitored, via dry contacts or an intermediary terminal interface provided by the fire alarm Contractor. See Specification Section 2.4.3.C.

E. The Contractor shall make all terminations necessary to monitor sprinkler supervisory devices.

F. The Contractor shall coordinate with the Sprinkler Contractor and the Owner for testing these devices. Documentation of sprinkler supervisory device testing shall be a part of this contract.

2.11 AUDIBLE/VISUAL NOTIFICATION APPLIANCES

2.11.1 DESCRIPTION OF EQUIPMENT

A. Signaling devices shall consist of fire alarm speakers and visual notification appliances.

B. Audible notification appliances shall be installed, spaced, and tapped so as to produce a sound output on alarm, which is clearly audible above the ambient noise level throughout the building. In no case shall the audible alarm be less than 15 decibels (dB) above the ambient room noise level as defined in *NFPA 72*.

C. Visual signaling appliances shall not be less than 15 candela (cd).

2.11.2 VISUAL NOTIFICATION APPLIANCES

A. Visual notification appliances shall consist of a vibration resistant solid-state flasher assembly that, upon actuation, illuminates a white or clear lens labeled "Fire" in red letters.

B. When mounted on the wall, the word "Fire" shall read from top to bottom. When mounted on the ceiling, the word "Fire" shall read left to right.

C. Visual notification appliances shall be installed in all public areas, public rest rooms, meeting rooms, dining rooms, ballrooms, and all common area corridors. Spacing and layout shall be as required by *NFPA 72*.

D. Visual notification appliances shall be listed in accordance with *UL 1971*.

E. Audible/visual appliances shall be placed in all public areas, assembly areas and in all areas having high ambient noise conditions.

2.12 CONDUCTORS AND RACEWAY

A. Except as otherwise required by the *Code* and/or these Specifications, the installation of fire alarm circuits shall conform to the requirements of *NEC*

Article 760 and raceway installation to the applicable sections of Chapter 3 of *NFPA 70, National Electrical Code*. Fire alarm circuit wiring shall include all circuits described in Section 760.1.

B. Non–power-limited fire alarm (NPLFA) circuits shall have overcurrent protection and be installed in conformance with Parts I and II of Article 760 and the applicable requirements of Chapters 1 through 4 of *NFPA 70*. The power sources for NPLFA circuits shall meet the requirements of Section 760.41.

C. Power-limited fire alarm (PLFA) circuits shall be installed in conformance with Parts I, III, and IV of Article 760. The power sources for PLFA circuits shall meet the requirements of Section 760.121 and the equipment supplying power shall be durably marked as required by Section 760.124.

D. Fire alarm circuits installed in locations other than ordinary indoor, dry locations shall be in conformance with 760.3(D) with conductor or cable types suitable for the installation conditions.

E. Separation of circuits shall be in conformance with Section 760.48 for NPLFA circuits and Section 760.136 for PLFA circuits.

F. NPLFA circuit conductor sizes, insulation types, and installation methods shall be as limited by Section 760.49.

G. PLFA circuit wiring material and installation methods shall be by either of two methods:

   1. Cable listed and marked in accordance with Section 760.154 or substitutions in accordance with 760.154(A), using NPLFA wiring methods in accordance with 760.53

   2. Conductors in raceway in accordance with Section 760.53

H. PLFA cable, where not installed in UL Listed metal conduit or raceway, shall be mechanically protected by building construction features:

   1. Installation shall be in areas not subjected to mechanical injury.

   2. Fire alarm circuits shall be supported by the building structure. Cable shall be attached by straps to the building structure at intervals not greater than 10 feet. Wiring installed above drop ceilings shall not be laid on ceiling tiles. Wire shall not be fastened in such a manner that puts tension on the cable.

   3. Cable shall be types FPLP, FPLR or FPL, or permitted substitutions, selected for the installation application as required by Section 760.135.

I. All cable runs shall be continuous between devices, without splices, wherever feasible. Where a continuous run is not feasible, connections shall be made using terminal blocks in a UL Listed metal electrical box. All other connections shall be to terminal blocks. Wire nuts shall not be permitted. Wires connected together shall have the same color insulation. All connections shall be accessible for inspection and servicing and shall be clearly identified on the contractor record drawing.

J. Wire and cable shall be sized, twisted, and shielded as recommended by the fire alarm system manufacturer and shall meet the requirements of the *National Electrical Code*, Article 760.

K. All conduit shall be grounded by approved ground clamps or other means in conformance with the *National Electrical Code*.

L. Where conduit is embedded in plaster or masonry, the Contractor shall use a type approved by the *National Electrical Code* for this use. All embedded conduit shall be liquid and gas tight. Continuous run of conduit without joints is preferred for embedding.

M. All electrical enclosures, raceways, and conduits shall contain only those electrical circuits associated with the fire detection and alarm system and shall not contain any circuits that are unrelated to the system.

N. All cable that is not enclosed by conduit shall be supported and anchored with nylon straps or clamps. The use of staples is prohibited. Other conduits shall not be used to support cables.

O. Cables and conductors having scrapes, nicks, gouges, or crushed insulation shall not be used.

P. The use of aluminum wire is prohibited.

Q. All electrical circuits shall be numerically identified at both ends with wire-taped numbers.

R. All system conductors, except grounding conductors, shall be solid copper.

S. All end-of-line resistors shall be mounted on terminal blocks or on device terminals.

T. Power-limited circuits that are reclassified as NPLFA circuits shall meet the requirements of Part II of Article 760, including overcurrent protection, and shall have the marking required by Section 760.121 eliminated. Reclassified circuits shall remain non–power-limited throughout their entirety.

U. All conduit, junction boxes and enclosures subjected to moisture shall be weatherproof (NEMA 4 or 4X) as required.

V. All junction boxes and terminal cabinets shall be red in color.

W. All conduit shall have red bands that are a nominal one-inch wide at spacings not to exceed 10 feet.

2.13 BOX LOCATION

All device back boxes, junction boxes, and pull boxes shall be accessible for inspection and maintenance. Junction pull boxes shall be installed on 100-foot centers maximum. Terminal cabinets installed outdoors or in areas subjected to moisture shall be weatherproof (NEMA 4 or 4X) and shall be installed no less than 18 inches above grade. In areas subject to lightning, terminal strips shall be isolated from the enclosure by non-metallic base plates to prevent arcing of contacts to enclosure. Boxes shall also be grounded using approved grounding rods.

3.0 EXECUTION

3.1 STARTING AND COMPLETION DATES

The schedule for the project will be established at the pre-construction meeting.

3.2 INSPECTION

The job site supervisor shall examine daily all areas in which the work will be performed on the day prior to beginning work. The supervisor shall immediately report unsatisfactory working conditions to the Owner for resolution. The Contractor shall not proceed with the work until all unsatisfactory working conditions have been corrected.

3.3 INSTALLATION

3.3.1 GENERAL

A. All holes made by the Contractor in any wall, ceiling, or floor shall be patched by the Contractor, restoring the wall, ceiling, or floor to its original condition, fire resistance, and integrity.

B. All piping and conduit shall be installed at a height so as not to obstruct any portion of a window, doorway, stairway, or passageway and shall not interfere with the operation of any mechanical or electrical equipment.

C. System riser(s) shall be installed in mechanical raceways or conduit, located to avoid physical harm.

D. Locations of all equipment, controls, and system components are subject to the approval of the Owner.

E. The Contractor is responsible for protecting both new and existing smoke detectors during construction. These detectors shall be covered during construction. Covers used for such protection shall be removed upon completion of work in that area and at the end of each work day.

3.3.2 CONCEALMENT

A. All wire, cable, conduit, raceways, junction boxes, and device back boxes shall be concealed in walls, ceiling spaces, electrical shafts, or closets in all finished areas. Conduit, raceways, junction boxes, and device back boxes may be exposed in unfinished areas, electrical, or mechanical equipment rooms.

B. Exposed conduit, raceways, junction boxes, and equipment back boxes shall be painted to be as inconspicuous as possible. The Owner shall approve the paint color selected. The Contractor shall prepare color samples for inspection by the Owner prior to painting. Exposed conduit, raceways, junction boxes, and other associated items related to the conduit network shall be provided with red bands every 10 feet with junction box covers labeled as fire alarm, unless specifically instructed otherwise.

3.4 INSPECTION AND TESTS

3.4.1 SYSTEM TESTS

The Contractor shall provide the Owner with written certification that all equipment:

A. Has been inspected and tested by a manufacturer's certified representative

B. Is installed in accordance with the manufacturer's recommendations and and UL Listings

C. Is in proper working order

D. The Contractor shall provide completed Inspection and Testing forms as outlined in *NFPA 72*.

3.4.2 ACCEPTANCE TESTING

Upon completion of each installation phase, the Contractor shall perform and document, in an approved format, system tests as required herein. All acceptance tests shall be performed in the presence of the Owner or a designated representative.

A. All conductors, including shielding conductors, shall be tested for continuity, shorts to ground and shorts between pairs.

B. All remote transmitting panel monitor points shall be functionally tested and monitor point identifications verified.

C. All alarm initiating devices shall be functionally tested. This includes all smoke detectors, including both Owner-supplied smoke detectors or vendor-supplied smoke detectors.

D. All supervisory functions of each initiating device and signaling line circuit shall be functionally tested.

E. Receipt of all alarm and trouble signals, initiated during the course of the testing, shall be verified at each annunciation device.

F. Correct labeling of all annunciation device LEDs shall be verified.

G. Sound level tests shall be performed in on each floor and all corridors to judge a minimum conformance with Section 2.12.

H. The system CPU and annunciators shall be load tested on standby battery power, according to NFPA 72, *National Fire Alarm and Signaling Code*.

I. Any additional tests, required by the referenced codes, standards, or criteria, or by the Owner, shall be performed. This documentation shall include:

   1. The date and time of each test

   2. A reference set of Contractor record drawings, numerically identifying the individual components and circuits tested, test locations, and indicating the measured sound level in each corridor and guest room location

   3. A description of each test performed

   4. A checklist of each device and circuit tested, indicating the results of each test

   5. The names and signatures of the individuals conducting and witnessing each test

J. The Contractor shall post suitable signs the day prior to, and shall maintain them during testing which indicate the date and time fire alarm testing is to occur. The signs shall be located in lobbies, elevator lobbies and other suitable locations so as to notify hotel occupants of the testing.

3.4.3 FINAL INSPECTION AND TESTS

A. The Contractor shall make arrangements with the Owner for final inspection and witnessing of the final acceptance tests. The Owner and the Fire Protection Consultant will conduct the final inspection and witness the final acceptance test. The Contractor shall be responsible for notification of the County of Stafford Fire Prevention Bureau and Building Department that the system is ready for testing. The Contractor shall arrange and schedule all tests.

B. All tests and inspections required by the referenced codes and standards, County of Stafford Fire Prevention Bureau and Building Department, and the Owner shall be performed by the Contractor.

   1. When local code authorities are required to witness tests, the Contractor shall be responsible for making all necessary arrangements with the code authorities and coordinating the work with the Owner.

   2. The Contractor shall be responsible for obtaining all test documents with necessary approval stamps and signatures of the code authorities. The Contractor shall submit one copy of each of these documents to the Owner.

C. The Owner's consultant will visit the job site to observe the work and witness final acceptance tests when the Consultant has been advised by the Contractor that the work is completed and ready for test. If the work has not been completed or the final acceptance tests are unsatisfactory, the Contractor shall be responsible for the Owner's extra expenses for re-inspection and witnessing the retesting of the work. Such extra fees shall be deducted from the payments made by the Owner to the Contractor.

D. The Contractor shall provide at least five working days' notice for all tests.

E. All Contractor testing shall conform minimally to the requirements of Chapter 14 of *NFPA 72*. Contractor testing shall include 100% of all devices, appliances, interfaced systems, and control unit functions.

F. The Contractor shall record the results of all tests and shall submit a completed copy of an Inspection and Testing form.

3.5 MATERIAL HANDLING

3.5.1 STORAGE

A. The Owner will provide the Contractor with a storage space for the Contractor's use during this project. The Contractor shall be responsible for the security of this space.

B. Overnight storage of materials is limited to the assigned storage area. Materials brought to the work area shall be installed the same day or returned to the assigned storage area, unless previously approved by the Owner.

3.5.2 RECEIVING AND HANDLING

A. The Contractor shall be responsible for all receiving, handling, and storage of materials at the job site.

B. Use of loading docks, service drives, and freight elevators shall be coordinated with the Owner.

C. All job planning shall conform to *NECA 100*.

3.5.3 RUBBISH REMOVAL

A. The Contractor shall remove rubbish and debris resulting from his or her work on a daily basis. Rubbish not removed by the Contractor will be removed by the Owner and back charged to the Contractor.

B. Removal of debris and rubbish from the premises shall be coordinated with the Owner.

END OF SECTION

Index

A

A-weighted scale, 90
Ablative firestopping products, 120
Acceptance and completion documentation, 35–37
Acceptance tests, 268, 316–317
Acoustically distinguishable space (ADS), 92, 194–195
Active multiplex transmission system, 221–222
ADA. *See* Americans with Disabilities Act (ADA)
ADAAG. *See* Americans with Disabilities Act Accessibility Guidelines (ADAAG)
Adaptive network system, 223
"ADDRESS 002: DEVICE MISSING/NO REPORT," 283
Addressable controls, 13–14
Addressable initiating devices, 15
Addressable monitor module, 131, 140
Adoption by reference, 24
Adoption by transcript, 24
Adoption of codes and standards, 23–24
ADS. *See* Acoustically distinguishable space (ADS)
Advanced detection topics, 156–182
   air sampling smoke detectors, 168–170
   detector selection, 163–164
   door release service, 179, 180
   factors affecting detector response, 165
   fire science, 158–163
   high ceilings, 170–171
   high-volume air-movement areas, 166–168
   HVAC systems, 175–178
   projected beam smoke detector, 170–171
   radiant energy smoke detector, 163–164, 171–173
   smoke control applications, 174–178
   suitable ambient conditions, 164–166
   time, 163
   video image smoke detection, 173–174
Aerosol smoke, 272
AFCI. *See* Arc-fault circuit interrupter (AFCI)
Agglomeration, 159
AHJ. *See* Authority having jurisdiction (AHJ)
Air sampling smoke detector, 65, 168–170, 273
Alarm capacity, 52
Alarm current, 53
Alarm initiating devices. *See* Initiating device
Alarm-initiating devices, 14–15, 58–59. *See also* individual initiating device
Alarm signal deactivation, 34
Alarm signals, 15, 37
Alarm silencing switch, 34
Amber alert, 186
Ambient environment, 164–166
American National Standards Institute (ANSI), 23
Americans with Disabilities Act (ADA), 8, 95–96
Americans with Disabilities Act Accessibility Guidelines (ADAAG), 95
Americium 241 radioactive material, 63
Ampere-hour capacity, 52
Annunciation and zoning requirements, 35, 38
ANSI. *See* American National Standards Institute (ANSI)
Arc-fault circuit interrupter (AFCI), 239
Architect's scale, 253, 254
As-built drawings, 252
ASME A17.1-2010/CSA B44-07, 143, 150
ASME A17.1-2010/CSA B44-10, Safety Code for Elevators and Escalators, 25, 26, 141
Audible notification appliances, 86, 88–94
Authority having jurisdiction (AHJ), 24–25, 255
Automatic alarm-initiating devices, 58. *See also* individual initiating devices
Automatic engine-driven generators, 51
Automatic-starting, engine-driven generators, 51
Automatic transfer switch, 53
Auxiliary alarm system, 9, 12
Average ambient sound pressure level, 91

B

Back box, 112
BACNet. *See* Building automation control network (BACNet)
Bathtub curve, 262
Battery, 51–53
Battery capacity, 53
Beam ceiling, 70
   heat detector, 71–72, 234
   smoke detector, 74–77, 234
Bells, 88
Bimetallic strip, 61
BIOS, changes to, 270
Blueprints, 248
Border lines, 254
Building automation control network (BACNet), 133
Building codes, 22
Building Fire Research Laboratory, 74
Burn-in failure, 262
Butterfly valve, 81

C

Cables, 109
Calibrated test methods, 271
Carbon dioxide (CO_2) sensor, 65
Carbon dioxide (CO_2) suppression system, 133, 139
Carbon monoxide (CO) detector, 59, 65
Cathode ray tube (CRT) monitor, 95
Ceiling-mounted appliances, 93, 97
Central Station Alarm Association (CSAA), 33
Central station alarm system, 9, 11, 209–213
   Code requirements, 212–214
   compliance, 212
   contract, 210–211
   delinquency signal, 213
   expense, 212
   operators, 214
   overview (figure), 211
   prime contractor, 211–214
   record retention, 214–215
   required elements, 210–211
   retransmission, 213, 214
   runner, 214
   supervisory signals, 214
   test signals, 214
   trouble signals, 214
   UL certificate, 212
   verification call, 214
Central station operators, 214
Certification, 32, 33, 263, 289–295
   after the exam, 296
   Code requirements and qualifications, 289–290
   exam strategies, 296–297
   IMSA, 290–292
   NICET, 293–294
   purpose, 289
Certified engineering technicians (CETs), 293
Changes to executive software, 270

Channing, William, 4, 204
Chemistry of fire, 158
Chimes, 86, 89
CI cable. *See* Circuit integrity (CI) cable
Circuit breaker shunt trip mechanisms, 149
Circuit classes, 45–50
Circuit identification, 109, 110
Circuit integrity (CI) cable, 116–117, 198–200
Circuit types, 16–19, 39–44
Civilian deaths (2011), 2
Class A and Class X circuit separation, 49, 117, 200–202
Class A and Class X circuit separation exceptions, 49–50, 201–202
Class A circuit, 45–46
Class A initiating device circuits, 280
Class B circuit, 46–47
Class B initiating device circuits, 274–280
Class B signaling line circuit, 44, 283
Class C circuit, 47–48
Class D circuit, 48
Class E circuit, 48
Class X circuit, 49
CO detector. *See* Carbon monoxide (CO) detector
Cocoanut Grove fire (Boston, 1942), 22
Code wheel pattern, 207
Coded alarm signals, 37
Coded EVACS signal, 193
Coded manual station, 193
Coded signals, 86, 193
Coded voice signal, 86
Coded walk test signal, 193
Codes and standards, 21–27
  adoption, 23–24
  building codes, 22
  enforcement, 24–25
  example of code requirements, 25, 26
  NFPA 72, 24, 26–27
  SDOs, 21, 22–23
Colonial America, 4
Combination burglary/fire alarm system, 230
Combination EVACS/ECS, 186
Combination systems, 128–130
Combined loop resistances, 119
Combustion, 158
Combustion products, 158–159
Compatibility, 34
Concept drawings, 250
Conceptual design drawings, 255
Continuity of power supplies, 53
Contracting methods, 257
Control equipment reacceptance tests, 270
Control equipment test methods, 271
Control unit, 12–14
Control unit diagrams, 251
Control valve supervisory signal initiating devices, 80–81
Conventional circuits, 39
Conventional controls, 14
Conventional initiating devices, 14–15

Copper conductor, 107, 113
CPVC piping, 169
CSAA. *See* Central Station Alarm Association (CSAA)

D

DAC system. *See* Digital alarm communications (DAC) system
DACT. *See* Digital alarm communicator transmitter (DACT)
DARS system. *See* Digital alarm radio system (DARS)
dBA. *See* Decibels, A-weighted (dBA)
Decibel, 90
Decibels, A-weighted (dBA), 90
Dedicated branch circuit, 50, 51
Delinquency signal, 213
Deluge sprinkler system, 20, 21
Derived local channel (DLC), 221
Design/bid/build format, 257
Design-build format, 257
Design/build specifications, 247
Design drawings, 249–250
Detection. *See* Advanced detection topics
Detection coverage requirements, 66–67
Diazo process, 248
Digital alarm communications (DAC) system, 219–221
Digital alarm communicator transmitter (DACT), 219–221
Digital alarm radio system (DARS), 223–224
Dimension lines, 254
Direct connect non-coded system, 222
Distortion, 194
Distributed recipient mass notification system (DRMNS), 12, 186, 189
Distributed system, 130
DLC. *See* Derived local channel (DLC)
Documentation, 35–37. *See also* Plans and specifications
Door and shutter release, 151–152, 179, 180
Door release magnet, 152
Door unlocking, 152
Drawing notes, 250, 254
Drawing scale, 252–255
Drawings. *See* Plans and specifications
DRMNS. *See* Distributed recipient mass notification system (DRMNS)
Dry contacts, 131
Dry-pipe sprinkler, 20, 78–79
Duck and cover bomb threat, 129
Duct smoke detector, 175–178, 273
Duplicate terminals and leads, 44
Dwelling fire alarm system. *See* Household fire alarm system
Dynamic load test, 161
Dynamic penetrations, 121

E

Early fire alarm systems, 4–6
ECS. *See* Emergency communications system (ECS)
Electrical. *See* Wiring and wiring methods

Electronic heat detector, 62–63
Electronic rolling scales, 254
Elevator safety functions, 141–151
Elevator shunt trip power supervision, 150
Elevator shutdown, 148–149
"EMERGENCY COMMUNICATIONS," 51
Emergency communications system (ECS), 38, 94, 184–204
  Class A and Class X circuit separation, 200–202
  DRMNS, 186, 189
  evacuation signals, 192–194
  fire command center, 191–192
  firefighters' telephones, 189–190
  in-building ECS, 186–189
  in-building MNS, 189
  intelligibility, 194–196
  one-way ECS, 186–189
  power supplies, 191
  survivability, 196–200
  two-way ECS, 189–191
  two-way radio communications enhancement system, 190–191
  wide-area MNS, 189
Emergency control functions, 14. *See also* System interfaces and safety control functions
Emergency forces notification, 209
End-of-life failure, 262
End-of-line (EOL) device, 17, 39
Engine-driven generators, 51–53
Engineering scale, 254
Environmental protection, 3
EOL device. *See* End-of-line (EOL) device
Equivalent resistances, 118
Eutectic materials, 60
EVACS. *See* In-building fire emergency voice alarm communications system (EVACS)
EVACS controls, 188
Evacuation signals, 192–194
Example specifications. *See* One Prosperity Place specifications
Exit marking audible appliances, 94
Exposed cable, 112

F

F rating, 120
Factory Mutual (FM), 33
Factory Mutual Global (FM Global), 212
Farmer, Moses, 4, 204
Fire, 158
Fire alarm and emergency communication system inspection and testing form, 267, 268
"FIRE ALARM CIRCUIT," 51, 110, 113
Fire alarm circuits, 16–19, 39–44
Fire alarm control unit, 12–14
Fire alarm controls, 12–14
Fire alarm cover, 80
"FIRE ALARM/ECS," 51
Fire alarm notification appliance. *See* Notification appliance
Fire alarm speaker, 188

Fire alarm symbols, 256
Fire alarm system
   circuit types, 16–19, 39–44
   codes and standards, 21–27
   control units, 12–14
   detection. *See* Advanced detection topics
   dwellings. *See* Household fire alarm system
   general requirements, 32–37
   goals, 2–3
   history. *See* Historical overview
   initiating devices. *See* Initiating devices
   inspection. *See* Inspection, testing, and maintenance
   monitoring for integrity, 38–50
   notification appliances. *See* Notification appliance
   plans and specifications. *See* Plans and specifications
   power supplies, 50–54
   purpose/importance, 2
   SDOs, 7–8, 21, 22–23
   signals and signal types, 15, 37–38
   suppression systems, 19–21
   types, 8–12
   wiring. *See* Wiring and wiring methods
Fire alarm system interfaces. *See* System interfaces and safety control functions
Fire alarm system reliability, 3
Fire alarm system wiring. *See* Wiring and wiring methods
Fire alarm testing. *See* Inspection, testing, and maintenance
Fire command center, 191–192
Fire modeling, 170
Fire plume development, 162
Fire Prevention Week (FPW), 3
Fire pump, 79, 137
Fire Resistance Directory, 199
Fire science, 158–163
   chemistry of fire, 158
   fire triangle, 158
   products of combustion, 158–159
   smoke detectors and standardized tests, 159–163
   stratification, 163
Fire service access elevator, 149–151
Fire statistics, 2
Fire triangle, 158
Firefighter's hat symbol, 144
Firefighters' telephone, 89–90, 93, 189–190
Firefighters' telephone stations, 189, 190
Firestopping, 120–122
Firestopping caulk, 122
Firestopping pillow, 121
Firewall, 256
Fixed temperature heat detector, 59–61
Flat scales, 253
Flying spices, 112
FM. *See* Factory Mutual (FM)
FM-200 system, 139

FM Global. *See* Factory Mutual Global (FM Global)
Foam system, 20
FPL power-limited fire general use, 113, 114
FPLP power-limited fire plenum use, 113, 114
FPLR power-limited fire riser use, 113, 114
Fuel-use calculations, 53
Fully-sprinklered hoistway and elevator machine room, 145
Functional test (smoke detectors), 270

G

Gamewell loop, 206
Gamewell public fire reporting system, 5
Gaseous suppression system, 20, 21
Gasoline-powered engine-driven generators, 54
Gateways, 132–133
Gemini Smoke Detector Analyzer, 271
General requirements, 32–37
   acceptance and completion documentation, 35–37
   alarm signal deactivation, 34
   compatibility, 34
   inspection and testing personnel qualifications, 32–33
   installer qualifications, 32
   listed equipment, 33–34
   operating conditions, 34–35
   protection of control equipment, 35
   supervising station operator qualifications, 33
   system designer qualifications, 32
   wiring, 35
   zoning and annunciation, 35
General Services Administration (GSA), 248
Giant voice, 186, 189
Graphic annunciator, 16, 17, 94–95, 192
Great Chicago Fire (1871), 4
Great Fire of London (1666), 3
Grounding, 108
GSA. *See* General Services Administration (GSA)
Guard's tour delinquency signals, 214

H

Hearing impaired persons, 86, 237. *See also* Visible signaling
Heat detector, 59–63, 67–73, 165
   beam ceilings, 71–72
   dwellings, 237–238
   electronic, 62–63
   fixed temperature, 59–61
   high ceilings, 72–73
   installation, 67–68
   joisted ceilings, 69–70
   line-type, 62
   listed spacing, 68
   rate-compensated, 61
   ROR, 62
   sample specifications, 311

   sloping and peaked ceilings, 73
   smooth ceiling spacing, 69
   spacing requirements, 68–73
   testing, 161–162, 272–273
Heat detector installation requirements, 68
Heat release, 159
Heating, ventilating, and air-conditioning (HVAC) systems
   shutdown, 151
   smoke control, 175–178
Heritage protection, 3
Hidden lines, 254
High ceilings, 72–73, 170–171
High-profile fires, 22
High-volume air-movement areas, 166–168
Historical overview, 3–7
   building codes, 22
   colonial America, 4
   early fire alarm systems, 4–6
   high-profile fires, 22
   modern equipment, 6–7
   SDOs, 21
Horns, 86, 89
Household fire alarm system, 226–243
   AC primary power source, 238–239
   audible notification, 236–237
   basement, 234
   combination burglary/fire alarm system, 230
   control units, 230–231
   escape plan, 231
   heat detector, 234–235
   multiple-level dwellings, 234
   non-rechargeable battery primary power source, 240
   non-supervised interconnected wireless alarms, 241–242
   occupant notification, 236–237
   permitted equipment, 228–231
   power supplies, 238–241
   purpose, 228
   rechargeable battery primary power source, 240–241
   secondary power source, 240
   smoke detection, 231–235
   sounder bases, 230
   supervising station connections, 241
   types of dwelling units covered, 228, 229
   visible appliances in sleeping areas, 238
   visible signaling, 237
   workmanship and wiring, 236–237
HVAC systems. *See* Heating, ventilating, and air-conditioning (HVAC) systems
Hydrogen-carbon bonds, 158
Hydrogen-hydrogen bonds, 158

I

IBC. *See* International Building Code (IBC)
IDC. *See* Initiating device circuit (IDC)
Impairments, 262–263

IMSA. *See* International Municipal Signal Association (IMSA)
IMSA interior fire alarm certification, 290–292
In-building fire emergency voice alarm communications system (EVACS), 9, 10, 186–189
In-building mass notification system (MNS), 12, 189
In-building two-way communication system, 89–90, 93
Inergen system, 139, 166
Initial/acceptance tests, 268
Initiating device, 14–15, 56–83
　alarm signaling devices, 14–15, 58–59
　detection coverage requirements, 66–67
　heat detector. *See* Heat detector
　installation and spacing requirements, 65–82
　manual fire alarm box, 79–80
　primary purpose, 56
　protection, 66
　recessing the devices, 66
　smoke detector. *See* Smoke detector
　supervisory signaling devices, 58, 59, 80–82
　theory, 59–65
　waterflow, 78–79
Initiating device circuit (IDC), 16–18, 39–41, 274–281
Inspection, testing, and maintenance, 262–285
　burn-in/end-of-life failures, 262
　changes to executive software, 270
　Class A initiating device circuits, 280-281
　Class B initiating device circuits, 274–281
　control equipment, 270
　documentation, 265–266
　EVACS, 273
　heat detectors, 273, 274
　IDCs, 275–281
　impairments, 262–263
　initial/acceptance tests, 268
　inspection checklist, 269
　maintenance, 274
　notification, 264–265
　notification appliance circuits, 281–283
　periodic inspections and tests, 268–269
　qualifications, 263–264
　reacceptance tests, 269–270
　releasing systems and interfaces, 266
　required form, 267
　responsibilities, 263
　safety control functions, 273
　sample specifications, 316–318
　signaling line circuits, 283–284
　smoke detectors, 270–272, 273
　special situations, 273
　sprinkler systems, 273
　tools, 265
　troubleshooting, 274–284
　visual inspection, 268

Inspection and testing personnel qualifications, 32–33
Installer qualifications, 32
Integrated burglary and fire alarm system, 129
Intelligibility, 91–92, 194–196
Interconnected fire alarm systems, 130–131. *See also* System interfaces and safety control functions
Interconnected non-fire protection system, 133
Interconnected suppression systems, 133–139
　fire pumps, 137
　general requirements, 134
　special hazard systems, 137–139
　sprinkler system attachments, 135–137
International Building Code (IBC), 22
International Mechanical Code (IMC), 25
International Municipal Signal Association (IMSA), 32, 290
Internet Protocol (IP) gateway, 219
Interstate Bank fire (Los Angeles, 1988), 190
Intumescent firestopping products, 120, 121
Ionization smoke detector, 63, 162, 164
IP gateway. *See* Internet Protocol (IP) gateway
Iroquois Theater fire (Chicago, 1903), 22

J

Jaeger, Walter, 6
Joisted ceiling
　heat detector, 69–70, 234
　smoke detector, 74–77, 234
Junction box cover, 109

L

L rating, 120
LAN. *See* Local area network (LAN)
Layout drawings, 249
LCD. *See* Liquid crystal display (LCD)
LED. *See* Light-emitting diode (LED)
Legacy systems, 218
Legs, 209
Level 0 survivability, 196–199
Level 1 survivability, 196–199
Level 2 survivability, 196–199
Level 3 survivability, 196–199
Life-safety drawings, 256
Life safety signals, 129
Light-emitting diode (LED), 95
Light obscuration smoke detection, 64
Light-scattering photoelectric smoke detector, 64
Lightning strikes, 107, 108
Line-type heat detector, 62, 164
Linear amplifier, 190
Linear smoke detection, 65
Lines and line weights, 254
Liquid crystal display (LCD), 94, 191
Listed BACNet gateways, 133
Listed BACNet portals, 133
Listed equipment, 33–34

Local area network (LAN), 47
Local licenses, 264
Local system, 204
Loop resistance calculations, 117

M

Magnet circuit, 152
Maintenance, 274–275
Managed facilities-based voice network (MFVN), 47
Manual fire alarm box, 58, 79–80
Manual-starting, engine-driven generators, 51
Manufacturer's calibrated test instruments, 271
Mass notification system (MNS), 8, 12, 94
Master box, 204, 208
MasterFormat system, 247, 248
Match lines, 254
McCulloh loop, 206
MFVN. *See* Managed facilities-based voice network (MFVN)
MGM Grand fire (Las Vegas, 1980), 22
MI cable. *See* Mineral-insulated (MI) cable
Microprocessor-based fire alarm system, 7
Microprocessor controlled (addressable) controls, 13–14
Mineral-insulated (MI) cable, 116, 199–200
Mini-horns, 89
Mission continuity, 3
MNS. *See* Mass notification system (MNS)
Monitor module, 39, 131
Monitoring for integrity, 38–50
　basic requirements, 38–39
　circuit classes, 45–50
　IDCs, 39–41
　NACs, 43–44
　SLCs, 42–43
　T-tapping, 44
Morse, Samuel, 4
Multi-criteria detector, 65
Multi-sensor detector, 65
Multiple-station smoke alarm, 228
Multiple-station smoke and heat alarm, 9, 12

N

NAC. *See* Notification appliance circuit (NAC)
National Electrical Code (NEC), 21, 106. *See also* Wiring and wiring methods
National Fire Protection Association (NFPA), 5, 21
National Institute for Certification in Engineering Technologies (NICET), 32, 293
National Institute of Standards and Technology (NIST), 74
NEC. *See National Electrical Code (NEC)*
NECA 100, Symbols for Electrical Construction Drawings, 256
NECA 305-2010, Standard for Fire Alarm System Job Practices (ANSI), 108

NFPA. *See* National Fire Protection Association (NFPA)
NFPA 13, *Standard for the Installation of Sprinkler Systems,* 136
NFPA 20, *Standard for the Installation of Stationary Fire Pumps,* 137
NFPA 70, *National Electrical Code,* 21, 106
NFPA 72, *National Fire Alarm and Signaling Code,* 24, 26–27
NFPA 72-G, *Recommended Practice for the Installation of Notification Appliances,* 8
NFPA 90A, *Standard for Ventilating Systems,* 26
NFPA 92A, *Standard for Smoke-Control Systems Utilizing Barriers and Pressure Differences,* 175
NFPA 92B, *Standard for the Installation of Air-Conditioning and Ventilating Systems,* 175
NFPA 101, *Life Safety Code,* 22, 26
NFPA 170, *Standard for Fire Safety and Emergency Symbols,* 256
NFPA 1221: *Standard for the Installation, Maintenance, and Use of Emergency Services Communications Systems,* 207
NICET. *See* National Institute for Certification in Engineering Technologies (NICET)
NICET fire alarm systems certification, 293–295
Non-addressable circuit, 39
Non-coded signals, 193–194
Non–power-limited cable types, 111–112
Non–power-limited fire alarm circuits, 111–113
Non–power-limited wiring methods, 112–113
Non-rechargeable battery primary power source, 240
Non-required coverage, 67
Non-restorable fixed temperature heat detector, 60
Non-smooth ceiling
   heat detector, 69–72
   smoke detector, 74–77
Non-sprinklered elevator machine and pit sprinklers, 147
Non-sprinklered hoistway and elevator machine room, 148
Non-supervised interconnected wireless alarms, 241–242
Non-voice coded signals, 86
Notification appliance, 15–16, 84–103
   ADA, 95–96
   audible signaling, 88–94
   bells, 88
   chimes, 89
   coded signals, 86
   corridor strobe spacing, 99–100
   defined, 84, 86
   ECS and MNS, 94
   exit marking audible appliances, 94
   firefighters' telephones, 89–90, 93
   horns, 89
   intelligibility, 91–92
   listing requirements, 87
   mechanical protection, 88
   mounting, 87, 93
   performance-based arrangement, 86
   physical construction, 88
   power booster panels, 86
   private mode signaling, 92
   public mode signaling, 90–92
   public mode visible signaling, 97–101
   sirens, 89
   sleeping areas, 92–93, 100–101
   sound transmission, 90–93
   spacing, 86
   speakers, 89
   strobes, 94–101
   tactile display, 86
   telephone appliances, 93
   textual display, 86
   visible signaling, 94–101
Notification appliance circuit (NAC), 19, 43–44, 281–283
NPL fire alarm cables. *See* Non–power-limited fire alarm circuits
NPLF non–power-limited fire general use, 111, 112
NPLFP non–power-limited fire plenum use, 111, 112
NPLFR non–power-limited fire riser use, 111, 112
Nuisance alarms, 58, 164

O

Object lines, 254
Occupant evacuation elevator, 151
Ohm's Law, 118, 119
OM manual. *See* Operations and maintenance (OM) manual
One Prosperity Place specifications, 299–318
   acceptance testing, 316–317
   annunciation, 306–308
   annunciation devices, 309–310
   approvals by code authority, 303
   audible notification appliances, 312
   batteries/secondary power sources, 306
   conductors and raceway, 312–315
   contractor record drawings, 303
   description of the system, 305
   final approval and acceptance, 304–305
   final inspection and tests, 317–318
   fire alarm control panel, 308–309
   fire alarm signals, 307
   general description, 299
   guarantee, 304
   heat detectors, 311
   IDCs/SLCs, 305
   indicating lamps displays, 309
   inspection, 315
   inspection and tests, 316–318
   installation, 315–316
   intent of specifications, 299
   manual fire alarm stations, 310
   material handling, 318
   NACs, 305
   notification appliance circuit power extender panel, 308–309
   operator acknowledgment signals, 308
   owner's representative, 301
   power supplies, 305–306
   qualifications of contractor, 302–303
   quality assurance, 301–303
   receiving and handling, 318
   remote annunciators, 309–310
   rubbish removal, 318
   smoke detectors, 310–311
   spare parts, 304
   special tools, 304
   sprinkler alarm supervisory equipment, 311–312
   starting and completion dates, 315
   storage, 318
   supervisory signals, 307
   system configuration, 305
   system tests, 316
   test record, 304
   trouble signals, 307
   visual notification appliances, 312
   work included, 300–301
   working conditions, 301
One-way emergency communication system, 186–189
One-way private radio system, 221
One-way radio system, 222
Open screw and yoke (OS&Y) valve, 81
Operations and maintenance (OM) manual, 265, 273
OS&Y valve. *See* Open screw and yoke (OS&Y) valve
Other protection systems. *See* System interfaces and safety control functions
Outdoor mass notification system (MNS), 12
Outside circuits, 107–108
Overview (figure)
   suppression systems, 20
   types of alarm systems, 9
Oxidation, 158

P

Paddle-type switch, 136
Paddle-type waterflow device, 78, 79
Paint spray booth, 166
Pan fire, 161, 162
Partial or selective coverage, 67
Particle counting (laser) smoke detector, 65
Pathway survivability, 196–200
"Pathways," 45
Peaked and shed ceilings
   heat detector, 73
   smoke detector, 78, 231
Pendant-mounted smoke detector, 175
Penetrations, 120–122
Performance-based arrangement, 86
Periodic inspections and tests, 268–269
Permitting process, 37

Photoelectric light obscuration, 64–65
Photoelectric light-scattering smoke detector, 64
Piezoelectric buzzer, 86
Pigtails, 44
Pit sprinklers, 147, 148
PIV. *See* Post indicator valve (PIV)
Plans and specifications, 244–258
   blueprints, 248
   contracting methods, 257
   design drawings, 249–250
   documentation requirements (Chapter 7), 249
   drawings, sizes, 248, 249
   firewalls and other barriers, 256
   government agencies, 248
   legend, 254
   lines and line weights, 254
   MasterFormat system, 246-247, 251
   notes, 250, 254
   plans check, 256
   plans review and approvals, 255–256
   record drawings, 252
   red lines, 252
   required contents, 251
   RFIs, 255
   sample specifications, 299–318. *See also* One Prosperity Place specifications
   scale, 252–254
   shop drawings, 250–252
   specification content, 247–248
   standard symbols, 256
   storage of system documentation, 256
   title block, 254
   working drawings, 248
Plenum, 108
Plotter, 248
Plugging out, 208
Plume, 162
Post indicator valve (PIV), 81, 136
Power booster panels, 86
Power-limited cable types, 113–114
Power-limited fire alarm circuits, 113–116
Power-limited wiring methods, 114–116
Power supplies, 50–54
   continuity, 53
   ECS, 191
   household fire alarm system, 240–243
   monitoring for integrity, 39
   primary supply, 50–51
   sample specifications, 305–306
   secondary supply, 51–54
Pre-action sprinkler system, 20, 139
Pressure supervisory signal initiating devices, 81
Pressure switch, 136
Primary power supply, 50–51
Prime contractor (central station service), 210–213
Printers, 95
Private microwave system, 222
Private mode signaling, 92
Products of combustion, 158–159
Professional associations, 290

Projected beam smoke detector, 7, 64, 162, 170–171
Property protection signals, 129
Proprietary supervisory station alarm system, 9, 11, 215–217
Protected premises (local) system, 8–10, 208
Protection of control equipment, 35
PSTN. *See* Public switched telephone network (PSTN)
Public authorities having jurisdiction, 25–26
Public emergency alarm reporting system, 9, 11–12, 206–209
Public mode occupant-notification appliances, 16
Public mode signaling, 90–92
Public mode visible signaling, 96–101
Public switched telephone network (PSTN), 47
Pyrolysis, 158

R

Radiant energy smoke detector, 163–164, 171–173
Radiated energy, 159
Radio frequency (RF) system, 222–223
Radio systems, 222–223
Raster scan software, 58
Rate-compensated heat detector, 61
Rate-of-rise (ROR) heat detector, 62
Reacceptance tests, 269–270
Recessing detector, 66
Rechargeable battery primary power source, 240–241
Record drawings, 252
Record of Completion, 36, 37
Record retention, 213–214, 266
Red lines, 252
Releasing system controls, 134, 135
Remote supervising station alarm system, 9, 11, 217–218
Request for information (RFI), 255
Restorable fixed-temperature heat detector, 60–61
Retard, 78, 136
Retard chambers, 79
Retransmission, 210, 211, 213
Return-side smoke detection, 175, 176
Reverberation, 194
Review process, 37
RF system. *See* Radio frequency (RF) system
RFI. *See* Request for information (RFI)
Riser diagrams, 251
Risk zone, 54
Rolling scales, 253, 254
ROR heat detector. *See* Rate-of-rise (ROR) heat detector
Runner, 216, 218

S

Sample specifications. *See* One Prosperity Place specifications
Scale, 252–254

SDOs. *See* Standards developing organizations (SDOs)
Secondary battery power supply capacity, 52
Secondary power calculations, 51–54
Secondary power supply, 51–54
Section 907, *IBC*, 25
Sensitivity test (smoke detectors), 270–272
Sensitivity test unit, 271, 272
Shielded conductor, 108
Shop drawings, 250–252
Shunt trip, 148
Shunt trip power supervision, 150
Shunt-type and local energy public emergency alarm reporting system, 208
Signal hierarchy, 129
Signal-to-noise ratio, 194
Signaling line circuit (SLC), 18, 42–43, 221, 283–285
Signals and signal types, 15, 37–38
Simplified loop resistances, 119
Single-station smoke alarm, 9, 12, 236
Sirens, 89
SLC. *See* Signaling line circuit (SLC)
Sleeping areas
   audible signaling, 92–93
   visible appliances (figure), 237
   visible signaling, 100–101
Sloped ceiling
   heat detector, 73
   smoke detector, 77–78, 231–232
Smoke alarm, 231
Smoke barriers, 256
Smoke control, 151
Smoke control systems, 174–175
Smoke density, 159
Smoke detector, 63–65, 73–78
   air sampling, 65, 168–170
   during construction, 74
   control units, and, 165
   detection process, 159
   door releasing service, 152, 179, 180
   duct, 175–178
   dwellings, 229–233
   effectiveness, 240
   elevator recall, 142
   fan shutdown, 151
   historical overview, 6
   HVAC systems, and, 175–178
   installation/spacing, 73–78
   ionization, 63
   light obscuration, 64–65
   non-smooth ceilings, 74–77
   particle counting, 65
   peaked and shed ceilings, 78
   permitted ambient conditions, 73–74
   photoelectric light-scattering, 64
   projected beam, 170–171
   quantity of smoke detected, 162
   radiant energy, 163–164, 171–173
   sample specifications, 310–311
   sloped ceilings, 77–78
   smooth ceilings, 74
   sprinkled hoistways, 144

testing, 159–161, 270–272, 273
underfloor spaces, 165, 166
video image smoke detection, 173–174
when not used, 164
"SMOKE DETECTOR SAMPLING TUBE — DO NOT DISTURB," 169
Smoke detector sensitivity testing, 270–272
Smoke door release, 151–152, 179, 180
Smoke particles, 159
Smooth ceiling
heat detector, 69
smoke detector, 74
Solenoid valve, 149
Solo True Test, 271
Sound pressure level (SPL), 90, 91
Sounder bases, 230
Spark/ember detector, 172
Speaker, 89
Speaker circuit, 111
Special hazard suppression systems, 137–139
Specification documents. See Plans and specifications
Speech intelligibility, 194–196
Speech transmission index (STI) meter, 195
Speech transmission index (STI) method, 195
SPL. See Sound pressure level (SPL)
Spot-type smoke detector, 63
Sprinkler system attachments, 135–137
Sprinkler waterflow alarm-initiating devices, 78–79
"SPRINKLER WATERFLOW AND SUPERVISORY SYSTEM," 138
Sprinkler zone, 78, 135
Sprinklered elevator machine room and non-sprinklered hoistway, 146
Stack effect test, 161
Stairway pressurization, 152
Stale gasoline, 54
Standards developing organizations (SDOs), 7–8, 21, 22–23
Standby battery, 51–53
Standby capacity, 52
Static firestopping applications, 121
Station Nightclub fire (Rhode Island, 2003), 22
STI meter. See Speech transmission index (STI) meter
STI method. See Speech transmission index (STI) method
Stratification, 163
Street box, 5, 206–208
Strobes, 94–101
Structure fires (2011), 2
Subsidiary supervising station, 209
Supervising station alarm system, 209–224
active multiplex communications, 221–222
central station service, 210–215
communication methods, 218–223
DAC systems, 219–222

DARS systems, 223–224
direct connect non-coded system, 222
DLC, 221
dwellings, 240
emergency forces notification, 209
IP gateway, 219
legacy systems, 218
overview (figure), 209
proprietary stations, 218–220
radio systems, 222–223
remote stations, 217–218
Supervising station operator qualifications, 33
Supervisory signal–initiating devices, 14, 58–59, 80–82
Supervisory signals, 15, 37–38, 214
Supervisory station alarm system, 9, 10–11
Supply-side smoke detection, 175, 176
Suppression system, 19–21
Suppression system actuation, 166
Survivability, 116, 196–200
Synchronization modules, 99
System defects and malfunctions, 262–263
System designer qualifications, 32
System interfaces and safety control functions, 126–154
combination systems, 128–130
connection methods, 131–133
door and shutter release, 151–152
door unlocking, 152
elevator safety functions, 141–151
elevator shutdown, 148–149
emergency control function interfaces, 139–152
fire pumps, 137
fire service access elevator, 149–151
HVAC shutdown, 151
interconnected and networked systems, 130–131
interconnected suppression systems, 133–139
occupant evacuation elevator, 151
smoke control, 151
special hazards systems, 135–137
stairway pressurization, 152

T

T rating, 120
T-tap/T-tapping, 44, 109
Ted Williams Tunnel (Boston), 58, 173
Telephone appliances, 93
Telephone jacks, 189-190
Telephone stations, 189
Temperature and humidity extremes, 34
Temperature supervisory signal–initiating devices, 82
Temporal-3 signal, 90, 192–193
Terminals, 109
Terror alerts. See Emergency communications system (ECS)
Testing. See Inspection, testing, and maintenance
Thermal lag, 60

Thermal paper integral printers, 95
Title block, 254
Tone coded signals, 193
Total (complete) coverage, 66–67
Tractor-feed, line printers, 95
Trades, 250
Traffic delays, 186
Trouble signals, 15, 38, 214
Troubleshooting, 274–284
Trunks, 209
Two-hour rated enclosure or shaft, 199
Two-way, in-building wired emergency communications system, 189–190
Two-way emergency communication system, 189–191
Two-way radio communications enhancement system, 190–191
Two-way RF multiplex system, 222
Type 4 system, 223
Type 5 system, 223
Type A system, 208
Type B system, 208
Type FPL cable, 114
Type FPLP cable, 114
Type FPLR cable, 114
Type NPLF cable, 112
Type NPLFP cable, 112
Type NPLFR cable, 112

U

UL. See Underwriters Laboratories (UL)
UL 217, 159
UL 268, 160
UL 827, Standard for Central-Station Alarm Services, 212, 213
UL 985: Standard for Household Fire Warning System Units, 240
UL 1730: Standard for Smoke Detector Monitors and Accessories for Individual Living Units of Multifamily Residences, and Hotel/Motel Rooms, 240
UL 1971-4-2002: UL Standard for Safety Signaling Devices for the Hearing Impaired, 96
UL Fire Protection Equipment Directory, 211
UL heat detector test setup, 161
UL smoke detector test setup, 160
ULC. See Underwriters Laboratories Canada (ULC)
Underfloor spaces, 165, 166
Underground circuits, 108
Underwriters Association of the Middle Department, 21
Underwriters Laboratories (UL), 5, 21, 33, 34, 159–162, 211
Underwriters Laboratories Canada (ULC), 162
Uninterruptible power supply (UPS), 53
United States Government General Services Administration (GSA), 248
UPS. See Uninterruptible power supply (UPS)

V

Valve laces, 81
Valve supervisory switches, 15, 59
Vane-type switch, 136
Vane-type waterflow device, 78, 79
Verification call, 214, 217
Very early smoke detection apparatus (VESDA), 6
Video image smoke detection, 58, 173–174
Virginia Tech shootings, 184
Visible signaling (visible notification appliances), 94–101
Visual inspection, 268
Voice intelligibility, 194–196
Voltage drop calculations, 118–120
Volunteer technical committee, 23

W

Waffle ceiling, 76, 77
Wall-mounted appliances, 93, 96
Water level supervisory signal–initiating devices, 81–82
Waterflow initiating devices, 78–79
Waterflow switch, 136
Wet and dry chemical suppression system, 20, 21
Wet pipe sprinkler, 20
Wide-area mass notification system (giant voice), 186, 189
Wireless public emergency alarm system box, 208–209
Wiring and wiring methods, 35, 39, 104–124
 access, 110
 cables, 109
 circuit integrity cable, 116–117
 Class A and Class X circuit separation, 117
 classified locations, 110
 corrosive/damp, wet locations, 110
 firestopping, 120–122
 general requirements, 106–107
 grounding, 108
 hazardous locations, 109–110
 identification, 109
 loop resistance calculations, 117
 non–power-limited fire alarm circuits, 111–113
 outside circuits, 107–108
 plenums and other air-handling spaces, 108
 power-limited fire alarm circuits, 113–116
 survivability, 116
 terminals, 109
 voltage drop calculations, 118–120
 workmanship, 108
Working drawings, 248

Z

Zone address module, 39
Zoning and annunciation, 35, 38